Nuclear
Fission
Reactors

Nuclear Fission Reactors

I. R. Cameron
University of New Brunswick
Saint John, New Brunswick, Canada

PLENUM PRESS · NEW YORK AND LONDON

Library of Congress Cataloging in Publication Data

Cameron, I. R.
  Nuclear fission reactors.

  Bibliography: p.
  Includes index.
  1. Nuclear reactors. 2. Atomic power. I. Title.
TK9202.C17 1982   621.48′3   82-18128
ISBN 0-306-41073-7

198009

© 1982 Plenum Press, New York
A Division of Plenum Publishing Corporation
233 Spring Street, New York, N.Y. 10013

Printed in the United States of America

Preface

This book is intended to provide an introduction to the basic principles of nuclear fission reactors for advanced undergraduate or graduate students of physics and engineering. The presentation is also suitable for physicists or engineers who are entering the nuclear power field without previous experience with nuclear reactors. No background knowledge is required beyond that typically acquired in the first two years of an undergraduate program in physics or engineering. Throughout, the emphasis is on explaining why particular reactor systems have evolved in the way they have, without going into great detail about reactor physics or methods of design analysis, which are already covered in a number of excellent specialist texts.

The first two chapters serve as an introduction to the basic physics of the atom and the nucleus and to nuclear fission and the nuclear chain reaction. Chapter 3 deals with the fundamentals of nuclear reactor theory, covering neutron slowing down and the spatial dependence of the neutron flux in the reactor, based on the solution of the diffusion equations. The chapter includes a major section on reactor kinetics and control, including temperature and void coefficients and xenon poisoning effects in power reactors. Chapter 4 describes various aspects of fuel management and fuel cycles, while Chapter 5 considers materials problems for fuel and other constituents of the reactor. The processes of heat generation and removal are covered in Chapter 6.

Following a general survey of the basic types of nuclear power reactor, separate chapters are devoted to each of the principal designs—the gas-cooled graphite-moderated reactor, the light-water-moderated reactor, the heavy-water-moderated reactor, and the fast reactor. Each chapter includes a discussion of the evolution of the design and a detailed description of one or more typical power plants.

In recognition of the importance which reactor safety has assumed, both within the nuclear industry and with the public at large, the remainder of the book is devoted to a study of the safety and environmental aspects of nuclear reactors. After a discussion of the biological effects of radiation, the potential hazards from both routine operation and accidents at nuclear power plants are evaluated in the light of the most recent studies. These are illustrated by reference to incidents such as that at the Three Mile Island reactor. Other environmental concerns, such as waste disposal and thermal effects, are also considered.

Both SI and English units are used in the text, the latter particularly in dealing with heat transfer, where the older system is still extensively used in North America. Problems using both systems are included at the end of the chapters where problem solving is appropriate.

My thanks are due to a number of individuals in the nuclear industry who co-operated readily in providing technical information, and in particular to Mr. G. Gibbons and Dr. J. E. Sanders of the United Kingdom Atomic Energy Authority. I would also acknowledge the invaluable help of Mr. W. Morris, who prepared most of the illustrations, and of Mrs. E. Duffley, who typed much of the manuscript. Finally, I record my gratitude to my wife and daughter, without whose sustained forbearance this book would not have been written.

Contents

5. Materials Problems for Nuclear Reactors

6. Heat Generation and Removal in Nuclear Reactors

7. General Survey of Reactor Types

8. The Gas-Cooled Graphite-Moderated Reactor

9. The Light-Water-Moderated Reactor

I

Basic Physics of the Atom and the Nucleus

1.1. The Nuclear Atom

The concept that matter is made up of minute individual entities originated with the philosophers of ancient Greece, but the idea did not gain general acceptance until it was put on a quantitative basis by the growth of modern science in the 19th century. In particular, the British chemist John Dalton, in 1808, explained the known laws of chemical combination on the assumption that any given element is composed of identical *atoms*, which differ from the atoms of any other element. Chemical compounds are then formed by the combination of one or more atoms of one element with atoms of another; for example, the combination of one atom of carbon with two of oxygen yields carbon dioxide.

The 19th century saw numerous extensions of the atomic theory, one of the most important being the kinetic theory of gases. The macroscopic behavior of gases may be explained by considering them to be composed of molecules in rapid random motion, with characteristic speeds of the order of several hundred meters per second (m/s). The continual collisions of the high-speed molecules with the walls constitute the pressure exerted by the gas on a container. The molecules exhibit a characteristic statistical spread of velocities, known as the Maxwell–Boltzmann distribution, which will be discussed in detail in a later chapter because of its relevance to the physics of neutrons in a scattering medium such as a reactor lattice.

1.2. The Structure of the Atom

Up to the end of the 19th century the generally accepted picture of the atom was that of a hard, indestructible sphere. In the last years of the century two discoveries provided evidence suggesting that atoms did in fact have an internal structure. J. J. Thomson, in Cambridge, showed that the so-called cathode rays, produced when an electrical discharge was passed through a gas at low pressure, always consisted of identical negatively charged particles of very small mass, about one two-thousandth part of the mass of the lightest atom known, that of hydrogen. This particle, the *electron*, appeared therefore to be a constituent common to the structure of all atoms. Further investigation yielded the following values for the mass and charge of this fundamental particle:

$$\text{mass of electron} = 9.1 \times 10^{-31} \text{ kilograms (kg)}$$
$$\text{charge of electron} = 1.6 \times 10^{-19} \text{ coulombs (C)}$$

Further evidence of the complex structure of the atom was provided in 1896 by the discovery, by the French physicist Becquerel, of the phenomenon of radioactivity. It was found that certain elements, such as uranium, spontaneously emitted radiation at an apparently constant rate. Ernest Rutherford showed that the emissions involved two types of particle, to which he gave the names *alpha* (α) and *beta* (β) rays. The first of these had a positive charge and a mass essentially equal to that of an atom of the element helium, while the second was found to be simply an electron. In 1903, Rutherford, in collaboration with Frederick Soddy, explained the process of radioactivity by the revolutionary hypothesis that the particle emission was a means whereby an atom of one element converted itself spontaneously into an atom of another.

The next step in elucidating the structure of the atom was taken by Rutherford in 1911, following the classic experiment of Geiger and Marsden on the scattering of α particles by thin metal foils. The α particles, striking the foil with relatively high momentum, were in some cases unexpectedly deflected through large angles from their original direction of motion. Rutherford concluded that the only explanation for this remarkable effect was that these α particles had been subjected to a very intense electrostatic field. It had, of course, been realized that the atom, being as a whole electrically neutral, must have a positive charge to cancel the negative charge associated with its electrons. The intense field required to cause the large-angle deflections implied that the positive charge of the atom was concentrated in a volume very much smaller than that of the atom itself. This led to the concept of the atom as consisting largely of empty space, with a small nucleus containing all the positive charge of the atom, and nearly all its mass, at the center, the negatively charged electrons being held in association with the nucleus by the Coulomb electrostatic force. The simplest atom was hydrogen, the nucleus of which consisted of a single positive charge (*proton*).

Alpha-particle scattering experiments on a number of elements showed that the charge on the nucleus was always a multiple of the charge of the proton. It was, therefore, reasonable to conclude that any particular element is characterized by having a certain fixed number of protons in the nucleus, this number (the *atomic number, Z*) being the same as the number of electrons in the atom (the latter determining the chemical behavior of the element). It was clear, however, that additional information was necessary to complete the picture, since, for the typical atom, the combined mass of the number of protons required to supply the known nuclear charge was somewhat less than one half of the observed mass of the nucleus.

With the discovery, by Chadwick in 1932, of the *neutron*, an uncharged particle of mass closely equal to that of the proton, it was possible to provide a model for the nucleus which is essentially that accepted today. The nucleus is considered to be composed of Z protons and N neutrons, the total number of particles ($A = N + Z$) being known as the *atomic mass number*. The term *nucleon* is used to describe both types of particle in the nucleus. The nucleus of beryllium, for example, contains four protons and five neutrons, and therefore has a mass approximately nine times that of the hydrogen nucleus.

It is found, however, that the majority of elements have nuclear masses which do not lie close to integral multiples of the mass of the hydrogen nucleus. The reason is the existence of *isotopes*, i.e., atoms whose nuclei contain the same number of protons, but different numbers of neutrons. The element uranium, for example, exists as a mixture of three isotopes, which are always found in the proportions given in Table 1.1.

Another example of importance in reactor physics is hydrogen, which occurs in nature as a mixture of two isotopes. The nucleus of ordinary, or light, hydrogen consists of a single proton, while heavy hydrogen, or deuterium, has a neutron as well as a proton in the nucleus; the latter isotope is relatively rare, occurring as only about one atom in 6000 of natural hydrogen. In addition to these two stable isotopes, a radioactive isotope of hydrogen, tritium, with two neutrons and a proton in the nucleus, can be

Table 1.1. Isotopic Ratios in Natural Uranium

Atomic number (Z)	Neutron number (N)	Atomic mass number (A)	Proportion ($\%$)
92	142	234	0.0057
92	143	235	0.720
92	146	238	99.275

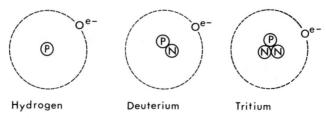

Hydrogen   Deuterium   Tritium

Fig. 1.1. Isotopes of hydrogen.

created artificially (see Fig. 1.1). This isotope is designated as $_1\text{H}^3$, where the number to the left of the chemical symbol indicates the number of protons and the number to the right the total number of nucleons in the nucleus (atomic mass number). Since the atomic number is implied in the element designation, it is frequently more convenient to omit the former, referring to the uranium isotopes, for example, as U-234, U-235, and U-238.

It is found from experiment that the radius of the nucleus is given approximately by the formula

$$r = r_0 A^{1/3} \tag{1.1}$$

where A is the atomic mass number and $r_0 = 1.35 \times 10^{-15}$ m. Based on this formula, the radius of the most massive naturally occurring nucleus, $_{92}\text{U}^{238}$, is calculated as 8.4×10^{-15} m.

1.3. Radioactivity

1.3.1. α, β, and γ Emission

The α and β particles discovered by Rutherford are now known to be emitted from the nucleus of the atom, thereby changing the atomic number and thus the identity of the element. In the process of α *emission*, the emitted particle consists of a nucleus of helium (two protons + two neutrons), leading to a reduction of the atomic number by two units. An example is the decay of U^{238} to form the *daughter nucleus* thorium-234:

$$_{92}\text{U}^{238} \rightarrow _{90}\text{Th}^{234} + _2\alpha^4$$

The process of β *emission* involves the spontaneous change of a neutron in the nucleus into a proton and an electron, the latter appearing as the emitted β particle, leading to an increase of one unit in the charge on the nucleus. The β decay of indium-116, for example, may be written as

$$_{49}In^{116} \rightarrow {}_{50}Sn^{116} + {}_{-1}\beta^0$$

(the mass of the β particle being effectively zero compared to that of a proton or neutron).

A much rarer form of radioactivity is *positron emission*, where a proton in the nucleus changes into a neutron, the unit of positive charge being removed by the emission of a positron, a particle with the same mass as the electron, but with a charge of opposite sign. An example is the decay of cobalt-56 to iron-56:

$$_{27}Co^{56} \rightarrow {}_{26}Fe^{56} + {}_{1}\beta^0$$

The emission of the alpha or beta particle is frequently followed almost instantaneously by the emission of a high-energy *photon* of radiation, called a *gamma* (γ) *ray*. Gamma rays are electromagnetic radiations of the same basic nature as the more familiar photons of light or x radiation, but are distinguished by their higher energy.

The application of quantum mechanics to the electrons surrounding the nucleus, notably by Schrödinger in 1927, showed that these electrons are limited in their possible energies to certain well-defined states or levels, characterized by quantum numbers which arise from the application of suitable boundary conditions to the so-called Schrödinger equation. In the normal atom, the electrons will occupy the states of lowest energy, but an electron may make a quantum jump to a higher level if the appropriate energy requirement is supplied by some external process, such as collision or the absorption of radiation of suitable frequency. An electron excited into a state above its normal, or ground, level can subsequently drop back to a lower level by emission of radiation of the appropriate frequency. If the electron jumps from an initial quantum state of energy, E_i to a final state of energy E_f, the frequency (ν) of the photon of radiation emitted is given by

$$h\nu = E_i - E_f \qquad (1.2)$$

where h is Planck's constant [6.626×10^{-34} joule seconds (J s)].

The energy levels of the simplest atom, that of hydrogen, are illustrated in Fig. 1.2. The energy unit used is the *electron volt* (eV), which is defined as the kinetic energy gained by a particle with a charge equal to that of

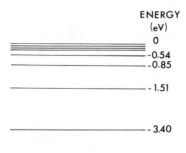

ENERGY
(eV)

Fig. 1.2. Energy levels of the hydrogen atom. The zero of the energy scale corresponds to the just-ionized state.

the electron $(1.6 \times 10^{-19}$ C) when it is accelerated through an electrostatic potential difference of 1 V. Commonly used multiples of the eV are the kilo-electron-volt (keV) $(10^3$ eV) and the mega-electron-volt (MeV) $(10^6$ eV). It will be seen that the possible energy states become much more closely packed together with increasing electron energy, tending to a limit at an energy of 13.58 eV above the ground state. If the electron in the hydrogen atom acquires an energy in excess of 13.58 eV, it will become completely dissociated from the atom, which is then said to be *ionized*. The quantity of energy required to ionize the atom is known as the *ionization potential*.

For atoms with more than one electron, the electrons are arranged in shells at varying distances from the nucleus, the electrons in the inner shells being the most closely bound to the nucleus. In this case, the outermost electrons are the ones which give rise to radiation in the optical region of the spectrum, while electrons jumping from one inner level to another emit x-ray photons which have considerably greater energies, up to more than 100 keV.

In a similar way, we find that the neutrons and protons inside the nucleus are restricted to occupying discrete energy levels. Because the forces holding nucleons together are much stronger than the electrostatic forces which bind the electrons to the nucleus, the average spacing between the levels in the nucleus is considerably greater than the spacing of the atomic energy levels.

The reason for the γ radiation accompanying α or β emission is that the daughter nucleus is frequently left in one of its excited states and will get rid of its excess energy by jumping, directly or by way of intermediate states, to the ground state. This process is illustrated for the decay of the α-active nucleus $_{92}U^{235}$ to $_{90}Th^{231}$ in Fig. 1.3.

The total amount of energy released when the disintegration takes place, which can appear in the form of kinetic energy shared between the α particle and the recoiling nucleus, may be calculated from a knowledge of the masses of the parent and the daughter nucleus. According to the mass–energy equivalence principle of Einstein, the energy associated with any mass m is of magnitude $E = mc^2$, where c is the speed of light (e.g., a 1-kilogram (kg) mass is equivalent to 9×10^{16} J of energy). For the decay

$$_{92}U^{235} \rightarrow {}_{90}Th^{231} + {}_2\alpha^4$$

the energy released is found by subtracting the sum of the masses of the Th^{231} nucleus and the α particle from the mass of the nucleus of U^{235}, and multiplying by the factor c^2. This quantity of energy will be the amount of kinetic energy shared between the α particle and the Th^{231} nucleus in the case where the disintegration leaves the daughter nucleus in its ground state.

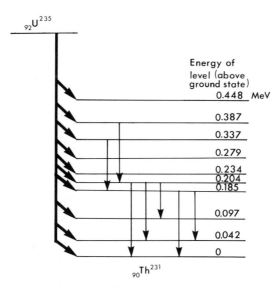

Fig. 1.3. The α decay of the nucleus $_{92}U^{235}$. The oblique arrows indicate the possible transitions by α emission to the levels of $_{90}Th^{231}$, while the vertical arrows represent the principal γ rays emitted in the de-excitation of the excited states of the daughter nucleus.

If the daughter nucleus is left in one of its excited states, the energy of this state above the ground level will have to be subtracted from the calculated energy release to find the kinetic energy given to the products. The energy spectrum of the α particles therefore consists of a series of discrete lines, each of the possible energy values corresponding to a transition from (usually) the ground state of the parent nucleus to one of the states of the daughter (see Fig. 1.4a).

In contrast, the β particles emitted from a β-active nucleus exhibit a continuous energy spectrum, including all energies from zero up to a maximum which corresponds to the value (E_{max}) predicted from the mass difference between the parent and daughter nucleus (see Fig. 1.4b). The average energy of the β particles is approximately equal to $0.3\,E_{max}$. The reason for the continuous spectrum is that the β-decay process involves the simultaneous emission of two particles, one being the β particle (electron) itself and the other a particle known as the *neutrino*. The available energy is therefore shared, in varying degree, between the β particle and the neutrino. The existence of the neutrino is of significance in reactor physics in that it is responsible for carrying away some of the energy released in the processes associated with nuclear fission, on account of the fact that neutrinos effectively escape from the reactor without losing any of their energy, because of their very weak interaction with matter.

1.3.2. The Systematics of Radioactive Decay

Each radioactive decay proceeds at a rate characteristic of the particular isotope involved. The basic law of radioactivity is that the rate at which disintegrations occur in a sample of a given radioactive isotope is proportional to the number of atoms (N) in the sample, that is,

$$\frac{dN}{dt} = -\lambda N \qquad (1.3)$$

where λ, which is a characteristic of the isotope involved, is known as the *decay constant*.

The decay rate, λN, is known as the *activity* of the sample. One unit of activity is the *curie* (Ci) defined as a rate of 3.7×10^{10} disintegrations per second. Commonly used subunits are the millicurie (mCi), equal to 10^{-3} Ci, and the microcurie (μCi), equal to 10^{-6} Ci. The curie is now being replaced as a unit of radioactivity by the *becquerel* (Bq), where 1 Bq is simply equal to a rate of one disintegration per second.

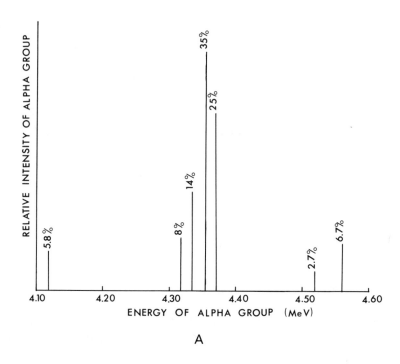

A

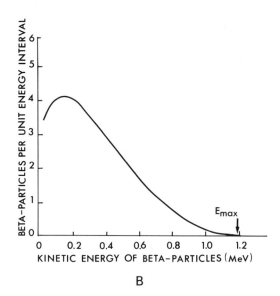

B

Fig. 1.4. (a) Energy spectrum of α particles emitted in the decay of $_{92}U^{235}$. (b) Energy spectrum of β particles emitted in the decay of a typical β-active nucleus.

Equation (1.3) may be integrated to yield an expression for the number of radioactive nuclei still remaining at time t, relative to N_0, the number present at $t = 0$:

$$N = N_0 \, e^{-\lambda t} \tag{1.4}$$

The activity of a given sample (which is proportional to N) therefore decays exponentially. The rate of decay is conventionally specified in terms of the *half-life*, defined as the time required for the activity to decrease by a factor of 2. It may readily be shown from equation (1.4) that the relation between the half-life ($T_{1/2}$) and the decay constant is

$$T_{1/2} = \frac{\ln 2}{\lambda} = \frac{0.693}{\lambda} \tag{1.5}$$

The naturally occurring radioisotopes, mainly associated with the complicated decay chains of the elements uranium and thorium, show a wide range of half-lives, from the 3×10^{-7} s of Po^{212} to the 1.4×10^{10} year (yr) half-life of Th^{232}. The very existence of the natural radioactive series today, despite the long time which has elapsed since the formation of the elements in some original cosmogenic event, is a consequence of the very long half-lives of U^{235}, U^{238}, and Th^{232}. The isotope U^{238}, for example, is the first member of a long chain comprising a total of 16 radionuclides, each produced by the α or β decay of the member preceding it. The chain is said to be in a state of *secular equilibrium*; since the half-life of the U^{238} (4.5×10^9 yr) is so much greater than that of any other member, the decay of the whole chain essentially proceeds at the same rate as the U^{238} which feeds it.

Example 1.1. A sample of pure $_{27}Co^{60}$ has an activity of 10^8 disintegrations per second. If the half-life of the isotope is 5.2 yr, calculate the mass of the sample.

The decay constant of $_{27}Co^{60}$ is

$$\lambda = \frac{0.693}{T_{1/2}} = \frac{0.693}{5.2 \times 3.154 \times 10^7} = 4.225 \times 10^{-9} \, \text{s}^{-1}$$

The activity is equal to λN, where N is the number of atoms in the sample. Hence

$$N = \frac{10^8}{4.225 \times 10^{-9}} = 2.367 \times 10^{16}$$

Since a mass of 60 grams (g) of $_{27}Co^{60}$ contains 6.022×10^{23} atoms (Avogadro's number), the mass of the sample is

$$M = \frac{2.367 \times 10^{16}}{6.022 \times 10^{23}} \times 60 = 2.36 \times 10^{-6} \, g \qquad \square$$

1.4. Mass Defect and Binding Energy

The close binding of a number of nucleons to form a nucleus makes it necessary to postulate a new kind of force, the *nuclear force*, since the gravitational force is many orders of magnitude too small, and the mutual electrostatic interaction between the protons in the nucleus is a disruptive rather than cohesive effect. The binding of nucleons is due to an extremely short-range force which arises from the continuous exchange of particles known as *pi* (π) *mesons* between the nucleons in the nucleus. While it is not necessary for the present purpose to consider the detailed nature of the nuclear force , a knowledge of the energy changes arising from rearrangement of the constituents of the nucleus is of fundamental importance in reactor physics. As an example, the effect of adding an extra neutron to the nucleus of U^{235} is very different from that produced by adding it to U^{238}.

For all stable nuclei, it is found that the mass of the nucleus is less than the sum of the masses of its constituent nucleons, considered independently. This difference is known as the *mass defect*. For a nucleus of mass number A, containing Z protons and N neutrons, the mass defect may be defined by the relation

$$\Delta M = Zm_p + Nm_n - {}_ZM^A \qquad (1.6)$$

where m_p and m_n are, respectively, the masses of an individual proton and neutron and $_ZM^A$ is the mass of the nucleus concerned. (It should be noted that the standard mass normally quoted for an isotope is the mass of the neutral atom; in order to obtain the mass of the nucleus, the combined mass of the electrons has to be subtracted.)

By the Einstein mass–energy equivalence principle, the mass defect is the mass equivalent of the work done by the nuclear force in bringing the nucleons together to form the nucleus or, alternatively, of the loss of mutual potential energy of the nucleons as a result of their accretion. The energy equivalent of the mass defect is called the *binding energy* of the nucleus:

$$\text{B.E.} = (Zm_p + Nm_n - {}_zM^A)c^2 \tag{1.7}$$

where c, as before, is the speed of light.

Another important parameter is the *binding energy per nucleon* for the nucleus, which is obtained by dividing by the number of nucleons:

$$\text{B.E./nucleon} = \frac{(Zm_p + Nm_n - {}_zM^A)c^2}{A} \tag{1.8}$$

This quantity represents the average energy which has to be supplied to remove a nucleon from the nucleus, or, alternatively, the average energy given to the nucleus when a free proton or neutron is absorbed into it.

Use of the relations given above to calculate mass defect or binding energy requires a very precise knowledge of the mass of the nucleus in question. Nuclear masses can be measured with high precision either by mass spectrometry or from a knowledge of nuclear reaction energies. While absolute masses can be substituted into the equations above, it is more convenient to use a system of units more directly related to nuclear physics. The mass unit employed is the *atomic mass unit* (u), which is defined as being one twelfth of the mass of the neutral atom of the isotope ${}_6C^{12}$.

Some useful conversion factors involving energy and mass units are summarized in Table 1.2; for a more detailed listing of fundamental constants and conversion factors see Appendix (Tables A.1 and A.2).

The use of nuclear masses in obtaining energy releases is illustrated by the two examples below, one for the case of the α-particle emission and the other for the calculation of the binding energy of a specific nucleus. A selection of the more important atomic masses is given in Appendix (Table A.3). It should be noted that the masses quoted are those of the neutral atom.

Example 1.2. Calculate the release of kinetic energy (of the α particle and the recoiling nucleus) which occurs in the decay

$$_{92}U^{235} \rightarrow {}_{90}Th^{231} + {}_2\alpha^4$$

when the transition takes place to the state of ${}_{90}Th^{231}$ which de-excites by the emission of a γ ray of energy 0.204 MeV (see Fig. 1.3).

The masses of the nuclei involved, and that of the α particle (nucleus of ${}_2He^4$), can be obtained by subtracting the appropriate number of electron

Table 1.2. Conversion Factors for Energy and Mass Units

1 u = 1.6605656 × 10^{-27} kg
Mass of proton = 1.00727647 u = 1.6726486 × 10^{-27} kg
Mass of neutron = 1.008665012 u = 1.6749544 × 10^{-27} kg
Mass of electron = 5.4858026 × 10^{-4} u = 9.109534 × 10^{-31} kg
1 MeV = 1.6 × 10^{-13} J
Energy equivalent of 1 u = 931.50 MeV

masses from the atomic masses given in Table A.3. Using the electron mass given above, the masses of the $_{92}U^{235}$ and $_{90}Th^{231}$ nuclei and the α particle are $_{92}U^{235}$, 234.99345586 u; $_{90}Th^{231}$, 230.98692634 u; $_2\alpha^4$, 4.00150609 u.

Subtracting the latter two masses from that of the $_{92}U^{235}$ nucleus gives the mass loss as

$$\Delta m = 0.0050233 \text{ u}$$

Hence, energy release = 0.0050233 × 931.5 = 4.679 MeV.

Since, in the transition being considered, 0.204 MeV is taken up in the γ-ray emission, the release of kinetic energy is 4.679 − 0.204 = 4.475 MeV.

(By applying conservation of momentum to the two recoiling masses, it can easily be shown that the α particle itself acquires 98.3% of the total kinetic energy released.)   □

Example 1.3. Calculate the binding energy per nucleon in the nucleus of $_{92}U^{235}$.

As in the previous example, the mass of the *nucleus* of $_{92}U^{235}$ is 234.99345586 u. The mass defect is, by equation (1.6),

$$\Delta M = Zm_P + Nm_N - {}_zM^A$$
$$= [(92 \times 1.00727647) + (143 \times 1.008665012) - 234.99345586] \text{ u}$$
$$= 1.91508 \text{ u}$$

The total binding energy is therefore equal to 1.91508 × 931.5 = 1783.9 MeV. The binding energy per nucleon is then 1783.9/235 = 7.59 MeV.   □

The binding energy per nucleon is a crucial parameter in considering the possible useful release of energy in nuclear processes, since it is this quantity that determines whether energy is released or consumed when nuclei are broken up or combined. The way in which the binding energy per nucleon varies with atomic mass number is illustrated in Fig. 1.5.

It will be seen that the greatest binding energies occur for mass numbers in the medium-mass range, the curve having a broad maximum around $A = 60$. It drops steeply on the low-mass side of the maximum, although there are a few nuclei (e.g., He^4, Be^8, C^{12}, O^{16}) which exhibit binding energies markedly above the general curve in this region. For mass numbers above the maximum, the binding energy per nucleon shows a steady decrease, the value for $_{92}U^{238}$, the heaviest naturally-occurring isotope, being some 1.22 MeV lower than the maximum of 8.79 MeV. The overall form of the curve suggests two mechanisms which should lead to the release of energy in nuclear processes, since any move towards a region of greater stability, i.e., towards nuclei of greater binding energies, will result in the liberation of energy. One such process is the break-up of a massive nucleus into two nuclei of intermediate mass (*fission*) and another is the combination of light nuclei (*fusion*). It is with the former process that we shall be concerned in the present book.

It is possible to derive an expression which describes the general behavior of the binding energy curve on the basis of a particularly simple

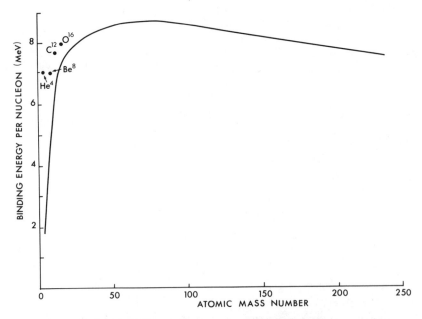

Fig. 1.5. Binding energy per nucleon as a function of the atomic mass number.

model of the nucleus. It is known that the nuclear force between the nucleons is a particularly short-range force, unlike the electrostatic force between the protons, which obeys the inverse-square law. The short-range nature of the force suggests an analogy between the nucleus and a liquid drop, in which the molecules are strongly attracted only by those others in their immediate vicinity. Accordingly, if we imagine nucleons being added successively to a nucleus, each nucleon added will form the same fixed number of bonds with those already present, since it is essentially limited to interacting only with the nucleons with which it is actually in contact. Since each nucleon in this way will therefore contribute the same amount to the binding energy of the nucleus, we may assume, as a first approximation, that the total binding energy is simply proportional to the total number of nucleons present, that is, to the atomic mass number, A, i.e.,

$$\text{B.E.} = a_1 A \qquad (1.9)$$

where a_1 is a numerical constant.

There are two principal corrections which can immediately be applied to compensate for obvious omissions in this oversimplified picture. The first of these is for the so-called *surface tension* effect, which is of particular importance for nuclei in the low-mass range. We have assumed that all nucleons in the nucleus have the same number of energy bonds. In practice, the nucleons at the surface of the nucleus, having fewer immediate neighbors than those in the interior, will form fewer bonds and therefore contribute less to the total binding energy. The latter will consequently be reduced by a term which will be proportional to the surface area of the nucleus. Since the nucleons are very closely packed, owing to the short-range nature of the nuclear force, the volume of the nucleus tends to vary directly with the number of nucleons which it contains. The radius of the nucleus is therefore proportional to the cube root of the atomic mass number, i.e., $r \propto A^{1/3}$, and the surface area will vary as $A^{2/3}$. The inclusion of the surface tension effect thus results in the addition of a correction term, which may be written as $-a_2 A^{2/3}$, to the right-hand side of equation (1.9), where a_2 is a (positive) numerical constant.

The second obvious correction takes account of the mutual electrostatic repulsion of the protons. According to the Coulomb force law, the mutual electrostatic potential energy of two protons, each having a charge of e coulombs, separated by a distance r, is proportional to e^2/r. To a reasonable approximation, we may consider each of the Z protons as interacting with a total of $(Z - 1)$ other protons, at an average distance equal to r, the radius of the nucleus. Hence each proton will contribute a term proportional to $(Z - 1)/r$, i.e., to $(Z - 1)/A^{1/3}$. Taking all of the Z protons into account, the total binding energy is reduced by a term which may be written as $-a_3 Z(Z - 1)/A^{1/3}$, where a_3, again, is a numerical constant.

Taking account of these two effects, the binding energy may be written, to a second approximation, in the form

$$\text{B.E.} = a_1 A - a_2 A^{2/3} - a_3 Z(Z - 1)/A^{1/3} \tag{1.10}$$

This expression reproduces reasonably well the behavior of the curve in Fig. 1.5. The rapid drop in binding energy for low mass number is associated with the second term in equation (1.10), and the more gradual reduction at large A with the third. Further terms can be added to include more subtle effects, such as the preference for equal numbers of protons and neutrons (e.g., He^4, O^{16}) and the effect of pairing of nuclear spins, which makes nuclei with even numbers of both protons and neutrons (even–even nuclei) particularly stable in comparison with others.

While the liquid drop model of the nucleus is successful in deriving the general form of the binding energy curve, there are features which can only be explained by adopting a more complex model. It was noted earlier that certain nuclei have binding energies per nucleon which lie considerably above the average in their region of the curve. It is found that nuclei are particularly stable if either the proton or neutron number (or both) is equal to one of a set of so-called *magic numbers* (2, 8, 20, 28, 50, 82, 126). This suggests, by analogy with the well-known explanation of the chemical stability of the inert gases in terms of the closure of electron shells, that some nuclear properties may be interpreted on the assumption that the nucleons also exhibit a shell structure. Neutrons and protons are assumed to occupy separate energy states defined by sets of quantum numbers as in the atomic electron case. The operation of the Pauli principle, that no two protons (or neutrons) can share the same set of quantum numbers, can be shown to lead to shell closure when either N or Z is equal to one of the observed values of the magic numbers.

Nuclei with atomic number $Z > 92$ do not exist in nature. The reason for the relatively high instability of such nuclei is the increasing importance of the Coulomb repulsion between the protons in reducing the binding energy. Such nuclei, which may be created artificially, have a strong tendency towards achieving greater stability by the emission of α particles, thereby reducing the nuclear charge. Another important result of the Coulomb repulsion is a steady increase in the neutron/proton ratio with increase of nuclear mass number (see Fig. 1.6).

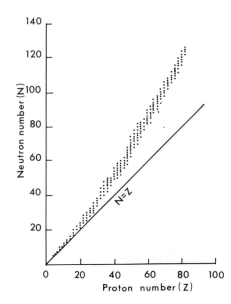

Fig. 1.6. Distribution of neutron and
proton numbers for the stable nuclei.

The distribution of stable nuclei follows a well-defined curve which
departs progressively from the line $N = Z$ as A increases. The narrow limits
of stability arise from two competing effects. While an increase in neutron/
proton ratio tends to reduce the mutual repulsion term, the additional
neutrons can only be added at the expense of forcing them into higher and
higher energy states. A relatively small departure from the observed line of
maximum stability in Fig. 1.6 therefore makes the nucleus unstable owing to
the reduction in binding energy due to having an excessive number of either
protons or neutrons.

1.5. Interactions of Neutrons with the Nucleus

As indicated earlier, the protons and neutrons inside the nucleus are
restricted to occupying discrete energy levels. A wealth of experimental data
on the energies and other characteristics of such levels has been obtained
from the study of nuclear reactions induced by bombarding elements with
charged particles (e.g., α particles or protons) of high energy (~ 1 MeV or
higher). The bombarding particle must have a high kinetic energy to
overcome the strong Coulomb repulsion of the positively charged nucleus.

Immediately following the discovery of the neutron, it was realized that
the new particle would be an ideal tool for inducing nuclear reactions, since

there would be no electrostatic repulsion acting to prevent it approaching the nucleus. Consequently, even neutrons of very low kinetic energy can interact readily with any nucleus with which they come into contact. The probability of the neutron being captured by the nucleus is a function of the kinetic energy with which it approaches it. On account of its importance in reactor physics, the main features of the interaction of neutrons with nuclei will be considered in outline.

A concept of great relevance in considering the interaction of neutrons of low or moderate energy with nuclei is that of the *compound nucleus*. When the neutron interacts with the nucleus, it is first captured by the nucleus (Z, N) to form the heavier nucleus $(Z, N + 1)$. The lifetime of this compound nucleus (typically 10^{-14} s) is long on the nuclear time scale, i.e., it is much longer than the time which the neutron would have taken to travel through a distance equal to the nuclear diameter, which is of the order of 10^{-21} s for a 1-MeV neutron (of velocity approximately 10^7 ms^{-1}) incident on a nucleus of diameter 10^{-14} m.

The absorption of the neutron involves work being done by the nuclear forces and this work appears as excitation energy of the compound nucleus; the amount of energy added in this way is equal to the binding energy of the neutron in the compound nucleus. The total excitation energy, E, of the compound nucleus above its normal, or ground state, is then equal to the sum of the binding energy of the neutron (E_B) and the kinetic energy due to the relative motion of the neutron and the target nucleus before the collision. The latter quantity has to be evaluated in the center-of-mass coordinate system, in which the neutron and the nucleus approach one another with equal and opposite momenta. Following the collision, the compound nucleus will therefore be at rest in the center-of-mass system, so that the total kinetic energy (E_K) of the colliding bodies relative to the center of mass will be totally converted into internal excitation energy of the compound nucleus.

The probability of the neutron being captured by the target nucleus increases sharply when the excitation energy $(E_B + E_K)$ happens to coincide with one of the intrinsic energy levels of the compound nucleus, i.e., for $E = E_1, E_2 \ldots$ (see Fig. 1.7). Consequently, as the kinetic energy of the bombarding neutron is steadily increased from zero, it will be found that the probability of its capture will pass through a series of maxima at the incident energies for which the above criterion is satisfied. This condition is known as *resonance*, by analogy with other familiar physical phenomena, such as the turning of an *RC* circuit.

In practice, the resonance condition for a given level can be achieved over a finite spread of incident energy values, since the energies of the

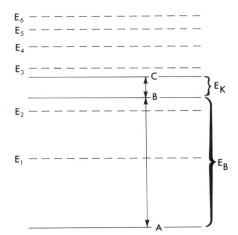

Fig. 1.7. Neutron resonance capture. The probability of capture is greatest when the sum of the binding energy of the neutron and the kinetic energy before the collision (this sum being represented by the line AC) coincides with one of the energy levels of the compound nucleus (dotted lines).

quantum levels of the nucleus are not sharply defined, but are characterized by a finite *level width*, Γ. The level width is linked through the Heisenberg uncertainty principle to the *mean lifetime*, τ, of the level, which is the average period of time for which the nucleus will remain in the excited state before de-excitation takes place by emission of either a particle or a photon of radiation.

The relation between τ and Γ is

$$\Gamma = \frac{\hbar}{\tau} = \frac{1.055 \times 10^{-34}}{\tau} \tag{1.11}$$

where $\hbar = h/2\pi$, h being Planck's constant. It is most convenient to express the level width Γ in eV, when the relation becomes

$$\Gamma = \frac{0.7 \times 10^{-15}}{\tau} \tag{1.12}$$

When the incident neutron is of low energy, the excitation energy of the compound nucleus arises almost entirely from the binding energy of the neutron (approximately 8 MeV for nuclei of medium or high mass). The typical separation of energy levels in the compound nucleus at this excitation energy is some tens of eV in the medium-mass range, decreasing to the order of a few eV for nuclei of very high mass. This is appreciably larger than the typical magnitude of the level width at these energies, and consequently marked resonance peaks will be observed at low incident neutron energies.

For higher neutron energies, (~ 1 MeV) the ratio of level width to level spacing in the compound nucleus increases to such an extent that the concept of sharp resonance peaks is no longer relevant.

Since the quantity Γ is inversely proportional to the time for which the compound nucleus exists before de-excitation, it is in fact a measure of the probability, per unit time, of the de-excitation taking place. In general, there will be a number of possible ways in which this may occur; for example, the compound nucleus may emit the same, or another neutron, or a charged particle, or it may lose its excitation energy by the emission of a quantum of radiation (γ ray). Each of these possible processes is characterized by its own *partial width*, Γ_i, which is proportional to the relative probability of the particular reaction taking place. The parameter Γ is then the *total level width*, and is the sum of all the partial widths corresponding to the possible modes of break-up of the compound nucleus, i.e.,

$$\Gamma = \sum_i \Gamma_i \qquad (1.13)$$

The way in which the probability of neutron capture and subsequent break-up of the compound nucleus varies with neutron energy for the various elements commonly present in the nuclear reactor is the basic information which one requires for practical calculation of the properties of a chain-reacting system. The quantitative measure of the probability of a given reaction is the appropriate nuclear cross section, which is dealt with in the following section.

1.6. Cross Sections for Neutron Reactions

In a nuclear reactor we have to deal with neutrons covering a wide range of kinetic energies, from about 10 MeV down to 0.001 eV. The neutrons are in continuous collision with the nuclei of the fuel and other materials present in the reactor. As the result of the collision of a neutron with an atomic nucleus, a variety of possible reactions may take place. We may illustrate by considering the collision of a neutron with a nucleus of Al^{27}, when the possible reactions are as follows (see Fig. 1.8):

$$_{13}Al^{27} + {}_0n^1 \rightarrow {}_{13}Al^{27} + {}_0n^1$$
$$_{13}Al^{27} + {}_0n^1 \rightarrow {}_{13}Al^{27*} + {}_0n^1$$
$$_{13}Al^{27} + {}_0n^1 \rightarrow {}_{13}Al^{28} + \gamma$$

$$_{13}Al^{27} + {_0}n^1 \rightarrow {_{12}}Mg^{27} + {_1}p^1$$
$$_{13}Al^{27} + {_0}n^1 \rightarrow {_{11}}Na^{24} + {_2}\alpha^4$$
$$_{13}Al^{27} + {_0}n^1 \rightarrow {_{13}}Al^{26} + 2\,{_0}n^1$$

The first of these reactions is known as *elastic scattering* (n, n). In this case, there is no change either in the identity of the struck nucleus, or in its internal energy, as a result of the collision. The process is simply analogous to an elastic collision between two bodies in classical physics, where the total momentum and total kinetic energy are both conserved, although there will in general be a transfer of kinetic energy from one body to the other. The process is of fundamental importance in nuclear physics because the primary mechanism by which the high-energy neutrons produced by the fission process can be slowed down is by elastic collision with the nuclei of the reactor constituents.

The process of elastic scattering can take place either with or without the formation of a compound nucleus. Elastic scattering without formation of a compound nucleus is known as *potential elastic scattering*, and can take place at any neutron energy. As explained in the previous section, elastic scattering involving the formation of a compound nucleus can take place only when the energy of the incident neutron is such as to produce resonance with one of the intrinsic levels of the compound nucleus; this process is consequently known as *resonance elastic scattering*. Since the neutron, once it has been absorbed to form the compound nucleus, spends a fairly long time, on the nuclear scale, colliding with the other nucleons, the neutron that is eventually ejected may not be the same as that originally absorbed; this, however, is of no importance so far as the overall effect of the scattering is concerned.

It is possible that, after ejection of the neutron from the compound nucleus, the residual nucleus, in this case Al^{27}, may be left in an excited state, rather than in the ground state from which it was raised by the neutron absorption. The excited nucleus, denoted by $Al^{27}*$, will then decay by γ-ray emission. Since the γ-ray energy is obtained at the expense of the scattered neutron, the latter will emerge with a lower kinetic energy than if the collision had been elastic. This process is known as *resonance inelastic scattering* (n, n'), and is the second listed above.

In order that the Al^{27} nucleus be left in an excited level, the initial kinetic energy of the colliding neutron and nucleus (measured in the center-of-mass system) has to be at least equal to the value of the excitation energy of the state. The neutron kinetic energy corresponding to excitation of the first level is known as the *threshold energy* for inelastic scattering from the particular nucleus. This situation is in contrast to that for elastic scattering, which can take place at any neutron energy. Since the level spacing decreases with increasing mass of the nucleus, the process is more important for the heavier nuclei. Inelastic scattering by U^{238}, for example, is important in nuclear reactors, since the threshold energy is only some 44 keV, while elastic scattering is much more important for carbon, where the first excited level of C^{12} is 4.4 MeV above the ground state.

The next three processes shown in Fig. 1.8 are the *neutron capture* reactions, designated as (n, γ), (n, p), and (n, α), respectively. In this type of reaction, the compound nucleus decays by the emission of either a γ photon or a charged particle. For neutrons of low energy, the first of these processes is usually the only one which is possible. The reason is that the Coulomb electrostatic barrier, which effectively prevents bombarding charged particles of low energy from getting into the nucleus, is equally effective in preventing the escape of charged particles such as a proton or α particle. At low incident neutron energies, processes involving the emission of a charged particle from the compound nucleus can occur only for certain light nuclei which combine a strongly exothermic reaction with a relatively low potential barrier. One of the most important examples is slow neutron absorption by the nucleus B^{10}, which constitutes some 19.8% of natural boron. The reaction in this case is an (n, α) reaction and may be written

$$_5B^{10} + _0n^1 \rightarrow _3Li^7 + _2\alpha^4$$

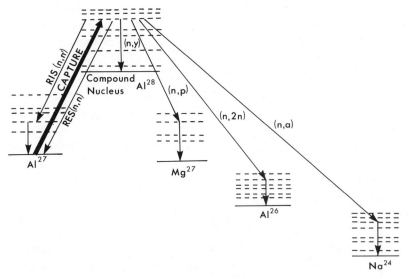

Fig. 1.8. Possible reactions due to a neutron collision with a nucleus of $_{13}Al^{27}$.

Example 1.4. Calculate the energy release in the reaction above, using the masses given in Table A.3.

Deriving the nuclear masses by subtracting the appropriate number of electrons, the masses involved are $_5B^{10}$, 10.010195098 u; $_0n^1$, 1.008665012 u; $_3Li^7$, 7.014358759 u; $_2\alpha^4$, 4.00150609 u.

Subtracting the two latter masses from the sum of the masses of the boron nucleus and the neutron gives a mass loss of

$$\Delta M = 2.9953 \times 10^{-3} \text{ u}$$

The energy release is therefore $2.9953 \times 10^{-3} \times 931.5 = 2.790$ MeV.

□

For all but light nuclei, the potential barrier is high enough that the compound nucleus formed by the capture of a slow neutron nearly always de-excites by the emission of a γ ray or of an elastically scattered neutron.

The nucleus produced by the (n, γ) reaction, which is known as *radiative capture*, is frequently radioactive. Since the result of the capture is to produce a nucleus with a neutron number greater by 1 than the neutron number of the stable isotope undergoing the reaction, the radioactive decay is generally by β emission, which essentially converts the additional neutron in the nucleus to a proton. The structural materials of a nuclear reactor, which are exposed continuously to bombardment by large numbers of neutrons, eventually acquire a high level of radioactivity, as a result of radiative capture reactions.

One other neutron-induced reaction, which is of importance in certain circumstances, is the $(n, 2n)$ reaction, where the capture of the original neutron leads to the emission of two neutrons from the compound nucleus. The most important example is that of the light element beryllium

$$_4Be^9 + _0n^1 \rightarrow _4Be^8 + _0n^1 + _0n^1$$

The reaction, which can only take place with incident neutrons of energy greater than 1.8 MeV, provides a significant enhancement of the neutron population in a reactor containing beryllium.

The last of the important neutron capture reactions, which is restricted in practice to nuclei at the very top end of the periodic table, is *nuclear fission*, where the excitation energy of the compound nucleus following neutron capture is sufficient to cause it to break up into two fragments of approximately equal mass. The detailed consideration of the fission process will form the subject of the next chapter.

It will be seen that, in a nuclear reactor where a considerable number of

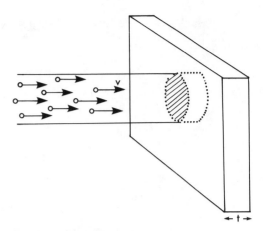

Fig. 1.9. Definition of nuclear cross section for neutron beam of density n neutrons per cm^2 and uniform neutron velocity v.

different elements are being subjected to bombardment by neutrons with a wide range of kinetic energies, it is essential to have some quantitative measure of the relative probability with which the various reactions described above will take place. The parameter used as a measure of the probability of a reaction is the *nuclear cross section*, which is defined as follows (see Fig. 1.9). Consider a neutron beam, where the particle density is n neutrons per cm^3, falling on a thin target of a given material. Let the target thickness be t, and the number of nuclei per cm^3 of the target material $= N$. Assuming that all the neutrons have the same velocity v, the number of neutrons striking unit area of the target per second is equal to the product nv. This quantity is known as the *neutron flux*, normally specified in neutrons per cm^2 per second. The number of nuclei per unit area of the target is equal to the product Nt. The *reaction cross section* for any nuclear reaction, σ, is then defined by the relation

$$R = Nt\sigma nv \qquad (1.14)$$

where R is the number of reactions of the type considered which take place per second per cm^2 of the target. Hence

$$\sigma = \frac{\text{reaction rate per unit area of target}}{\text{nuclei per unit area of target} \times \text{no. of neutrons striking unit area of target/second}} \qquad (1.15)$$

From equation (1.14), it is seen that R/nv, which represents the probability that any incoming neutron will react with one of the target nuclei, is equal to the product of σ and the total number of nuclei per unit area of the target. Hence, σ may be visualized as being the "target area" presented by each nucleus to the incoming neutrons. It is obvious that σ will not in general be equal to the physical cross-sectional area of the nucleus, since it will be different for each of the possible scattering or absorption reactions which the neutron can undergo. It will also be subject to very rapid variations depending on the velocity of the neutron, particularly in the region of a resonance.

For gold-197, for example, the (n, γ) cross section at the peak of the resonance at 4.9-eV neutron energy is in excess of 3×10^{-20} cm^2, while the "geometrical" area, πr^2, presented by the gold nucleus, is only of the order of 2×10^{-24} cm^2, so that the reaction cross section is vastly greater than the physical cross section of the nucleus. It is, in fact, only for very high neutron energies that the reaction cross section tends to be of the same order as the geometrical cross section.

Because of the very small areas involved, a special unit, the *barn* (b) is used for specifying the nuclear cross section. This is defined as

$$1\ \text{b} = 10^{-24}\ \text{cm}^2$$

The reaction cross sections of interest lie mostly in the range from 10^{-3}–10^4 b. Cross sections at the lower end of this range are frequently specified in millibarns (1 mb $= 10^{-3}$ b).

Each possible type of reaction which a neutron may undergo with the nucleus is associated with a specific cross section. The most important cross sections are σ_{se}, elastic scattering; σ_{si}, inelastic scattering; σ_{γ}, radiative capture; σ_f, fission; σ_p, (n, p) reaction; σ_α, (n, α).

The sum of the cross sections for all the possible reactions is known as the *total cross section*, defined as

$$\sigma_t = \sigma_{se} + \sigma_{si} + \sigma_{\gamma} + \sigma_f + \sigma_p + \sigma_\alpha + \cdots \qquad (1.16)$$

Another useful quantity is the *absorption cross section* (σ_a), which is the sum of the cross sections for all reactions which lead to absorption of the incoming neutron, i.e.,

$$\sigma_a = \sigma_{\gamma} + \sigma_f + \sigma_p + \sigma_\alpha + \cdots \qquad (1.17)$$

For the fissile nuclides, it is often useful to introduce a *capture cross section* (σ_c), incorporating all the absorptions which do not lead to fission, i.e.,

$$\sigma_c = \sigma_\gamma + \sigma_p + \sigma_\alpha + \cdots \tag{1.18}$$

Then, for a fissile nucleus

$$\sigma_a = \sigma_c + \sigma_f \tag{1.19}$$

In a nuclear reactor, we are usually concerned with the rate at which reactions of various kinds are taking place within some particular element of volume. For this purpose we introduce the *macroscopic cross section*, which, for a given nuclide, is defined as the product of the nuclear (or *microscopic*) cross section for the process involved and the number of atoms (N) of the particular nuclide per cm^3. The macroscopic cross section, Σ, is therefore

$$\Sigma = N\sigma \tag{1.20}$$

Where the region under consideration contains only a single element, the quantity N is equal to $\rho N_0/A$ where ρ is the density, N_0 is Avogadro's number, and A is the atomic mass number of the element. Hence the macroscopic cross section is

$$\Sigma = \frac{\rho N_0}{A}\sigma \tag{1.21}$$

Since Avogadro's number has the numerical value of 0.6×10^{24}, Σ may be conveniently calculated from the relation

$$\Sigma = 0.6\frac{\rho}{A}\sigma \tag{1.22}$$

where σ is the microscopic cross section in *barns*. The units of Σ are cm^{-1}.

As an example, the macroscopic total scattering cross section of U^{238}, for which the density is 19 g cm^{-3} and the nuclear cross section for scattering of slow neutrons is 13.8 b, is

$$\Sigma_s = \frac{0.6 \times 19}{238} \times 13.8 = 0.66 \text{ cm}^{-1}$$

Where the unit volume contains a mixture of nuclear species, the overall macroscopic cross section for any given process is specified by an extension of equation (1.20) as

$$\Sigma = N_1\sigma_1 + N_2\sigma_2 + \cdots N_n\sigma_n \qquad (1.23)$$

where $N_1, N_2, \ldots$ are the atom densities of the various constituents.

Example 1.5. Calculate the total macroscopic absorption cross section of a uniform mixture of U^{235} and graphite, for which the C/U^{235} atomic ratio is 10^4.

Since the number density of the U^{235} is so much lower than that of the carbon, the latter is effectively the same as for pure graphite and, using the data in Table A.4, we have for the macroscopic absorption cross section of the carbon (of density $1.6\ \text{g cm}^{-3}$)

$$\Sigma_{aC} = \frac{0.6 \times 1.6}{12} \times 3.4 \times 10^{-3} = 0.00027\ \text{cm}^{-1}$$

The number density of U^{235} atoms is less than that of carbon by a factor of 10^4 and hence the macroscopic cross section is

$$\Sigma_{aU} = 10^{-4} \times \frac{0.6 \times 1.6}{12} \times 681 = 0.00545\ \text{cm}^{-1}$$

The total macroscopic absorption cross section of the mixture is then

$$\Sigma_a = 0.0057\ \text{cm}^{-1} \qquad \Box$$

The number of absorption reactions taking place per second in $1\ \text{cm}^3$ of the volume in which the neutron flux is nv is

$$R = N\sigma_a nv = \Sigma_a nv \qquad (1.24)$$

as may be seen by letting the thickness of the target in equation (1.14) be 1 cm (with the assumption that the neutron absorption rate is low enough that the flux is not significantly depleted during the passage through the target).

An important quantity, which is related in a simple way to the macroscopic cross section, is the *mean free path* of a neutron which is moving through a medium. The mean free path is the average distance traveled by the neutron between the interaction events to which the cross section Σ refers. Consider a neutron flux of magnitude $(nv)_0$ incident on the surface of a medium for which the absorption cross section is Σ_a. Assume that, after the neutron beam has traveled a distance x into the medium, the magnitude of the flux has been reduced by absorption to the value nv (see Fig. 1.10).

From equation (1.14), the number of absorptions per unit cross-sectional area of the beam which take place per second in the volume element of thickness dx is $N\sigma_a nv\, dx$ and so the change in flux over the distance dx is

$$d(nv) = -N\sigma_a nv\, dx = -\Sigma_a nv\, dx$$

or

$$\frac{d(nv)}{nv} = -\Sigma_a\, dx \qquad (1.25)$$

The magnitude of the ratio $d(nv)/nv$ is simply the probability that a neutron is absorbed within the element dx. Integrating equation (1.25) yields

$$nv = (nv)_0\, e^{-\Sigma_a x} \qquad (1.26)$$

The probability that a neutron will penetrate to a depth x without being absorbed is given by the ratio of the remaining flux at that depth to the original flux $(nv)_0$, and is therefore equal to $e^{-\Sigma_a x}$.

The probability $p(x)\, dx$ that a neutron will penetrate to a depth x and will then be absorbed within the element dx is the product of the probabilities $e^{-\Sigma_a x}$ and $\Sigma_a\, dx$, i.e.,

$$p(x)\, dx = e^{-\Sigma_a x}\, \Sigma_a\, dx \qquad (1.27)$$

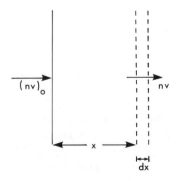

Fig. 1.10. Decrease of neutron flux in an absorbing medium.

The absorption mean free path is defined by the relation

$$\lambda_a = \frac{\int_0^\infty x p(x)\, dx}{\int_0^\infty p(x)\, dx} \tag{1.28}$$

Hence, from equation (1.27),

$$\lambda_a = \frac{\int_0^\infty x\, e^{-\Sigma_a x}\, \Sigma_a\, dx}{\int_0^\infty e^{-\Sigma_a x}\, \Sigma_a\, dx} = \frac{1}{\Sigma_a} \tag{1.29}$$

In a similar way, the mean free path for scattering, λ_s, can be shown to equal $1/\Sigma_s$, while the total mean free path for all events is $\lambda = 1/\Sigma_t$.

1.7. Cross Sections of Particular Importance in Reactor Physics

With a few exceptions, the scattering cross sections of elements for neutrons of low energy lie in the range from 1 to 10 b, while at very high energies the cross section reduces to a value more nearly related to the geometrical cross section of the nucleus. The typical behavior of the scattering cross section for elements of low atomic mass is illustrated for the cases of oxygen and carbon in Figs 1.11 and 1.12.

The scattering cross section at any energy arises from a combination of the slowly varying cross section due to potential scattering and a pronounced peak structure due to resonance scattering. For oxygen (O^{16}), the scattering cross section is approximately constant over the range from 0.1 eV to 0.4 MeV, with a few widely spaced resonances in the region from 0.5 up to 10 MeV. The resonance levels in the compound nucleus, O^{17}, corresponding to the peaks, are also shown in Fig. 1.11. These may be obtained by adding the kinetic energy of the neutron in the center-of-mass (c.m.) system to the binding energy of the neutron in the compound nucleus (obtained by subtracting the total binding energy of O^{16} from that of O^{17}).

The scattering cross section for carbon (C^{12}) shows a similar behavior, again characterized by the relatively widely spaced resonances typical of a low-mass nucleus. Above about 20 MeV the peaks are no longer visible, since

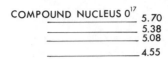

COMPOUND NUCLEUS O^{17}

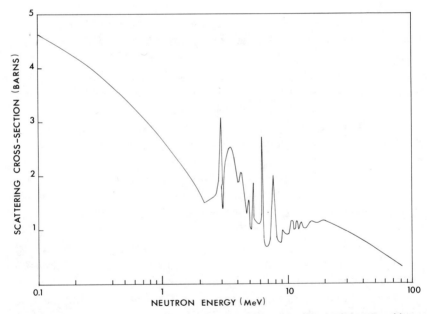

Fig. 1.11. Scattering cross section of oxygen [Based on *BNL 325*, Second Edition, Brookhaven National Laboratory (1964)].

Fig. 1.12. Scattering cross section of carbon [Based on *BNL 325*, Third Edition, Brookhaven National Laboratory (1976)].

at high energies the resonances of the compound nucleus tend to overlap and, in addition, the resolving power of the measuring equipment is usually no longer adequate to distinguish the individual resonances.

The closer-spaced resonance structure of the compound nuclei formed at the high-mass end of the periodic table leads to a pronounced difference in the form of the cross-section curve. The characteristic features of the absorption cross section of a heavy nucleus are illustrated in Fig. 1.13, which shows the (n, γ) absorption cross section of Au^{197}. The most prominent feature in the low-energy region is the strong resonance at 4.9 eV, followed by a succession of smaller peaks with an average spacing of about 15 eV. The energy above which the resonance structure disappears is now as low as 500 eV. Above this energy the absorption cross section decreases steadily,

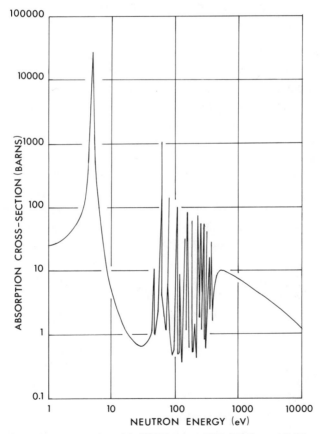

Fig. 1.13. Absorption cross section of Au^{197} [Based on *BNL 325*, Second Edition, Brookhaven National Laboratory (1966)].

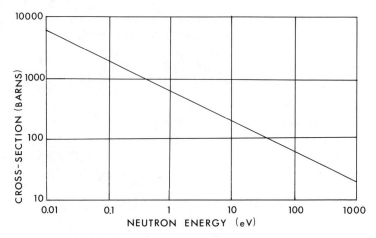

Fig. 1.14. Absorption cross section of boron.

reaching about 100 mb at 1 MeV. The low radiative capture cross section of elements at high neutron energy is a consequence of the high excitation energy of the compound nucleus, leading to an increased probability of decay by neutron emission (inelastic scattering) rather than γ emission.

A feature found in the absorption cross section of nearly all nuclei is the so-called "$1/v$ region" at low neutron energy, where the cross section varies inversely with the neutron velocity. When plotted on a logarithmic scale, the low-energy cross section of a "$1/v$ absorber" shows a linear variation with energy. An example of an element with a particularly extensive $1/v$ region is boron (Fig. 1.14). The absorption in this case is due almost entirely to the B^{10} isotope, which constitutes 19.8% of natural boron. Owing to its large cross section, boron is a useful material where high absorption is desirable, as for the absorber rods required for control of a nuclear reactor.

Another element of importance because of its high absorption cross section at low neutron energies is cadmium. The total cross section of cadmium at low energies is shown in Fig. 1.15. Owing to the presence of an unusually low-lying resonance at 0.18 eV, the absorption rises steeply below 1 eV and remains high down to zero energy. This feature has led to the use of cadmium as a selective filter when one wishes to expose elements to neutrons while excluding lower-energy neutrons.

The total cross section of U^{238}, which is of fundamental importance for virtually all reactors, is illustrated in Fig. 1.16. The intermediate energy region is characterized by the presence of a large number of resonances, with an average spacing of about 17 eV. An interesting feature is the clearly de-

fined minima which frequently occur at the low end of the resonances; these
are associated with the scattering contribution to the cross section, and arise
because of destructive interference between potential and elastic scattering
waves. The capture of neutrons in the resonances of U^{238} and Th^{232} is a
major factor in the neutron economy of reactors containing these isotopes.

One final cross section to which attention might be drawn is the
scattering cross section of hydrogen, which is important because of the
common use of water as a coolant and as a means of slowing down neutrons
in fission reactors. The behavior of the cross section of hydrogen in the form
of H_2O is modified by chemical binding effects when the neutron energy
becomes comparable with the binding energy of the hydrogen in the water

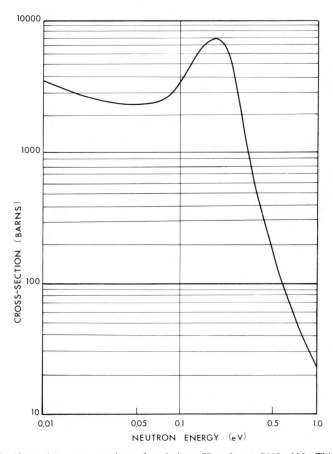

Fig. 1.15. Absorption cross section of cadmium [Based on *BNL 325*, Third Edition,
Brookhaven National Laboratory (1976)].

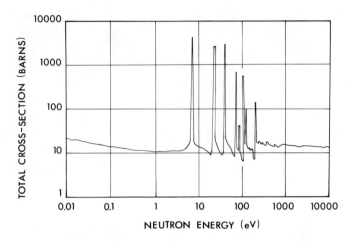

Fig. 1.16. Total cross section of $_{92}U^{238}$ [Based on *BNL 325*, Second Edition, Brookhaven National Laboratory (1965)].

molecule. At energies above 1 eV, the scattering cross section of hydrogen in the form of water is constant and has a value of around 20 b (see Fig. 1.17), but it increases steeply in the energy region below 1 eV. It can be shown that, in the general case of an element of atomic mass number A, the chemical binding effect causes the scattering cross section to increase at low energies by a factor of $[1 + (1/A)]^2$.

Absorption and scattering cross sections for the elements of importance in reactor physics are summarized in Table A.4. For reasons to be discussed later, the values given are for a neutron energy of 0.0253 eV, corresponding to a neutron velocity of 2200 ms^{-1}. Attention may be drawn to some of the particularly low absorption cross sections, such as that of zirconium, which is consequently valuable as a structural material in reactors. The reason for zirconium, and also lead, having such a low cross section is that the main isotope in both cases has a nucleus with a "magic number" of neutrons or protons, or both (see Section 1.4).

1.8. Experimental Techniques

In this section we consider some of the techniques which are used to obtain data relevant to the operation of a nuclear reactor. Some of the methods described are employed in the measurement of basic data, such as

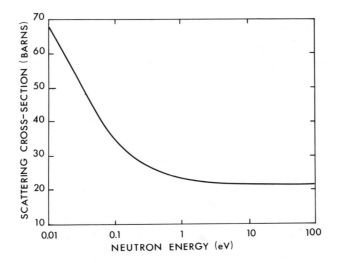

Fig. 1.17. Scattering cross section of hydrogen in the form of water, illustrating increase at low energies due to chemical binding effects.

scattering and absorption cross-sections, while others are useful in determining the characteristics of a particular reactor, for example, the way in which the neutron flux varies as a function of position in the system.

1.8.1. Neutron Sources

A nuclear reactor is, of course, itself a copious source of neutrons, and many experiments have been carried out using intense beams of neutrons extracted from an operating reactor. For many applications, however, it is desirable to have a cheap and compact source of neutrons with a defined energy distribution.

One of the most convenient mechanisms for producing neutrons is the bombardment of beryllium with α particles from a radioactive source. The reaction involved is

$$_4\text{Be}^9 + _2\alpha^4 \rightarrow _6\text{C}^{12} + _0\text{n}^1$$

This process is exothermic, the sum of the masses of the beryllium nucleus and the α particle exceeding the sum of the product masses by an amount equivalent to an energy of 5.7 MeV.

A common form of (α, n) source is a finely divided mixture of a radium salt with powdered beryllium. Because of the long half-life of Ra^{226}

(approximately 1600 yr), the neutron emission rate remains effectively constant over long periods of time. The α particles from Ra^{226} and its daughters are emitted at a few specific energies within the range from 4.6 to 7.7 MeV; since most of the α particles will have been considerably slowed down by collision before interacting with the beryllium, however, the neutrons emitted have a continuous spectrum of energies up to a maximum of 13 MeV, with an average of around 5 MeV. The neutron production rate is about 1.5×10^7 neutrons per second for each gram of radium.

One of the disadvantages of using radium as the source of α particles is the undesirably strong γ-ray emission associated with Ra^{226} and its daughters. An alternative is Po^{210}, which is free from γ-ray emission, but has the disadvantage of a rather short half-life [138 days (d)]. Another is Pu^{239}, which has the advantages of long half-life and freedom from γ-ray emission, but gives a rather low yield of neutrons per gram of plutonium.

Where it is desirable to have a monoenergetic source, i.e., one where all the neutrons are emitted with the same energy, use can be made of the (γ, n) reaction, in which the target nucleus is excited by the capture of a γ ray, and subsequently decays by the emission of a neutron (*photoneutron* source). The only practical targets are beryllium or deuterium (in the form of heavy water), in both of which the neutron binding energy has a particularly low value. The reactions are

$$_4Be^9 + \gamma \rightarrow {}_4Be^8 + {}_0n^1$$
$$_1D^2 + \gamma \rightarrow {}_1H^1 + {}_0n^1$$

Since both reactions are endothermic, the γ ray must have an energy above a threshold value (1.67 MeV for beryllium and 2.23 MeV for deuterium). The energy released in the reaction is equal to the amount by which the γ-ray energy of the chosen emitter exceeds the threshold energy; consequently, the source neutron energy may be varied to some extent by a suitable choice of γ emitter.

One of the most common photoneutron sources, which produces neutrons of relatively low energy (26 keV) is a combination of antimony-124 and beryllium (Sb–Be source). In contrast to the (α, n) source, where the short range of the α particles necessitates intimate mixing of the reactants, the antimony can be separate from the beryllium, and the source can be "switched off" by separating the two. The half-life of Sb^{124} is rather short (61 d), but the source can be reactivated by exposure to the high neutron flux in a reactor, when neutron capture in Sb^{123} leads to the formation of the active isotope.

Disadvantages of photoneutron sources include their relatively low yield and strong γ-ray emission.

For applications where compactness and low cost are not overriding factors, particle accelerators can be used as high-intensity sources which have the advantage of providing neutrons of well-defined and continuously controllable energy. A number of neutron-producing reactions may be utilized by bombarding suitable light-element targets with protons or deuterons of relatively modest energy, which can be achieved in a simple accelerator such as a Van de Graaff or Cockcroft–Walton generator. Some of the most useful reactions are listed below. The Q value quoted after each is the energy released in the reaction, derived from the difference in masses of the initial and product nuclei:

$$_1H^3 + {}_1H^1 \rightarrow {}_2He^3 + {}_0n^1 \qquad (Q = -0.76 \text{ MeV})$$
$$_3Li^7 + {}_1H^1 + {}_4Be^7 + {}_0n^1 \qquad (Q = -1.64 \text{ MeV})$$
$$_1H^2 + {}_1H^2 \rightarrow {}_2He^3 + {}_0n^1 \qquad (Q = 3.27 \text{ MeV})$$
$$_1H^3 + {}_1H^2 \rightarrow {}_2He^4 + {}_0n^1 \qquad (Q = 17.6 \text{ MeV})$$

The two proton-induced reactions are endothermic, while both of the deuteron-induced reactions are exothermic, the second strongly so.

Tritium targets are normally prepared by allowing tritium gas to be absorbed into a thin layer of titanium on a suitable base such as copper. Using this reaction, miniature accelerator tubes have been designed which are small enough to be completely inserted into a reactor. These devices can be operated in the pulsed mode; the rate of decay of the neutron flux following a rapid injection of neutrons into the reactor can be measured and used to obtain information on parameters such as the effectiveness of the reactor control system.

1.8.2. Detection of Charged Particles and γ Rays

When a charged particle, such as an α particle, proton, electron, or fission fragment travels through matter, it loses energy mainly through electromagnetic interactions with the electrons of the atoms along its path, causing the latter to become excited or ionized. Among the most widely used methods for detecting charged particles are the *proportional counter* and *Geiger–Müller (GM) counter*, where the charge carriers (electrons and ions) produced by the passage of the particle through a gas are separated by the application of an electrostatic field, giving an electrical pulse which may be amplified and recorded by a suitable circuit.

In its simplest form, a gas proportional counter or GM counter consists of a cylindrical chamber containing a gas such as argon or methane, with a thin central wire (anode) at a high potential relative to the chamber wall, which acts as the cathode. The movement in the electrostatic field of the electrons and ions produced by the passage of the charged particle induces a rapid change of voltage on the anode. If the anode potential is sufficiently high, the field in the neighborhood of the anode is strong enough that the original electrons acquire enough energy to cause further ionization. This phenomenon, known as *gas multiplication*, gives a large increase in the size of the electrical pulse produced by the counter. In the proportional counter, the anode potential is such that the pulse size is proportional to the initial ionization caused by the particle, thus allowing one to discriminate between particles of differing ionization. In the GM counter the field is still higher, and all dependence on the initial ionization is lost, the output pulse magnitude being the same irrespective of the initial ionization produced by the particle.

The GM counter may also be used to detect γ rays, but in this case it is necessary that the γ ray first interact with the material of the counter to produce one or more energetic charged particles which in turn produce the ionization to operate the detector. There are three types of interaction by which some or all of the energy of the γ ray can be converted into the kinetic energy of an electron. In *photoelectric absorption*, the γ ray loses all of its energy to an orbital electron which is then ejected from the atom. In *Compton scattering*, the γ ray is scattered by collision with an electron, and emerges with reduced energy, the energy lost appearing as kinetic energy of the electron. The third process, *pair production*, can take place only for γ rays with energies greater than about 1 MeV. The γ ray interacts with the field of a nucleus to produce a positron–electron pair, some of the energy (1.02 MeV) being used to provide the rest masses of the two particles, and the remainder appearing as kinetic energy shared between them. Both particles produce ionization and excitation as they slow down, and the positron eventually combines with an electron (*annihilation*) to give two γ rays, each of energy 0.51 MeV, which may produce further ionization.

The efficiency of the GM counter for the detection of γ radiation is low (about 1%), owing to the indirect method of detection. A more efficient device is the *scintillation counter*, where the ionization and excitation due to the electrons produced by one or more of the processes outlined above is converted into fluorescent radiation from a suitable scintillating material. The commonest scintillator is a single crystal of sodium iodide with a small amount of thallium added to it. The scintillator is optically coupled to a photomultiplier, where the light photons produced cause emission of electrons by the photoelectric effect; these electrons are multiplied by secondary emission in the electrodes of the photomultiplier, to give an electrical output pulse. Since the size of pulse produced is proportional to the amount of energy deposited in the scintillator by the γ-ray, it is possible to use the detector to measure both the intensity and energy of the γ radiation. Since large-volume sodium iodide detectors have a high counting efficiency, they are often used for the identification and measurement of radioisotopes in low-activity samples, for example, in environmental monitoring.

In order to analyze the energies of the γ rays emitted from a sample, it is necessary to determine the distribution of the pulse heights produced by the detector over the count time of the sample. The normal way of doing this is to use a *multichannel analyzer* (*MCA*), which measures the height of each output pulse and displays the information in the form of a spectrum of the number of counts falling within sequential pulse height intervals. A typical MCA record of the pulse height spectrum from a source of Co^{60}, which emits two γ rays, of energies 1.17 and 1.33 MeV, respectively, is shown in Fig. 1.18. Each γ ray energy is characterized by a peak (the *photopeak*) which repre-

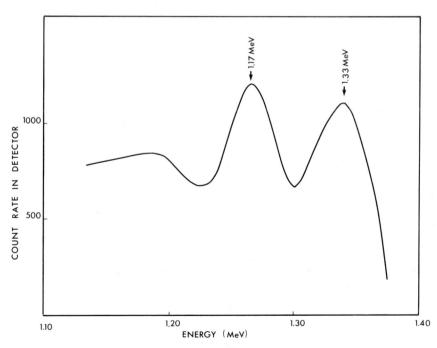

Fig. 1.18. The 1.17- and 1.33-MeV γ rays of Co^{60} as measured by a sodium iodide detector.

sents the accumulation of all the events where the incoming γ rays of that energy have interacted with the scintillator by the photoelectric effect, leading to the deposition of all the energy of the γ ray. The photopeaks are superimposed on a general background arising from the pulses produced by the Compton interaction of γ rays of higher energy, which leads to a deposition of only part of the energy of the γ ray, the remainder escaping in the form of the scattered (lower-energy) γ.

In practice, the mechanism for conversion of γ-ray energy into photons absorbed at the photocathode is rather inefficient, and so the γ-ray energy dissipated for each photoelectron liberated is relatively high. As a result, the statistical spread of the number of photoelectrons is such that, even if γ rays of only a single energy are being counted, one has a fairly wide spread of output pulse amplitudes, so that a broad peak is produced in the pulse height spectrum. Owing to the rather poor resolution of the scintillation detector, therefore, it is difficult to analyze a spectrum where many γ rays are present because of the overlapping of the peaks.

A device which gives a resolution for γ-ray spectra which is much better than that of the scintillation crystal is the *semiconductor detector*, which operates on the same principle as the gas ionization counter, but with the gas replaced by a solid dielectric, across which an electric field is maintained. In this case the charge carriers produced by the interaction of the γ ray are holes and electrons, and because the energy required to produce a hole–electron pair is much less than that required to produce a photoelectron at the cathode of a scintillation detector, the statistics of the process are much more favorable, leading to lower fluctuations and hence much greater resolution. The peak produced by γ rays of a given energy is much narrower than for the scintillation counter (see Fig. 1.19), making it possible to resolve γ rays of closely similar energy without the complication of overlapping peaks.

The only semiconducting materials readily available to the exacting standard of purity required are silicon and germanium, the latter being the one used for γ-ray measurements. In order to reduce the leakage current, it is necessary to compensate for the small amounts of acceptor impurity present (mostly boron). A technique has been developed for diffusing lithium ions into the germanium in sufficient concentration to neutralize the acceptor impurities, and thus greatly decrease the numbers of charge carriers, over volumes as high as $60 \, \text{cm}^3$ or more. The detector produced in this way is known as a lithium-drifted germanium, or Ge(Li) detector. It must be maintained at liquid nitrogen temperature to prevent redistribution of the lithium in the crystal.

A variety of special circuits can be used to improve the performance of

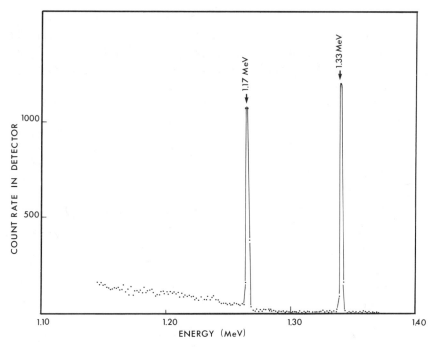

Fig. 1.19. The 1.17- and 1.33-MeV γ rays of Co^{60} as measured by a Ge(Li) detector.

nuclear counting systems, depending on the specific application required. An example of the simplest type of γ-counting circuit might be a single scintillator connected through an amplifier to a *single-channel analyzer* (*SCA*), a unit which gives a standard output pulse for each input pulse which lies within the photopeak width of the γ ray being counted. The output pulses from the SCA are recorded cumulatively in a device known as a *scaler* (Fig. 1.20a). One of the disadvantages of such a system is the inability to distinguish the pulses due to the γ ray under study from pulses in the same amplitude range which are created by Compton scattering of higher-energy rays which may be present either in the sample or in the general background, arising, for example, from radioactive impurities in the walls of the counting enclosure.

One way of improving the discrimination against background is the *coincidence method*, which can be used when the radionuclide being measured emits two or more γ rays essentially simultaneously (as, for example, when an excited level decays to ground in a cascade process involving several intermediate levels). A typical coincidence arrangement is shown in Fig.

1.20b. The sample is placed between two separate scintillation detectors, and
the pulses from each are amplified and fed to single-channel analyzers. The
output pulses from one detector are transmitted to a multichannel analyzer,
but on the way have to pass through a gate which is normally closed and can
only be opened by receiving simultaneous pulses from the SCAs of both
detectors. If several coincident γ rays are present, the SCA of detector 1 can
be set to cover only the limits of the photopeak corresponding to one of the

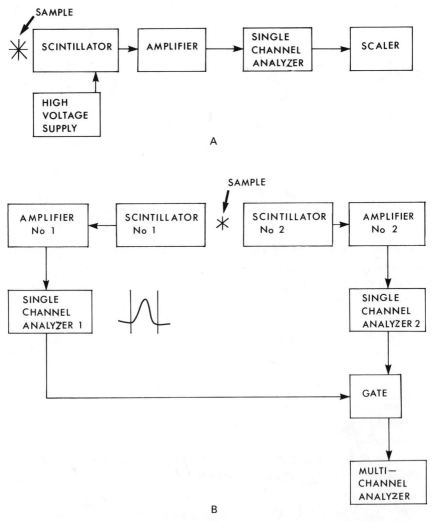

Fig. 1.20. Gamma-ray counting: (a) single-channel (b) coincidence system.

coincident γ rays, while the SCA of detector 2 can be set to cover the complete range of pulse heights from zero up to the maximum expected for the γ-ray spectrum of the source. When set up as in Fig. 1.20b the MCA will record the pulse height distribution of all pulses from detector 2 which are in coincidence with pulses from detector 1 which lie within the limits of the peak of the selected γ ray. The elimination of noncoincident γ rays greatly improves the selectivity of the system for the desired radionuclide.

1.8.3. Neutron Detectors

Since neutrons are uncharged, they do not produce direct ionization as does a charged particle. The detection of neutrons, therefore, depends on their first being made to take part in some reaction which produces charged particles, which may then be detected in the conventional manner. One of the commonly chosen reactions is the production of α particles following neutron capture in B^{10}:

$$_5B^{10} + {_0}n^1 \rightarrow {_3}Li^7 + {_2}\alpha^4$$

The Q value is 2.8 MeV, but most of the reactions leave the Li^7 nucleus in an excited level, so that the capture of very-low-energy neutrons leads, in 94% of the cases, to the release of 2.3 MeV, which is shared between the α particle and the recoiling Li^7 nucleus. The boron may be incorporated as a thin layer on the walls of a gas-filled proportional counter, or the counter gas may contain a proportion of gaseous boron trifluoride (BF_3). Since it is only the B^{10} isotope, which comprises 19.8% of natural boron, which has the high cross section for neutron capture (3840 b at a neutron energy of 0.0253 eV), the detector efficiency can be increased by using boron enriched in the B^{10} isotope. The B^{10} cross section falls off markedly at higher neutron energies and, in order to detect fast neutrons, the counter must be surrounded by a quantity of a material such as paraffin wax, which rapidly slows the neutrons down to low energies.

Scintillation methods offer the possibility of high-efficiency detectors with a more rapid time response than the BF_3 counter. As mentioned in the previous section, the basis of the scintillation detector is the conversion, in a suitable crystal, such as thallium-activated sodium iodide, NaI(Tl), of the kinetic energy of the charged particle to light, which can be amplified by a photomultiplier tube to provide an electrical pulse. Again, the neutron has to interact to produce either a charged particle or a γ ray, the latter of which may in turn interact to produce ionizing particles.

Scintillators which have hydrogen as a constituent, such as organic liquids for example, may be used for fast neutron detection, since the protons produced by fast neutron collisions create the ionization required to operate the detector. In order to adapt a sodium iodide scintillator for the detection of slow neutrons, a small concentration of boron may be distributed in the crystal, giving α particles on neutron capture as discussed above. Alternatively, it is possible to add a neutron absorber which emits γ rays following the (n, γ) capture reaction. Another possibility is the use of lithium iodide (LiI) which, in addition to its own suitability as a scintillator, interacts with neutrons through the reaction

$$_3Li^6 + {_0}n^1 \rightarrow {_1}H^3 + {_2}\alpha^4$$

The efficiency of the detector may be improved by increasing the isotopic ratio of the Li^6, which is present as only 7.4% in natural lithium. Owing to its sensitivity to γ radiation, a scintillation detector is not generally suitable where a strong background of high-energy γ rays is present.

Where it is required to measure neutrons in the presence of a significant γ-ray background, it is possible to make use of the properties of certain scintillators to distinguish the pulses produced by neutrons from those due to γ rays. This is known as a *pulse-shape discrimination* (PSD) system. In stilbene and some organic liquid scintillators, the pulse rise time for the fluorescence caused by the secondary electrons from a γ-ray interaction is considerably shorter than that due to the recoil protons produced by neutron scattering. By the use of fast timing discriminators, it is possible to separate the pulses caused by neutrons from those due to the γ rays.

A neutron detector of particular value for measurements in nuclear reactors is the *fission counter*. As will be discussed in Chapter 2, the break-up of a fissile nucleus such as U^{235} produces fission fragments which fly apart with a very large kinetic energy, and produce intense ionization over a rather short path length. The fissile material is generally distributed in a thin layer on one of the electrodes of a standard gas-filled proportional counter. The large pulses produced by the recoiling fission fragments are easily distinguished from the much smaller effects produced by γ rays, and so the counters are ideally suited for working in the high-γ-ray background inside a reactor. Another advantage is that, by using fission counters with different fissile isotopes, such as U^{235} or Pu^{239}, one obtains a count rate which is directly related to the reaction rate of the corresponding isotope in the reactor fuel, a quantity which is of primary importance for comparison with the theoretical model of the reactor performance.

The fission counter may be used to measure only the flux of high-energy neutrons by incorporating a coating which is only fissionable by neutrons above a certain threshold energy. Suitable coatings include U^{238} (threshold energy 1.45 MeV) and Th^{232} (1.75 MeV).

A small and simple instrument for measuring slow neutron fluxes in operating reactors is the self-powered, or Hilborn, detector. This consists of a length of rhodium wire (the emitter) separated by a layer of insulating material from a surrounding cylindrical metal tube (the collector). Slow neutron capture in Rh^{103} leads to the production of the short-lived β emitter Rh^{104}, which decays with a half-life of 43 s to Pd^{104}. Most of the energetic β particles emitted ($E_{max} = 2.4$ MeV) travel through the insulator to the collector, building up a negative charge there and a positive on the emitter. A direct current, of magnitude proportional to the neutron flux, is therefore generated when the two are connected through an external circuit.

A very convenient technique for obtaining information about both the magnitude and the energy spectrum of neutrons in a reactor is the foil activation method. This involves exposing thin foils of some suitable material to the reactor flux for a period of time and then measuring the radioactivity induced by the (n, γ) or some other reaction. Among the advantages of the activation foil technique are minimum perturbation to the system on which measurements are being made, accessibility to regions where it would be difficult or impossible to introduce a BF_3 or fission counter, and the possibility of making measurements on the energy distribution of the neutron flux, by using a range of foils with large resonances in the reaction cross section at different neutron energies.

Consider a thin foil, of mass M, made up of an element which, on exposure to neutrons of suitable energy, undergoes a reaction leading to the formation of a radioactive isotope of decay constant λ (see Section 1.3). Assume that the foil is exposed for some time to a neutron flux of magnitude nv n cm^{-2} s^{-1} (neutrons per cm^2 per s) (see Section 1.6). Let the macroscopic cross section of the foil material for the reaction considered be Σ_a for neutrons of velocity v.

The rate at which neutrons undergo the reaction is $\Sigma_a nv$ per unit volume and hence the rate of formation of the active isotope in the foil is

$$\frac{M}{\rho} \Sigma_a nv$$

where ρ is the density of the foil material.

If the number of active nuclei present at some time during the irradiation is N, the expression for the net rate of formation of the active isotope is obtained by subtracting from this the rate of decay, which is equal to λN, i.e.,

$$\frac{dN}{dt} = \frac{M}{\rho} \Sigma_a nv - \lambda N \tag{1.30}$$

The solution of this equation, giving the number of radioactive nuclei, $N(t)$, present at a time t after the start of the irradiation, is

$$N(t) = \frac{M\Sigma_a nv}{\rho\lambda} (1 - e^{-\lambda t}) \tag{1.31}$$

It will be seen that, when the period of irradiation is considerably greater than the half-life of the radioisotope produced, i.e., for $t \gg T_{1/2}$, or $t \gg 1/\lambda$, the number of active nuclei reaches a *saturation value*, given by

$$N_s = \frac{M\Sigma_a nv}{\rho\lambda} \tag{1.32}$$

After irradiation, the sample is removed and its activity measured with a suitable detector, such as a Geiger or scintillation counter. If the efficiency of the detector (counts recorded per disintegration in the foil) is ε, the count rate immediately after the foil is removed from the reactor will be obtained by multiplying ε by the activity of the foil, λN_s. Then

$$C = \frac{\varepsilon M\Sigma_a nv}{\rho} \tag{1.33}$$

Knowing the mass, density and reaction cross section of the foil material, and the efficiency of the counting system, the neutron flux, nv, can be derived. Corrections, of course, have to be made for decay of the foil activity during any time lapse between removal from the reactor and start of the counting sequence.

It has implicitly been assumed that the insertion of the foil does not itself result in an alteration of the neutron flux in the measuring region, i.e., that all the atoms of the foil are subjected to the same neutron flux which would be present in the absence of the foil. In practice, the average flux seen by an atom of the foil will be reduced below the undisturbed value because the flux in the foil falls off with increasing distance from the surface, owing to absorption in the outer layers, and also because the overall effect of introducing the foil is to cause a depression of the total flux in that region of the reactor. Even when the foils used are very thin, a correction for flux depression effects may have to be applied.

Example 1.6. A thin film of $_{49}In^{115}$, of mass 60 mg, is exposed for 20 minutes (min) to a flux of 2200-ms^{-1} neutrons of magnitude 10^6 n cm^{-2} s^{-1}. The $_{49}In^{115}$, which has a capture cross section of 157 b, is converted by neutron capture into the β emitter $_{49}In^{116}$, which has a half-life of 54 min. If, immediately after exposure, the foil is removed and placed in a detector system which has a counting efficiency of 10% for the β particles of $_{49}In^{116}$, what will the initial count rate of the detector be?

From equation (1.31), the activity of the sample immediately after removal will be

$$\lambda N = \frac{M\Sigma_a nv}{\rho}(1 - e^{-\lambda t})$$

The count rate in the detector is therefore

$$C = \varepsilon\lambda N = \frac{eM\Sigma_a nv}{\rho}(1 - e^{-\lambda t})$$

where the detector efficiency $\varepsilon = 0.1$. The macroscopic absorption cross section of $_{49}In^{115}$ is, from equation (1.22),

$$\Sigma_a = \frac{0.6\rho}{A}\sigma_a$$

so that

$$C = \frac{0.6\varepsilon M\sigma_a nv}{A}(1 - e^{-\lambda t})$$

The decay constant of $_{49}In^{116}$ is

$$\lambda = \frac{0.693}{T_{1/2}} = \frac{0.693}{54 \times 60} = 2.139 \times 10^{-4}\,s^{-1}$$

So that the factor in parentheses is equal to 0.2264 for $t = 20 \times 60 = 1200$ s. The initial count rate is therefore

$$C = \frac{0.6 \times 0.1 \times 0.060 \times 157 \times 10^6 \times 0.2264}{115} = 1113 \text{ counts/s} \qquad \square$$

Problems

(The data required for the solution of the problems are given in Tables A.1–A.4, Appendix I.)

1.1. Using equation (1.1) for the radius of an atomic nucleus, calculate the density (in $kg\ m^{-3}$) of a nucleus of $_{92}U^{235}$.

1.2. Calculate the number of atoms in 1 kg of $_{13}Al^{27}$.

1.3. Calculate the atomic mass of natural uranium (see Table 1.1 for composition).

1.4. Calculate the mass of a sample of $_{90}Th^{226}$, which has an activity of 10^8 Bq. The half-life of $_{90}Th^{226}$ is 31.2 min.

1.5. Calculate the activity of 1 g of $_{88}Ra^{226}$ (half-life 1622 yr).

1.6. A sample of pure $_{88}Ra^{227}$, of mass 4×10^{-14} kg, has an activity of 3×10^7 Bq. Calculate the half-life of $_{88}Ra^{227}$.

1.7. Calculate the frequency of the photon emitted when a hydrogen atom drops from its first excited state, of energy -3.40 eV, to the ground state (of energy -13.58 eV).

1.8. Calculate the energy release in the α decay of $_{92}U^{238}$.

1.9. Calculate the maximum energy of the β particles emitted in the decay

$$_{49}In^{116} \rightarrow {}_{50}Sn^{116} + {}_{-1}\beta^0$$

1.10. The binding energy of the nucleus of $_{26}Fe^{56}$ is 492.3 MeV. Calculate the mass of the nucleus in atomic mass units.

1.11. The mass of the neutral atom of $_{13}Al^{27}$ is 26.9815413 u. Calculate the binding energy of its nucleus.

1.12. Calculate the energy which has to be supplied to remove a neutron from the nucleus of $_{92}U^{235}$.

1.13. Calculate the energy of the γ ray emitted when a slow neutron is absorbed by a nucleus of $_{49}In^{115}$ in a (n, γ) reaction.

1.14. What is the minimum energy which a photon must have if it is to cause an α particle to break up into a proton and a tritium nucleus $(_1H^3)$?

1.15. Using the given nuclear masses, check the correctness of the Q values given for the four proton- or deuteron-induced reactions quoted near the end of Section 1.8.1.

1.16. Calculate the scattering mean free path of 2200-ms^{-1} neutrons in beryllium.

1.17. Calculate the macroscopic absorption cross section of natural uranium for 2200-ms^{-1} neutrons, using the relative atomic abundances of the isotopes given in Table 1.1.

1.18. The macroscopic absorption cross section of a material of density 13.3 g cm^{-3} and atomic mass number 178.5, for neutrons of a particular energy, is equal to 4.2 cm^{-1}. Calculate the ratio of the absorption cross section to the geometrical cross section of the nucleus at this neutron energy (use equation (1.1) for estimating the geometrical cross section).

1.19. The composition (in percent *by weight*) of a typical stainless steel is 69% iron, 17% chromium, 12% nickel, and 2% molybdenum. Calculate the absorption cross section of this stainless steel for 2200-ms^{-1} neutrons.

1.20. A monoenergetic beam of neutrons falls on a boron foil of thickness 0.1 millimeter (mm). The absorption cross section of boron at the energy of the neutrons in the beam is 200 b. What fraction of the beam is absorbed in the foil (atomic mass number of boron = 10.8).

1.21. A uniform neutron beam with a density of 10^4 neutrons per cm^3 and a neutron velocity of 2200 ms^{-1} is directed perpendicular to a manganese foil of thickness 0.1 mm. What is the absorption rate of neutrons per cm^2 of the foil (atomic mass of manganese = 54.9)?

1.22. A foil of $_{25}$Mn^{55}, of area 1 cm^2 and thickness 0.005 cm, is exposed for several days to a flux of 2200-ms^{-1} neutrons of magnitude 2.5 × 10^{10} n cm^{-2} s. If the half-life of $_{25}$Mn^{56} is 2.57 hours (h), calculate the activity of the foil immediately after removal from the flux.

Nuclear Fission and the Nuclear Chain Reaction

2.1. Nature of the Fission Process

Following the discovery of the neutron, detailed measurements of the effect of bombarding elements with neutrons were carried out, notably by Enrico Fermi and his collaborators in Rome. Among the large number of elements investigated was the heaviest naturally occurring element, uranium. As with many of the other elements, the bombardment of uranium led to the production of induced β activity. At least four different β-active isotopes, of different half-lives and disintegration energies, were detected.

Fermi concluded that the activities arose following the capture of a neutron in the most abundant uranium isotope, U^{238}, to form U^{239}:

$$_{92}U^{238} + _0n^1 \rightarrow _{92}U^{239}$$

The β-decay of U^{239} would then lead to the creation of the element of atomic number $Z = 93$:

$$_{92}U^{239} \rightarrow _{93}X^{239} + _{-1}\beta^0$$

which in turn could possibly decay by a further β emission to form the element of atomic number 94. Those unstable, artificially created elements of atomic number $Z > 92$ were given the name of *transuranic elements*. The element with $Z = 94$ (plutonium) is of particular importance in the operation of the nuclear reactor.

While the process postulated by Fermi and his coworkers must indeed have taken place in their experiments, they failed to detect the simultaneous occurrence of a much more dramatic and unexpected phenomenon. The

198009

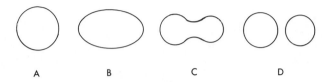

A   B   C   D

Fig. 2.1. Nuclear fission; oscillation of liquid drop or nucleus, leading to breakup.

discovery of the process of *nuclear fission* was not made until five years later, in 1939, by the German radiochemists Hahn and Strassmann. Following bombardment of uranium with slow neutrons, they had isolated an active product which, because it was precipitated using barium carrier, they first assumed to be an isotope of uranium, which is chemically similar to barium. Further evidence, however, forced them to the conclusion that the active product was actually barium itself rather than a heavier element of similar chemistry. This hypothesis was advanced reluctantly because of the re-volutionary nature of the conclusion that the bombardment of a nucleus of uranium ($Z = 92$) with neutrons could produce a nucleus of approximately half the mass of the original (Z for barium $= 56$).

The physical explanation of fission was provided very shortly afterwards by Frisch and Meitner. The process may most conveniently be considered in terms of the liquid drop model of the nucleus (see Section 1.4). Nuclei of high mass number tend to be unstable owing to the high number of protons and the consequent reduction of the nuclear binding energy due to the magnitude of the Coulomb term [see equation (1.10)]. The addition to an already unstable nucleus of the energy gained from the binding energy of the added neutron, together with any kinetic energy it may have had prior to the collision, may be enough to set the nucleus into a state of violent oscillation.

A physically similar situation may be set up by allowing a drop of water to rest on a hot horizontal surface, such as a heated metal plate. If the surface is sufficiently hot, the drop will essentially float on an insulating layer of steam, in a friction-free situation. It is then possible for the drop to be set into oscillation between its original state of circular symmetry and an ellipsoidal shape as shown in Fig. 2.1. The oscillation represents a state of dynamic equilibrium between the inertia of the moving matter in the drop and the surface tension which attempts to maintain it in a symmetrical configuration.

If the surface tension force is sufficiently large, the process of elongation will be reversed before break-up of the drop can take place. If there is sufficient kinetic energy associated with the inertia of the moving matter, however, the drop can attain a dumbbell shape, as in part C of Fig. 2.1, and continuation of the outward movement will result in its splitting into two parts, as in part D. The process in the nuclear case is similar, with the addition of the Coulomb repulsion as an additional factor operating against the internucleon forces which tend to keep the nucleus together.

The condition for fission to take place may be considered in terms of a potential energy diagram, where the mutual potential energy of the two nuclei produced by the fission (the *fission fragments*) is plotted as a function of the distance between their centers (see Fig. 2.2).

We may start by considering the fission process in reverse, i.e., with the two fragments some distance apart. If the two nuclei are now forced to move together, there will be an increase in their mutual potential energy arising from the external work required to overcome the electrostatic force between the two positive charges (section AB of curve in Fig. 2.2). As the two nuclei approach the point B, where they come into contact, the strong short-range nuclear force overcomes the electrostatic repulsion, the two nuclei are drawn together, and there is a corresponding decrease in the potential energy of the system (section BC of curve).

Considering the break-up of a nucleus, the difference in potential energy between the points B and C (CD in diagram), represents the minimum energy which has to be supplied to the uranium nucleus to cause it to undergo fission. If the combination of the binding energy of the neutron and the initial kinetic energy of the approaching neutron and the nucleus in the center-of-mass system exceeds this critical energy, nuclear fission will take place. In practice, there is no unique threshold energy for the fission of a heavy nucleus, since there are a large number of ways in which the nucleus can split up to yield products whose combined masses add up to give the mass of the original nucleus.

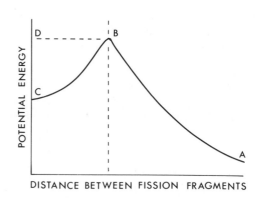

Fig. 2.2. Mutual potential energy of fission fragments as a function of separation.

It is found that for a few very heavy nuclei, the energy contributed by the addition of the neutron is in itself sufficient to cause fission. As we go to nuclei of progressively smaller mass, the bombarding neutron has to contribute an increasingly greater quantity of kinetic energy for fission to take place. It can readily be seen, by a simple qualitative argument, that the readiness of a nucleus to undergo fission depends on the value of the parameter Z^2/A. The reason for this is that the disruptive Coulomb energy term is proportional to $Z(Z-1)$, which is approximately equal to Z^2 for heavy nuclei (see Section 1.4), while the basic contribution to the binding energy from the saturation-type nuclear forces between the nucleons is proportional to A, the number present. The quantity Z^2/A, of course, increases steadily as one progresses through the periodic table.

It turns out that, of the two uranium isotopes, U^{235} and U^{238}, it is only the former that will undergo fission following capture of a slow neutron. The basic reason for this is the spin-pairing effect mentioned in Section 1.4. The increase in nuclear binding energy on adding a neutron to the 143-neutron nucleus U^{235} is greater than for the 146-neutron nucleus U^{238}, since in the first case one is completing a neutron pair while, in the second, one is adding an unpaired neutron. This difference in the spin-pairing effect leads to U^{238} having a critical fission energy more than 1 MeV greater than U^{235}. Consequently, fission of U^{238} can take place only with neutrons possessing a kinetic energy in excess of 1 MeV.

Isotopes which undergo fission as a result of absorbing a slow neutron are described as *fissile*. The most important examples of fissile materials are the isotopes U^{233}, U^{235}, Pu^{239}, and Pu^{241}. Of these, only U^{235} occurs naturally, the others having to be created as a result of neutron capture in thorium or uranium. Isotopes such as U^{238} and Th^{232}, for which the critical fission energy is larger than the binding energy of the added neutron, are described as *fissionable*.

A phenomenon of some importance for the uranium and transuranic isotopes is *spontaneous fission*. In this case, the heavy nucleus undergoes fission spontaneously, and not as a result of neutron absorption. The process is a quantum mechanical one, similar to the emission of α particles through a potential barrier which, according to classical physics, is too high for them to surmount. In general, the rate of spontaneous fission is extremely low; for U^{238}, for example, the half-life for spontaneous fission is of the order of 10^{16} years, while it is even longer for U^{235} and Pu^{239}. Spontaneous fission is of some importance in providing a background source of neutrons in reactors containing large amounts of U^{238}, or where a significant proportion of the fuel is Pu^{240}, as in a fast breeder reactor.

2.2. The Products of Fission

There are, of course, many possible combinations of fragment pairs which may be formed by the disintegration of the compound nucleus. It is found that the preferred mode of break-up for fission caused by slow neutrons is a highly unsymmetrical one, where the most probable products are in the mass regions around 95 and 140. This is illustrated in Fig. 2.3, which shows the percentage yield of fission products from the fission of U^{235}, as a function of mass number. Similar bimodal distributions are obtained for the fission of the isotopes U^{233} and Pu^{239}, which also undergo fission with slow neutrons.

As noted already (see Fig. 1.6), the ratio of neutrons to protons in the nucleus of an element increases steadily as we move through the periodic table. The neutron/proton ratio for U^{235}, for example, is 1.55, while for a nucleus of around one half of the mass of U^{235}, the stable isotopes have neutron/proton ratios of around 1.30. The nuclei formed by fission are therefore neutron rich, with neutron/proton ratios which are appreciably higher than those characterizing the stable isotopes of the same species. The return of the fission product nuclei to the region of stable neutron/proton ratio may be accomplished, in the first place, by the emission of one or more neutrons and then, more slowly, by the mechanism of β emission (which is essentially the conversion of one of the neutrons in the nucleus to a proton).

It is this phenomenon of the emission of the so-called *fission neutrons*

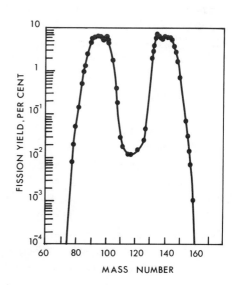

Fig. 2.3. Yield of fission products from slow neutron fission of U^{235} as a function of mass number (from ANL-5800, *Reactor Physics Constants*, 2nd ed. 1963. Courtesy Argonne National Laboratory).

that gives fission its practical importance, since it opens up the possibility of a *chain reaction*. The number of fission neutrons emitted depends on the identity of the fission fragments formed; in some cases no neutron may be emitted, while the number can be as high as eight. The important quantity is the *average number of neutrons per fission*, which is given the symbol v (nu). For fission of U^{235} by slow neutrons for example, v is equal to 2.42.

To illustrate the complete process, we may consider the typical fission

$$_{92}U^{235} + _0n^1 \rightarrow _{56}Ba^{140} + _{36}Kr^{96}$$

The Ba^{140} decays to form stable Ce^{140} by successive β emissions:

$$_{56}Ba^{140} \xrightarrow[12.8\,d]{\beta} _{57}La^{140} \xrightarrow[40\,h]{\beta} _{58}Ce^{140}$$

while Kr^{96} emits two neutrons to form Kr^{94}, which decays to form stable Zr^{94}:

$$_{36}Kr^{96} \rightarrow _{36}Kr^{94} + 2_0n^1$$

$$_{36}Kr^{94} \xrightarrow[1.4\,s]{\beta} _{37}Rb^{94} \xrightarrow[2.9\,s]{\beta} _{38}Sr^{94} \xrightarrow[1.3\,min]{\beta} _{39}Y^{94} \xrightarrow[20.3\,min]{\beta} _{40}Zr^{94}$$

When fission takes place in a nuclear reactor, approximately 99 % of the fission neutrons are emitted within a very short time interval ($\sim 10^{-17}$ s) of the fission taking place; these are known as *prompt neutrons*. The remainder, the *delayed neutrons*, are emitted for periods of up to several minutes after fission. If a sample of uranium is subjected to a high flux of slow neutrons for some time, and then removed, it is possible to study the way in which the intensity of delayed neutrons varies with time after the fission events have taken place. The output of delayed neutrons decreases in a manner similar to the decay of a mixture of radioactive isotopes of different half-lives. The reason for the delayed emission is that the neutrons are not emitted from the direct products of the fission, but from nuclei which are formed by subsequent β decay of these products.

Delayed neutron emission is illustrated in Fig. 2.4 by consideration of the delayed neutrons emitted following decay of the fission product Br^{87}. Br^{87} is a β-emitter with a half-life of 55 s. The decay of Br^{87} leads to the formation of Kr^{87}. If the latter is formed in its ground state, it will subsequently decay through two successive β emissions to stable Sr^{87}. It is also possible, however, for the decay of Br^{87} to lead to a level of Kr^{87} with an

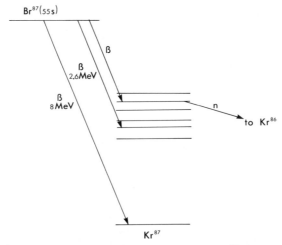

Fig. 2.4. Delayed neutron emission following the decay of 55-s Br^{87}. In addition to the two principal β-decay branches of 2.6 and 8 MeV, a small number of the decays lead to excited levels at energies greater than the binding energy of a neutron of the Kr^{87} nucleus. In this case, the β decay is followed by neutron emission, leading to Kr^{86}.

excitation energy in excess of the binding energy (about 5.5 MeV) of the last neutron in the Kr^{87} nucleus. A neutron can therefore be emitted from the excited Kr^{87} level. The production of secondary neutrons by this process will obviously be controlled by the decay characteristics of Br^{87}, and their intensity will decrease with the same half-life of 55 s. Isotopes such as Br^{87} are referred to as *delayed neutron precursors*.

Although the fraction of fission neutrons which is delayed is only 0.0065 for U^{235}, and is even smaller (0.0021) for Pu^{239}, it will be seen later that the existence of delayed neutrons is of crucial importance for the control of nuclear reactors, since it is only the relatively long time constant associated with the delayed neutrons that slows the dynamic response of the reactor sufficiently to make it controllable by the insertion of removal of mobile absorbers.

The kinetic energy of the fission neutrons accounts on average for about 5 MeV of the energy released in a fission event. The prompt neutrons are emitted with a wide range of energies (see Fig. 2.5), from quite low values up to more than 10 MeV. The energy spectrum has a peak in the neighborhood of 0.7 MeV, but the *average* prompt neutron energy is nearly 2 MeV. The curve shown, which applies approximately to U^{233}, U^{235}, and Pu^{239}, can be represented by the empirical formula

$$P(E) = Ae^{-BE} \sinh (2E)^{1/2} \qquad (2.1)$$

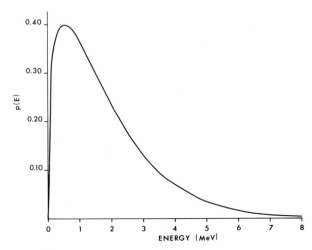

Fig. 2.5. Energy spectrum of prompt fission neutrons; the function $p(E)$ gives the relative probability of emission of a neutron of energy E.

where $p(E)$ is the probability per unit energy interval that the neutron will be emitted with energy E (in MeV), and A and B are constants fitted for the particular fissile isotope.

In contrast to the prompt neutrons, the delayed neutrons have a discontinuous energy spectrum, as each delayed precursor emits neutrons of a fairly clearly defined energy. The average energy of emission of the delayed neutrons is appreciably lower, with a range of values running from about 250–500 keV.

2.3. Energy Release in Fission

The total release of energy due to the fission process may be calculated from a knowledge of the masses before and after the fission takes place.

Example 2.1. For the fission discussed in the previous section, calculate the total energy release (in MeV).

Starting with a nucleus of $_{92}U^{235}$ and the neutron which it absorbs, the end products of the fission are two prompt neutrons, six β particles, and the nuclei of the stable atoms $_{58}Ce^{140}$ and $_{40}Zr^{94}$. Using the atomic mass data in Table A.3, and subtracting the appropriate number of electron masses to find the masses of the nuclei, the mass balance obtained is as given below:

Total mass before fission:

$_{92}U^{235}$ nucleus	234.993454 u
$_{0}n^1$	1.008665 u
	236.002119 u

Total mass after fission:

$_{58}Ce^{140}$ nucleus	139.873623 u
$_{40}Zr^{94}$ nucleus	93.884375 u
$2 \times _{0}n^1$	2.017330 u
$6 \times _{-1}\beta^0$	0.003292 u
	235.778620 u

Mass reduction = 0.2235 u

Energy release = $0.2235 \times 931.5 = 208$ MeV   □

A similar value may be found by a less precise calculation, based on the average binding energies for the fission products compared with that of the U^{235}. The average binding energy per nucleon for U^{235} is 7.6 MeV, while the average for Ce^{140} and Zr^{94} is around 8.5 MeV. The energy released is then approximately equal to the product $(8.5 - 7.6) \times 235 = 212$ MeV.

When averaged over all the modes of disintegration, the energy release from slow neutron fission of U^{235} is approximately 205 MeV. The way in which this release is distributed between the various processes involved is shown in Table 2.1.

It will be noted that 178 MeV, or 87% of the total, is released immediately, either as kinetic energy of the recoiling fission fragments or the fission neutrons, or in the form of γ radiation from the product nuclei. The remainder is released gradually in the decay chains of the radioactive fission products. The energy of the neutrinos accompanying β emission is lost to the reactor, since they escape without interacting with the system. The kinetic energy of the fission fragments is dissipated locally in the fuel, since their range is very short owing to the very rapid loss of energy by ionization for particles of such high charge. The fission products are brought to rest in a

Table 2.1. Breakdown of Energy Release
in Slow Neutron Fission of U^{235} (MeV)

Kinetic energy of fission fragments:	167
Kinetic energy of fission neutrons:	5
Energy of prompt γ rays:	6
β-decay energy of fission products:	8
γ-decay energy of fission products:	7
Neutrino energy:	12
Total:	205

time of the order of 10^{-12} s, after traveling a distance of a few micrometers (μm).

The energy generated by fission product decay in a nuclear reactor is of particular importance for the analysis of plant safety, since the presence of this persistent source of heating means that the reactor must be provided with a reliable cooling supply even after the fission reaction itself has been shut down. As an example, for a reactor with an electrical output of 1000 MW, which has been running for a long time at full power, the residual heating rate due to fission product decay will typically be as great as 15 MW a full day after shutdown.

Of the component energies given in Table 2.1, all but the 12 MeV associated with the neutrinos is potentially recoverable as useful heat. The loss of the neutrino energy is partly compensated for by the capture γ rays produced as a result of a proportion of the fission neutrons being absorbed in (n, γ) reactions. Consequently, the amount of recoverable energy per fission is close to 200 MeV.

Owing to the relatively large mass fraction which is converted to energy when fission occurs, the amount of energy produced per unit mass of fuel consumed is many orders of magnitude greater than in chemical reactions such as the burning of fossil fuels. The total energy produced by the fission of the atoms in 1 g of U^{235} is

$$E = \frac{1}{235} \times \underbrace{6.02 \times 10^{23}}_{\text{Avogadro's number}} \times 200 \times \underbrace{1.6 \times 10^{-13}}_{\text{MeV to joules}} \text{J} = 8.2 \times 10^{10} \text{ J}$$

This is equivalent to 23 megawatt-hours (MW h), or nearly 1 megawatt-day (MW d). The production of an equivalent amount of energy from fossil fuel would involve the burning of some three *tons* of coal.

2.4. The Nuclear Chain Reaction

The fact that the fission process involves the emission of secondary neutrons leads immediately to the possibility of setting up a chain-reacting system. We start by considering the problem of designing a nuclear reactor in which the fuel is natural uranium. The criterion for a successful chain reaction is the following: starting with a certain number of fission events taking place per unit time, it is necessary that the fraction of the secondary neutrons produced in fission which survive to cause further fissions should be sufficient to maintain the fission rate in the system at a constant level.

This condition may be described quantitatively by introducing a parameter known as the *neutron multiplication factor*, k. Consider a reactor in which at some instant a large number of fission events occur simultaneously. These events give rise to a number, N_0, of fission neutrons (which are all assumed to be prompt neutrons for the purposes of the present argument). Some of these secondary neutrons will be lost to the chain reaction by various processes which will be considered in detail later, and a certain fraction will survive to cause further fissions. Let the number of secondary neutrons produced as a result of these fissions be N. The neutron multiplication factor, k, may be defined by the relation

$$k = N/N_0 \qquad (2.2)$$

i.e., it is the ratio of the number of neutrons in one generation to the number in the preceding generation. If k is equal to unity, the fission rate in the reactor will remain at a constant level, and the system is said to be *critical*. For values of k less than unity, the fission rate will decay, and the reaction cannot be maintained, giving a *subcritical* condition. For k greater than unity, the reactor is said to be *supercritical*, and the rate of energy generation will increase at a steady rate.

In order to study the neutron balance in a reactor, we have to consider the processes which compete with fission absorption by U^{235} for the available neutrons. These processes include (i) leakage of neutrons out of the reactor volume; (ii) capture by elements other than uranium, e.g., structural materials, moderator, coolant, and fission products; (iii) capture by U^{238} to form U^{239}, leading to production of Pu^{239}; (iv) nonfission capture by U^{235}, leading to the formation of U^{236}.

The magnitude of the leakage will obviously depend on the size of the reactor. It is because of leakage that even an assembly of pure U^{235} will not attain a multiplication factor of unity until a certain mass of the material, known as the *critical mass*, is present. The calculation of the leakage from a reactor of given size and shape will be dealt with in Chapter 3. For the present we shall separate out the effect of leakage by writing the multiplication factor in the form

$$k_{\text{eff}} = k_\infty \mathscr{L} \qquad (2.3)$$

where $\mathscr{L}$ is the *nonleakage probability*, i.e., the probability that the average neutron does not leak out of the reactor during its lifetime. k_{eff} is the *effective multiplication factor*, that which applies to the finite reactor, while k_∞ is the value which the multiplication factor would have if the reactor were infinitely large, when the leakage would, of course, be zero. The advantage of separating the factors as in equation (2.3) is that the quantity k_∞ is a function only of the composition of the reactor and can in principle be calculated from a knowledge of the cross sections of the relevant materials for the various processes involved.

For the natural uranium reactor, it is found that even an infinitely large volume of pure uranium will not form a critical assembly. The reason for this becomes apparent when we consider the history of one of the fast secondary neutrons produced following the fission of a U^{235} nucleus. A neutron with, say, 2 MeV of energy is capable of causing fission in either U^{235} or U^{238}. On account of the much greater proportion of U^{238} present, the so-called *fast fission effect* is due almost exclusively to this isotope. The fast neutrons lose energy by a combination of elastic and inelastic scattering collisions with the uranium nuclei, the first reaction resulting in a small decrease in kinetic energy, and the second a large decrease, because energy is taken up in raising the uranium nucleus to an excited level. In general the result of an inelastic scattering will be to reduce the energy of the neutron to below the fission threshold of U^{238}. In practice the inelastic scattering cross section is such that less than 10% of the original fission neutrons would cause fast fission in U^{238} in an assembly of pure uranium.

Below about 100 keV, the inelastic scattering cross section of U^{238} drops steeply, and below an energy of about 45 keV, corresponding to the first excited state of U^{238}, the neutrons can lose energy only by elastic scattering from the uranium atoms. The small energy loss per collision in this process, owing to the large mass of the struck nucleus, leads to the neutron spending a relatively long time in the intermediate energy region where the probability of resonance capture by U^{238} is very high (see Fig. 1.16). The result is that the overwhelming majority of neutrons are captured in this way, with the eventual formation of Pu^{239}, while very few survive to the low-energy region (< 1 eV) where they have a high probability of being captured by U^{235}. It therefore turns out that, owing to the combination of inelastic scattering and resonance absorption in U^{238}, even an infinitely large mass of natural uranium would have a multiplication factor well below unity ($k_\infty \sim 0.25$).

The solution to this problem is to add to the uranium a material of low atomic mass, called a *moderator*. The average energy loss of a fast neutron in an elastic collision with a light moderator nucleus is much greater than in a similar collision with uranium. The result is that the neutrons are slowed down much more rapidly, through the region of high U^{238} resonance absorption, to the energy range below about 1 eV, where the absorption by

U^{235} is predominant. In this way it is possible to construct a critical assembly using natural uranium as fuel. Such a system is known as a *thermal reactor*, since the neutrons which cause the bulk of the fissions have energies which are approaching thermal equilibrium with the vibrational energies of the atoms in the moderator lattice. The details of the energy spectrum of these *thermal neutrons* will be covered later.

The criteria for a good moderator will be discussed in a subsequent chapter. For the moment, we may note that the dual requirement, that the moderator nuclei be of low mass and have a low absorption cross section, places a severe restriction on the possible choice of material. For a natural uranium reactor, it turns out that the only practicable moderators are carbon (graphite), deuterium oxide (heavy water), and beryllium. It is unfortunate that the absorption cross section of hydrogen is such that it is not possible to design a critical natural uranium reactor using ordinary water alone as moderator; the so-called light water reactors, with ordinary water as moderator, utilize fuel which has been artificially enriched in the U^{235} isotope as a means of enhancing fission capture at low energies.

We shall now consider the problem of calculating the infinite multiplication factor for a reactor made up of natural uranium with a suitable moderator. For this purpose we start by considering a reactor of infinite size and therefore zero leakage, a condition which, while strictly unrealistic, is approximately realized in practice by actual natural uranium reactors. The effect of finite leakage will be put in at a later stage. We also assume that the system is homogeneous, i.e., that the fuel and moderator are uniformly mixed throughout the reactor. It will be seen later that it is advantageous to introduce heretogeneity by having the uranium in the form of fuel rods separated by regions of moderator.

The first parameter of importance is the *average number of prompt neutrons emitted per thermal fission event in the fuel* (v). The values of this quantity for the fissile isotopes U^{233}, U^{235}, Pu^{239}, and Pu^{241} are given in Table 2.2.

Since only a proportion of the thermal neutrons absorbed in the fuel gives rise to fissions, the *number of fast neutrons emitted per thermal neutron absorbed in the fuel*, η (*eta*), is lower than v. If the fuel consists of a single fissile isotope, η is given by the relation

$$\eta = v \frac{\sigma_f}{\sigma_a} \tag{2.4}$$

where σ_f and σ_a are, respectively, the microscopic fission and absorption cross sections for the isotope. The values of η for the fissile isotopes are also given in Table 2.2.

When the fuel consists of a mixture of isotopes, η is defined as the *average* number of fast neutrons emitted per thermal neutron absorbed in fuel. In this case, we have

$$\eta = v \frac{\Sigma_f}{\Sigma_a} \tag{2.5}$$

where Σ_f is the macroscopic cross section for fission by thermal neutrons and Σ_a the total absorption cross section (including both fission and nonfission events). The value of η for natural uranium fuel, for example, is lower than for U^{235} because of nonfission capture in U^{238}.

Example 2.2. Calculate the value of η for natural uranium.

Taking the proportions of U^{235} and U^{238} as 0.72% and 99.28%, respectively, we have

$$\eta = v \frac{0.72\sigma_{f5}}{99.28\sigma_{a8} + 0.72\sigma_{a5}}$$

where v is the number of fast neutrons per fission of U^{235}, σ_{f5} the microscopic thermal fission cross section of U^{235}, σ_{a8} the microscopic thermal total absorption cross section of U^{238}, and σ_{a5} is the microscopic thermal total absorption cross section of U^{235}.

Using the cross sections given in Table A.5, we find that $\eta = 1.34$ for natural uranium.   □

Let us start by assuming that at some instant a number N_0 of thermal

Table 2.2. Values of v and η for Thermal Fission of Various Fuels[a]

Fuel	v	η
U^{233}	2.49	2.285
U^{235}	2.42	2.07
Pu^{239}	2.87	2.11
Pu^{241}	2.93	2.145

[a] From S. F. Mughabghab and D. I. Garber, *Neutron Cross Sections*, BNL-325, 3rd ed., Vol. 1, Brookhaven National Laboratory, Upton, New York (1973).

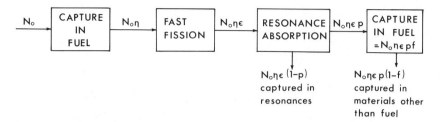

Fig. 2.6. Diagrammatic representation of fission chain reaction.

neutrons are captured in the natural uranium fuel (see Fig. 2.6). The capture of these neutrons will result in the emission of $N_0\eta$ fast neutrons as a result of fission. These fast neutrons will slow down, mainly as the result of collisions with the moderator atoms, but a small proportion will cause fast fission in U^{235} or U^{238}. Because of the high ratio of U^{238} to U^{235}, and the existence of a fission threshold at around 1 MeV for U^{238}, this effect will not be large. Its magnitude is described by *the fast fission factor, ε*, which may be defined as the number of fast neutrons appearing just below the U^{238} fast fission threshold to the number initially present. For natural uranium fuel in a homogeneous mixture with moderator, ε differs by less than 1 % from unity, while in a practical heterogeneous reactor it is approximately 1.03.

We now have a total of $N_0\eta\varepsilon$ neutrons of intermediate energy slowing down into the region of U^{238} resonance absorption. The resonance absorption process is treated in terms of the *resonance escape probability, p*, which is the fraction of the neutrons which escape capture while slowing down through the resonance region. A total of $N_0\eta\varepsilon p$ neutrons therefore appear below this region. These neutrons will diffuse throughout the fuel–moderator mixture until they are captured, either by a nucleus of uranium or one of the nonfuel components of the reactor.

The fraction of these thermal neutrons which is absorbed in the fuel is known as the *thermal utilization, f*. It is given in terms of the macroscopic absorption cross section of the fuel, Σ_{aU}, and of the combined fuel–moderator mixture as a whole, Σ_a, by the relation

$$f = \frac{\Sigma_{aU}}{\Sigma_a} \tag{2.6}$$

It follows that the number of thermal neutrons captured in fuel at the end of the cycle is $N = N_0\eta\varepsilon pf$. The infinite multiplication factor, i.e., the ratio of the number of neutrons in one generation to the number in the preceding generation [see equation (2.2)] is then

$$k_\infty = \eta \varepsilon p f \qquad (2.7)$$

This equation is referred to as the *four-factor formula*.

The condition for a critical reactor of infinite size is then that

$$\eta \varepsilon p f = 1 \qquad (2.8)$$

For the homogeneous natural uranium reactor, ε is close to unity. Taking $\eta = 1.34$, it follows that for a critical system, the product pf must be equal to 0.75. It turns out that, for a homogeneous mixture of uranium and a moderator such as graphite, the resonance escape probability is too low to permit this condition to be met. It will be shown later that "lumping" the fuel results in an increase in p which enables a critical uranium–graphite system to be attained. The other alternative is the enrichment of the U^{235} isotope in the fuel, which increases η and f to the point where the four-factor product can reach unity.

Example 2.3. Calculate the value of the thermal utilization for a mixture of U^{233}, Th^{232}, and graphite in the ratios (by atom) of $1:10:10^4$.

The macroscopic cross section of the carbon is given by

$$\Sigma_{aC} = (\text{number of carbon atoms/cm}^3) \times \sigma_{aC}$$

$$= \rho \frac{N_0}{A} \sigma_{aC}$$

where ρ, the graphite density, is 1.60 g cm^{-3} and $\sigma_{aC} = 3.4 \times 10^{-3} \text{ b}$ (Table A.4). N_0 is Avogadro's (6.022×10^{23}) and $A = 12$. Hence,

$$\text{number of carbon atoms/cm}^3 = \frac{1.60 \times 6.022 \times 10^{23}}{12} = 8.03 \times 10^{22}$$

Hence,

$$\Sigma_{aC} = 8.03 \times 10^{22} \times 3.4 \times 10^{-3} \times 10^{-24} = 2.73 \times 10^{-4} \text{ cm}^{-1}$$

Similarly, for Th^{232},

$$\Sigma_{a2} = \text{number of thorium atoms/cm}^3 \times \sigma_{a2}$$

$$= \frac{8.03 \times 10^{22}}{10^3} \times 7.4 \times 10^{-24} = 5.94 \times 10^{-4} \text{ cm}^{-1}$$

and for U^{233},

$$\Sigma_{a3} = \frac{8.03 \times 10^{22}}{10^4} \times 578.8 \times 10^{-24} = 4.648 \times 10^{-3} \text{ cm}^{-1}$$

Hence, by equation (2.6),

$$f = \frac{\Sigma_{a3}}{\Sigma_{a3} + \Sigma_{aC} + \Sigma_{a2}} = \frac{4.648}{4.648 + 0.273 + 0.594}$$
$$= 0.843 \qquad \square$$

For a reactor of finite size, the condition for criticality is that $k_{\text{eff}} = 1$. From equation (2.3),

$$k_\infty \mathscr{L} = \eta \varepsilon p f \mathscr{L} = 1 \qquad (2.9)$$

Since $\mathscr{L}$ is always less than unity for a reactor of finite size, the value of the product $\eta \varepsilon p f$ must be correspondingly increased for criticality. In addition, any practical reactor must start off with a certain amount of excess multiplication in hand, to allow for control and to compensate for the gradual decrease in k_∞ as the fuel is burned up and fission product poisons accumulate.

For a natural uranium–graphite system, even with optimum design of the fuel rod dimensions and spacing, and careful choice of low-absorption structural materials, the maximum attainable value of k_∞ is in the region of 1.07. The small margin in k_∞ means that the nonleakage probability must be close to unity. Consequently, a reactor using natural uranium as fuel must necessarily be very large to minimize leakage; the first-generation reactors of the British graphite-moderated design, for example, had a graphite structure up to 17 m in diameter and 9 m high. One way of reducing leakage is to surround the reactor with a *reflector*, a region of moderating material which will scatter back into the reactor some of the neutrons which would otherwise have escaped.

Control of the reactor is effected by using *control rods* of some material of suitably high neutron absorption cross section, such as cadmium or boron.

The size and composition of the reactor are chosen to be such that the system would be supercritical in the absence of the control rods. The fuel is loaded with the control rods inserted and, on the completion of loading, the control rods are slowly raised, increasing the thermal utilization until criticality is attained. If the control rods are then withdrawn by a further small amount, increasing k_{eff} to a value slightly above unity, the system will be supercritical and the neutron density in the reactor, and hence the fission rate, will increase continuously. Once the desired power level has been attained, the rods are readjusted to restore the system to the critical condition, and the power will remain steady at the chosen level. As has been mentioned earlier, the only reason why the rate of power increase can be held to a practicable value is the existence of delayed neutrons; the dynamics of reactor control will be discussed in detail in Chapter 3.

2.5. The Role of Plutonium in Nuclear Reactors

The neutrons absorbed by U^{238} in a natural uranium reactor are not wasted, but lead eventually to the production of Pu^{239}, which is also fissile. The sequence for production of Pu^{239} is as given below:

$$_{92}U^{238} + {}_0n^1 \longrightarrow {}_{92}U^{238}$$

The plutonium is formed as the result of two successive β decays with the half-lives indicated:

$$_{92}U^{239} \xrightarrow[23 \text{ min}]{} {}_{93}Np^{239} + {}_{-1}\beta^0$$

$$_{93}Np^{239} \xrightarrow[2.3 \text{ d}]{} {}_{94}Pu^{239} + {}_{-1}\beta^0$$

The production of Pu^{239} from U^{238} is known as *conversion* and U^{238} is called a *fertile* material. Another fertile material of importance is thorium, where the Th^{232} is converted to fissile U^{233} by the following sequence of events:

$$_{90}Th^{232} + {}_0n^1 \longrightarrow {}_{90}Th^{233}$$

$$_{90}Th^{233} \xrightarrow[22 \text{ min}]{} {}_{91}Pa^{233} + {}_{-1}\beta^0$$

$$_{91}Pa^{233} \xrightarrow[27.4 \text{ d}]{} {}_{92}U^{233} + {}_{-1}\beta^0$$

While the production of Pu^{239} will take place automatically in any uranium-fueled reactor in which the fuel is not pure U^{235}, the production of U^{233} from thorium requires that the latter be deliberately introduced into the reactor.

The buildup of Pu^{239} in a reactor fueled with natural or low-enrichment uranium is of value in extending the life of the fuel charge. As the U^{235} content decreases due to burn-up, the corresponding buildup of Pu^{239} reaches the point where more than half of the fissions taking place in the reactor are occurring in the plutonium. The prediction of the effect of Pu^{239} buildup is complicated by the fact that irradiation of the Pu^{239} formed leads eventually to the formation of another fissile material, Pu^{241}, through the following sequence:

$$_{94}Pu^{239} + {_0}n^1 \longrightarrow {_{94}}Pu^{240}$$

$$_{94}Pu^{240} + {_0}n^1 \longrightarrow {_{94}}Pu^{241}$$

The effects of the production of Pu^{239} and the higher isotopes of plutonium in a high-power thermal reactor will be discussed in detail in Chapter 4. For the present, we will note the importance of the *conversion ratio*, which is defined as

$$C = \frac{\text{number of fissile atoms produced from the fertile isotope/second}}{\text{number of fissile atoms of fuel destroyed (by fission or capture)/second}}$$

$$(2.10)$$

To illustrate, consider the case of a natural uranium reactor, for which the processes leading to plutonium formation are shown in Fig. 2.7. For simplicity, the reactor is assumed to be infinite, so that leakage is negligible. The sequence is normalized to the absorption of one thermal neutron in the uranium. On average, the absorption of this neutron leads to the destruction of $\Sigma_{a5}/(\Sigma_{a5} + \Sigma_{a8})$ atoms of U^{235}, where Σ_{a5} and Σ_{a8} are the thermal (macroscopic) cross sections for the two uranium isotopes, and to the release of η_U fast neutrons, where η_U is the value appropriate to natural uranium fuel. Noting that $\eta_U \varepsilon p f = 1$ (critical condition) we have

$$C = \frac{\varepsilon \eta_U (1 - p) + \Sigma_{a8}/(\Sigma_{a5} + \Sigma_{a8})}{\Sigma_{a5}/(\Sigma_{a5} + \Sigma_{a8})}$$

$$= \varepsilon (1 - p)\eta_U \frac{\Sigma_{a5} + \Sigma_{a8}}{\Sigma_{a5}} + \frac{\Sigma_{a8}}{\Sigma_{a5}}$$

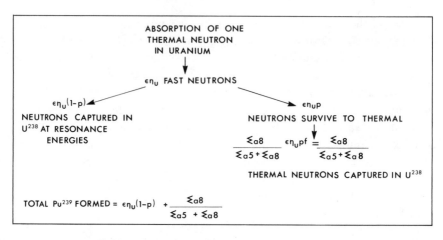

Fig. 2.7. Conversion ratio in a natural uranium reactor.

Now,

$$\eta_U = v \frac{\Sigma_{f5}}{\Sigma_{a5} + \Sigma_{a8}}$$

where Σ_{f5} is the (macroscopic) fission cross section of U^{235}. The corresponding value of η for U^{235} alone is

$$\eta_5 = v \frac{\Sigma_{f5}}{\Sigma_{a5}}$$

Hence

$$\eta_U \frac{\Sigma_{a5} + \Sigma_{a8}}{\Sigma_{a5}} = \eta_5$$

and we have

$$C = \varepsilon(1 - p)\eta_5 + \frac{\Sigma_{a8}}{\Sigma_{a5}} \tag{2.11}$$

Example 2.4. Calculate the conversion ratio for an infinite, graphite-moderated reactor, with $\varepsilon = 1.03$ and $p = 0.90$.

Taking the thermal absorption cross sections of U^{235} and U^{238} as 680.8 and 2.70 b, respectively (see Table A.5), we have

$$\frac{\Sigma_{a8}}{\Sigma_{a5}} = \frac{99.28 \times 2.70}{0.72 \times 680.8} = 0.547$$

Hence,

$$C = (1.03 \times 0.10 \times 2.07) + 0.547 = 0.760$$

(The value calculated is strictly the initial conversion ratio, since it applies only at the start of life of the reactor core, before any depletion of the U^{235} content has taken place.) □

While, in the case considered, the plutonium is being formed at a rate which is less than that at which the U^{235} is being consumed, it is possible to design a reactor in which the new fissile material will be produced at a rate which exceeds the fuel consumption rate. Such a reactor is known as a *breeder*. The only thermal reactor which offers the possibility for breeding is one where the fissile material is U^{233} and the fertile is Th^{232}.

As shown in Table 2.2, U^{233} has the highest η value of the fissile materials listed, with $\eta = 2.285$. To keep the chain reaction going, one of these neutrons, on average, must be captured eventually in U^{233}, to yield a further fission. Let us assume that another is captured by Th^{232} to form a replacement for the original U^{233} atom which underwent fission. Then we have 0.285 neutrons left over, which could, in principle, result in the U^{233} generation rate exceeding the rate at which it is being burned up. Since we have to allow for leakage and parasitic capture in moderator, coolant, and structural materials, however, it is only by expressly designing the reactor for good neutron economy that it is possible to achieve breeding with a thermal system. The only reactor which appears to make thermal breeding a practical possibility is the heavy-water-moderated and cooled system such as the Canadian CANDU reactor.

While the U^{233}–Th^{232} cycle is the only one which offers the possibility of breeding in a thermal reactor, the situation is different at high neutron energies due to the marked increase in the η values of the fissile plutonium isotopes, caused by an improvement in the ratio of σ_f/σ_c as the energy increases. For Pu^{239}, for example, the η value rises from 2.11 for thermal neutrons to 2.40 for neutrons of around 100-keV energy. This fact has led to considerable interest in the successful development of a *fast breeder reactor*, where the objective is to include the minimum amount of moderating material so that the mean energy of the neutrons causing fission remains high enough that a favorable value of η is achieved. The typical reactor of this type has a central core region containing Pu^{239} as fuel, surrounded by a so-called *blanket* of natural uranium, in which conversion of U^{238} to Pu^{239} can take place at a rate which allows breeding to occur. What such a reactor does is essentially to allow the whole of the natural uranium to be used as fuel, rather than simply the U^{235} component, and it therefore opens the road to a greater fuel supply for nuclear reactors.

2.6. Nuclear Cross Sections of the Fissile Isotopes

The fission cross section for U^{235} is shown in Fig. 2.8.

The intermediate energy region shows the closely spaced resonance structure to be expected for a massive even Z-odd N nucleus, where the addition of a single neutron is sufficient to raise the compound nucleus to an excitation energy where the level spacing is only a few eV. In the thermal energy region the capture cross section is about one sixth of the fission cross section, and, like the latter, exhibits a $1/v$ behavior. The fission cross section of U^{238}, which is significant only above 1 MeV, is shown for comparison in Fig. 2.9.

The behavior of U^{233} is generally similar to that of U^{235}, but the Pu^{239} and Pu^{241} cross sections are characterized by the presence of a prominent resonance at around 0.3 eV (see Fig. 2.10). As will be seen later, the

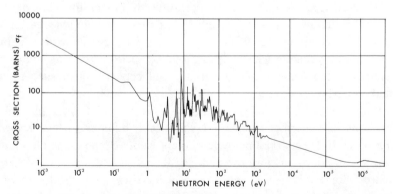

Fig. 2.8. Fission cross section of U^{235} [Based on *BNL 325*, Second Edition, Brookhaven National Laboratory (1965)].

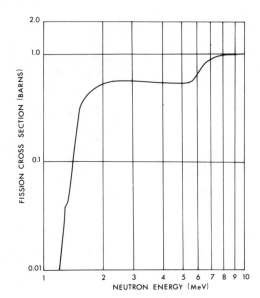

Fig. 2.9. Fission cross section of U^{238} [Based on *BNL 325*, Second Edition, Brookhaven National Laboratory (1965)].

absorption and fission rates in both these isotopes are sensitive to factors, such as the reactor temperature, which produce a change in the energy distribution of the thermal neutrons in the system.

Table 2.3 lists the values of the absorption and fission cross sections for the fissile isotopes, for neutrons of energy 0.0253 eV.

The parameter α, quoted in the last column, is the ratio of the capture and fission cross sections, i.e.,

$$\alpha = \sigma_c/\sigma_a \tag{2.12}$$

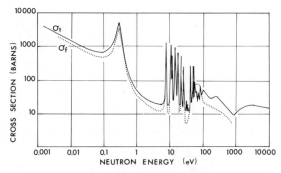

Fig. 2.10. Total and fission cross section of Pu^{239} [Based on *BNL 325*, Second Edition, Brookhaven National Laboratory (1965)].

Table 2.3. Nuclear Cross Sections for the Fissile Isotopes[a]

Isotope	Cross section at 0.0253 eV (b)		Capture-to-fission ratio (α)
	σ_a	σ_f	
U^{233}	579	531	0.090
U^{235}	681	582	0.170
Pu^{239}	1011	742.5	0.362
Pu^{241}	1377	1009	0.365

[a] From S. F. Mughabghab and D. I. Garber, *Neutron Cross Sections*, BNL-325, 3rd ed., Vol. 1, Brookhaven National Laboratory, Upton, New York (1973).

Hence, using equation (2.4),

$$\eta = v\frac{\sigma_f}{\sigma_a} = v\frac{\sigma_f}{\sigma_f + \sigma_c} = \frac{v}{1 + \alpha} \qquad (2.13)$$

It is, of course, the relatively high value of α for the plutonium isotopes that results in their thermal η values being of roughly the same magnitude as those for U^{233} and U^{235}, despite the higher number of neutrons per fission in the former case (see Table 2.2). The consequences of the variation of α with neutron energy will be considered in detail in later chapters.

Problems

(The data required for the solution of the problems are given in Tables A.1–A.5).

2.1. A nucleus of $_{94}Pu^{239}$ captures a slow neutron and undergoes fission to form the fission products $_{43}Tc^{107}$ and $_{51}Sb^{133}$. Calculate the energy release in the fission.

2.2. A nucleus of $_{92}U^{238}$ undergoes spontaneous fission to yield two nuclei of $_{46}Pd^{119}$. One of the fission fragments emits four neutrons to form $_{46}Pd^{115}$, which then undergoes four successive β-particle emissions to form stable $_{50}Sn^{115}$. The other emits five neutrons to form $_{46}Pd^{114}$, which undergoes two successive β-particle emissions to form stable $_{48}Cd^{114}$. Calculate the total energy released in the complete process described.

2.3. Calculate the total energy (in MW d) released by the fission of all the $_{92}U^{235}$ atoms in 1 kg of natural uranium.

2.4. Calculate the value of k_∞ for a uniform homogeneous mixture of uranium and carbon with the following atomic ratios: $U^{238}/U^{235} = 9.0$ and $C/U^{235} = 20,000$. (Take $p = 0.95$ and $\varepsilon = 1.00$).

2.5. A finite reactor is composed of a uniform mixture of U^{235}, beryllium, and boron in the following atomic ratios: $Be/U^{235} = 10,000$, $B/U^{235} = 0.5$. If the reactor is just critical, calculate the nonleakage probability.

2.6. Calculate the value of the initial conversion ratio for a reactor with the characteristics given in Problem 2.4.

3

Elements of Nuclear Reactor Theory

3.1. The Neutron Energy Spectrum in a Nuclear Reactor

A knowledge of the physical theory of nuclear reactors is necessary in order to understand their design and operating characteristics. For example, the calculation of the rate of power generation at different points in the core requires a knowledge of the neutron density and neutron energy distribution at each of these points. Other operating characteristics dependent on an understanding of the physical processes in the reactor include the response of the plant to changes in control rod position or operating power level and the effects of the gradual burn-up of the fuel and the accumulation of fission products.

The power generation rate at any point in a reactor is determined by the value of a quantity known as the *neutron flux* at that position. The concept of neutron flux was introduced, for the simple case of a beam of mono-energetic neutrons, in Section 1.6. In that case, when the density of neutrons in the beam is n neutrons per cm^3, and all the neutrons are traveling in the same direction with a common speed v, the neutron flux at any point in the beam is defined as

$$\phi = nv \qquad (3.1)$$

The units of neutron flux are neutrons per cm^2 per second. The description of the situation at any point in a reactor requires an extension of the simple definition given above to take account of the fact that the neutrons there have a wide range of energies and may be traveling in any direction. The most general description of the neutron distribution throughout the reactor will be by a function of the form

$$n(r, E, \Omega, t)\, dr\, dE\, d\Omega$$

which is defined as the number of neutrons at time t in the volume element dr

at the position r, whose energies lie in the range E to $E + dE$ and whose directions of motion lie within the solid angle range between Ω and $\Omega + d\Omega$. The corresponding neutron flux is given by multiplying by the neutron speed, i.e.,

$$\phi(r, E, \Omega, t) = v(E)n(r, E, \Omega, t) \qquad (3.2)$$

It was noted in Section 1.6 that the number of neutron absorptions taking place per unit volume per second in a region where the neutron flux is described by the product $\phi = nv$ is given by

$$\text{absorption rate} = \Sigma_a nv = \Sigma_a \phi \qquad (3.3)$$

where Σ_a is the macroscopic absorption cross section in the region. By an extension of the above, the absorption rate of neutrons in the volume element dr at position r is

$$\text{absorption rate} = \int_0^{4\pi} \int_0^{E_{max}} \Sigma_a(r, E)\phi(r, E, \Omega, t)\, dr\, dE\, d\Omega \qquad (3.4)$$

where $\Sigma_a(r, E)$ is the macroscopic absorption cross section at energy E at the position r, and E_{max} is the maximum neutron energy.

The quantity $\Sigma_a(r, E)$ is calculable provided that the composition of the reactor at the position r is given, and the nuclear absorption cross sections of the materials involved are known as a function of energy. The calculation of the quantity $\phi(r, E, \Omega, t)$, however, is an extremely complex one, as may be seen by considering the processes which the neutrons undergo in the reactor. High-energy neutrons are produced by fission in the fuel and then follow complicated zigzag paths, losing energy in a discontinuous fashion as a result of collisions with the nuclei of the moderator and other materials. During this process, some of the neutrons will be absorbed and others will escape from the reactor altogether. The situation is further complicated since the composition of the reactor is nonuniform because the fuel is in general concentrated in fuel rods separated by regions of moderator and because there are concentrations of absorbing material, such as control rods, which produce large local variations in the neutron flux.

The complexity of the problem makes it necessary to use sophisticated computer analysis to produce a complete description of the variation of neutron flux in the reactor. The most rigorous approach, which takes account of the directional properties of the neutron flux, i.e., the dependence on the variable Ω, is known as *transport theory*. It is possible, however, by introducing a number of simplifying assumptions, and by restricting the complexity of the systems treated, to produce reasonable approximations to the energy and spatial variations of the neutron flux.

We shall firstly simplify the problem by considering only the *steady state critical reactor*, i.e., a system in which the production rate of neutrons by fission is exactly balanced by the loss of neutrons by absorption and leakage. At this stage we shall also consider the reactor to be a completely *homogeneous* and *isotropic* mixture of fuel and moderator, and to be *infinite* in extent. Since all points of the medium are indistinguishable, this means that we have removed any dependence of the neutron flux on either r, or Ω, so that the only variable left is the neutron energy, E. Since the medium is homogeneous and isotropic, the sources of fission neutrons are distributed uniformly throughout the medium. For simplicity, we assume that all the fission neutrons are produced at the same energy, E_0, and are then slowed down by elastic collisions with the nuclei of the moderator. In this way we can obtain an expression for the variation of the neutron flux as a function of the energy variable alone.

The typical elastic scattering process is shown in Fig. 3.1. The neutron collides with a nucleus of one of the atoms of the medium (assumed to be at rest before the collision) and rebounds at an angle θ to its original direction of travel with reduced speed, while the struck nucleus is set in motion with a kinetic energy equal to that lost by the neutron. The kinetic energy lost by the neutron, ΔE, is a function of the angle θ, but it can be shown that, for neutrons scattered through a given angle θ, the *fractional loss of energy is independent of the initial energy of the neutron*, E, i.e.,

$$\frac{\Delta E}{E} = \text{const} \tag{3.5}$$

Now $\Delta E/E$ is equal to $\Delta(\ln E)$, the difference in the logarithms of the initial and final energies (E_f and E_i). The value of $\Delta(\ln E)$ in any collision will vary, depending on the angle of scattering, θ, but for a given moderator we can define a quantity known as the *average logarithmic energy decrement per collision* (ξ). We have

$$\xi = [\ln E_i - \ln E_f]_{\text{av}} = [\ln(E_i/E_f)]_{\text{av}} \tag{3.6}$$

The larger the value of ξ, the more effective the moderator is in slowing down neutrons. Values of ξ for scattering in some typical reactor materials

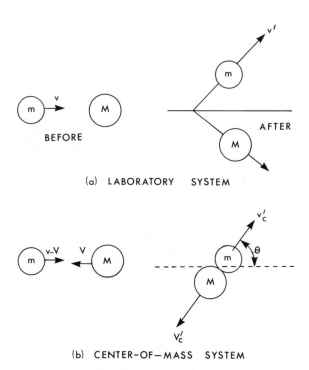

Fig. 3.1. Elastic scattering of neutron by nucleus of mass M, viewed (a) in the laboratory coordinate system, where the nucleus is at rest before the collision and (b) in the center-of-mass system, in which the center of mass of the two colliding bodies is assumed to be at rest.

are given in Table 3.1. These values can be used to calculate the average number of collisions which a neutron has to make with nuclei of a given moderator in order to reduce its energy by a specified amount.

Example 3.1. Find the number of collisions required to reduce the energy of a neutron from 2 MeV to thermal energy E (approximately 0.025 eV) when the moderator is graphite.

Since ξ is the average logarithmic energy decrement, the number of collisions is

$$\frac{\ln(E_0/E)}{\xi} = \frac{\ln(2 \times 10^6/0.025)}{0.158} = 115 \qquad \square$$

The effectiveness of a moderator in slowing down neutrons depends not

Table 3.1. Values of Average Logarithmic Energy Decrement

Element	Mass number	ξ
Hydrogen	1	1.000
Deuterium	2	0.725
Beryllium	9	0.207
Carbon	12	0.158
Uranium	238	0.0084

only on the value of ξ but also on the probability per unit volume that scattering does take place. Quantitatively then, the effectiveness may be described by the product $\xi\Sigma_s$, where Σ_s is the macroscopic scattering cross section of the moderator. This product is known as the *slowing-down power*. It is not in itself a sufficient indication of the suitability of the material as a moderator, since one also has to take into consideration the absorption cross section. Boron, for example, which has a good slowing-down power, is impractical as a moderator owing to its very high absorption for thermal neutrons. We can obtain an overall index of suitability by calculating the *moderating ratio*, which is defined as $\xi\Sigma_s/\Sigma_a$, where Σ_a is the macroscopic absorption cross section of the element.

Table 3.2 gives values of slowing-down power and moderating ratio for the commonly used moderators. The slowing-down power is calculated for scattering in the region from several keV down to 1 eV; since Σ_s will vary somewhat over this range, the values are approximate only. The moderating ratio has been obtained by dividing by the thermal neutron absorption cross section.

It may be shown fairly readily that, for a uniform, infinitely large mix-

Table 3.2. Slowing-Down Power and Moderating Ratio

Moderator	Slowing-down power (cm^{-1})	Moderating ratio
Water (H$_2$O)	1.40	70
Heavy water (D$_2$O)	0.175	6000
Beryllium	0.16	140
Graphite	0.060	220

ture of fuel and moderator, the energy dependence of the neutron flux is given by the expression

$$\phi(E) = \frac{S_0}{\Sigma_s \bar{\xi} E} \tag{3.7}$$

where $\phi(E)$ is the flux per unit energy at the energy E, Σ_s the macroscopic scattering cross section of the mixture, $\bar{\xi}$ the average logarithmic energy decrement of the mixture (calculated by weighting the ξ values of the fuel and moderator by their respective macroscopic scattering cross sections), and S_0 is the number of neutrons produced per second as a result of fission.

It is clear from equation (3.7) that the flux at any particular energy E is proportional to the inverse of the energy, i.e., that

$$\phi(E) \propto \frac{1}{E} \tag{3.8}$$

This relation applies over a wide range of energy, but ceases to apply at low neutron energies. As the neutrons slow down to low speeds, they eventually come into thermal equilibrium with the atoms of the moderator, and the neutron is as likely to gain energy as to lose energy in a given collision.

At this stage, it is useful to introduce a new quantity called the *neutron lethargy*. The lethargy of a neutron is closely related to its energy, the lethargy (u) of a neutron of energy E being defined as

$$u = \ln(E_0/E) \tag{3.9}$$

where E_0 is the energy of the source neutrons. It is obvious that the lethargy of the neutron increases as its energy decreases.

The change in lethargy between the two energies E_1 and E_2 is

$$u_2 - u_1 = \ln[E_1/E_2] \tag{3.10}$$

Also, we see from equation (3.9) that

$$du = -d(\ln E) = -\frac{dE}{E} \tag{3.11}$$

Since the neutron flux in a given lethargy interval must be the same as that in the corresponding energy interval,

$$\phi(u)\, du = -\phi(E)\, dE \qquad (3.12)$$

the negative sign taking account of the fact that changes in energy and lethargy must be in the opposite direction.

Combining (3.11) and (3.12), we have

$$\phi(u) = E\phi(E) \qquad (3.13)$$

The equivalent of equation (3.7) in lethargy units is

$$\phi(u) = \frac{S_0}{\Sigma_s \bar{\zeta}} \qquad (3.14)$$

Hence, when measured in lethargy units, the neutron flux is constant over most of the range of energies (or lethargies) from the source energy down to an energy somewhat above thermal. This is illustrated in Fig. 3.2, which shows the energy spectrum of a typical light water reactor. In practice, the fission neutrons are emitted over a range of several lethargy units, giving rise to a complex peak at high energies. The scattering cross section Σ_s of hydrogen drops appreciably at high energy, leading to a lower rate of energy loss, and increasing the peak due to the supply of source neutrons of low lethargy. The peak at high lethargy arises since the neutrons on reaching

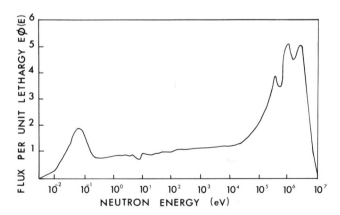

Fig. 3.2. Typical neutron spectrum for a light water reactor [from C. G. Campbell *et al.*, *J. Brit. Nucl. Energy Soc.* **I**, 364 (1962). Courtesy of the British Nuclear Energy Society].

thermal energies are no longer subject to a steady loss of energy and accumulate until they are absorbed.

For a well-moderated reactor, such as a dilute uranium–graphite type, the energies (or velocities) of the *thermal* neutrons are distributed according to the well-known *Maxwell–Boltzmann distribution*, which also occurs in the kinetic theory of gases. The number of neutrons per cm^3 per unit velocity interval at velocity v is

$$n(v) = 4\pi(m/2\pi kT)^{3/2}n_0 v^2 \, e^{-mv^2/2kT} \qquad (3.15)$$

where m is the neutron mass (1.675×10^{-27} kg), k is the Boltzmann constant (1.3805×10^{-23} J/K), T is the moderator temperature (in K), and n_0 is the total number of neutrons per cm^3.

The form of the Maxwellian distribution is shown in Fig. 3.3. For specifying parameters such as the absorption cross sections of reactor materials for neutrons of thermal energies, it is found convenient to characterize the neutron distribution in terms of the *most probable speed*, i.e., that corresponding to the peak of the curve, which is obtained by taking

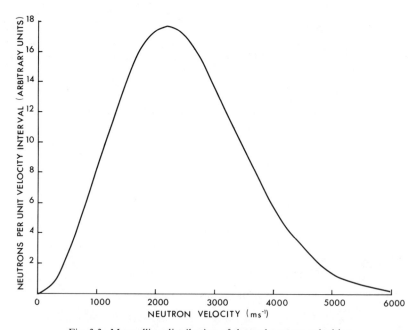

Fig. 3.3. Maxwellian distribution of thermal neutron velocities.

$dn(v)/dv = 0$ in equation (3.15). This speed turns out to be

$$v_0 = (2kT/m)^{1/2} \tag{3.16}$$

For neutrons in equilibrium with moderator at room temperature (293 K), $v_0 = 2200$ ms^{-1}, corresponding to a kinetic energy for the neutron of 0.0253 eV. Thermal neutron cross sections are customarily quoted for neutrons of this energy. For example, the neutron absorption rate per unit volume of any constituent of the reactor will be given by

$$R = N \int_0^\infty \sigma_a(v)n(v)v \, dv \tag{3.17}$$

where N is the number of atoms of the constituent per cm^3 and $\sigma_a(v)$ is the microscopic absorption cross section at neutron velocity v.

The calculation of this reaction rate becomes a complex matter when the cross section of the material varies rapidly with energy over the region of the thermal neutron spectrum. Fortunately, as mentioned in Section 1.7, most materials of interest have cross sections which have a $1/v$ dependence, or an approximate $1/v$ dependence, in this region. The absorption cross section of a $1/v$ absorber may be written as

$$\sigma_a(v) = \sigma_{a0} \frac{v_0}{v} \tag{3.18}$$

where σ_{a0} is the cross section at $v_0 = 2200$ ms^{-1}.

Substituting in equation (3.17), we have

$$R = N\sigma_{a0}v_0 \int_0^\infty n(v) \, dv \tag{3.19}$$

The integral is simply the total density of thermal neutrons, n_0, so that

$$R = N\sigma_{a0}(n_0v_0) = N\sigma_{a0}\phi_0 = \Sigma_{a0}\phi_0 \tag{3.20}$$

Hence the absorption rate is obtained simply by multiplying the macroscopic cross section at a neutron velocity of 2200 ms^{-1} by the quantity $\phi_0 = n_0v_0$, which is known as the *2200-meter-per-second flux*. It is for this reason that the cross sections are conveniently quoted at a neutron energy of 0.0253 eV.

For a nucleus whose cross section varies only approximately as $1/v$ in the thermal region, it is possible to calculate an empirical correction factor, $g(T)$, know as the *Westcott g factor*, based on the known variation of the cross section with energy. In this case, the absorption rate is calculated from the expression

$$R = g_a(T)\Sigma_{a0}\phi_0 \tag{3.21}$$

where the g factor is a function of the temperature T. Values of Westcott g factors for some of the more important reactor constituents are given in Appendix III.

Example 3.2. Calculate the power generation in a dilute homogeneous graphite-moderated reactor where the fuel content is 10 kg of pure U^{235}. The reactor is operating at a temperature of 600°C and the average 2200-ms^{-1} flux in the core is $\phi_0 = 2 \times 10^{13}$ n cm^{-2} s^{-1}.

If the volume of the core is V cm^3, the total fission rate is

$$F = g_f(T)\,\Sigma_{f0}\phi_0 V \tag{3.22}$$

where $g_f(T)$ is the Westcott g factor. Since the core is at a temperature of 600°C, $g_f(T)$ is equal to 0.9229.

Since the macroscopic fission cross section is given by equation (1.21) as

$$\Sigma_{f0} = \frac{\rho N_0}{A}\,\sigma_{f0}$$

where ρ is the mass of U^{235} per cm^3, and σ_{f0} the 2200-ms^{-1} fission cross section of U^{235} (582 b), we have

$$F = g_f(T)\frac{MN_0}{A}\,\sigma_{f0}\phi_0 \tag{3.23}$$

where $M = \rho V$ is the total mass of U^{235} in the core.

The amount of recoverable energy per fission, χ, for U^{235} is approximately 200 MeV, or 3.2×10^{-11} J. Hence, the thermal power of the reactor, or rate of release of recoverable energy, is

$$\text{power} = g_f(T)\frac{MN_0}{A}\,\sigma_{f0}\phi_0\chi$$

$$= \frac{0.9229 \times 20 \times 10^3 \times 6.022 \times 10^{23} \times 582 \times 10^{-24} \times 2 \times 10^{13} \times 3.2 \times 10^{-11}}{235}$$

$$= 17.6 \times 10^6 \text{ W} = 17.6 \text{ MW} \tag{3.24}$$

$\square$

3.2. The Spatial Dependence of the Neutron Flux in a Reactor

In the previous section we isolated the energy dependence of the neutron flux by considering an infinite reactor, thereby eliminating any variation of flux with position. In a finite reactor, the neutron flux will be a function of the spatial coordinates as well. Figure 3.4 illustrates the way in which the neutron flux will vary with position in the particular case of a reactor which is in the form of a slab of thickness a, the reactor being assumed to be infinitely long in the other two dimensions. The flux has a maximum at the central plane and decreases steadily as one moves out towards the edge, where the flux has a zero, or approximately zero, value.

In order to work out the flux distribution in a finite reactor, we have to consider the process of *neutron transport* from one part of the reactor to another. A neutron which is born as a result of a fission process will follow a zigzag path, traveling in a straight line between collisions with the nuclei of the reactor material, until it is eventually absorbed or escapes from the reactor. The simplest way of treating the problem is to regard the neutron motion as a *diffusion* process, in which neutrons have a tendency to diffuse from regions of high flux (or neutron density) to regions of lower flux as a result of a *flux gradient*.

In elementary diffusion theory, we assume that the neutron flux is a function of position only, and ignore any angular dependence. Thus, for example, we assume that neutron scattering is an *isotropic* process, i.e., that if we consider a large number of scattering events, the numbers of neutrons scattered into a given solid angle about any particular direction will be the same as the number scattered into the same solid angle at any other direction. This is not strictly true, particularly for the scattering of fast neutrons by a

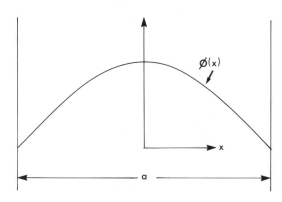

Fig. 3.4. Variation of neutron flux in an infinite slab reactor of thickness a.

light nucleus such as hydrogen, but for most practical cases it turns out to be an acceptable approximation.

We shall also simplify the problem by assuming that all the neutrons are *monoenergetic*. This, of course, is obviously far from the real situation, where the neutrons have energies ranging from several MeV (for fission neutrons) down to a small fraction of an eV (for neutrons in thermal equilibrium). The additional complication of the energy variation, however, can be dealt with at a later stage by dividing the neutron energy spectrum into a number of groups of effectively monoenergetic neutrons.

The basic equation of diffusion theory may be derived by considering the neutron balance in a small element of volume in the reactor. We assume steady state conditions, i.e., that the neutron density remains constant as it would in a just-critical reactor. The neutron balance in the element considered involves the following components: (a) the number of neutrons, S, which are created per second in the element by fissions occurring there, (b) the number of neutrons $(\Sigma_a \phi)$ which are absorbed in the element per second, (c) the number of neutrons scattered into the element per second as a result of collisions which have occurred in other parts of the reactor, and (d) the number of neutrons scattered out of the element per second due to collisions occurring within it.

The difference between the last two components, i.e., the number scattered out of the element minus the number scattered in, is known as the net *leakage* from the element. It can be shown that the expression for the net leakage out of a unit volume per second is

$$\text{net leakage} = -D\nabla^2\phi \qquad (3.25)$$

where D is a parameter known as the *diffusion coefficient* of the medium in which the scattering takes place. To a good approximation, for the majority of scattering materials, the diffusion coefficient is given by

$$D = \frac{1}{3\Sigma_s} = \frac{1}{3}\lambda_s \qquad (3.26)$$

where λ_s is the *scattering mean free path*, the average distance traveled by a neutron between scattering events.

We can now write down the neutron balance equation for the element considered since, in the steady state, we must have

$$\text{production rate} - \text{net leakage rate} - \text{absorption rate} = 0$$

Hence,

$$S + D\nabla^2\phi - \Sigma_a\phi = 0 \tag{3.27}$$

or

$$D\nabla^2\phi - \Sigma_a\phi + S = 0 \tag{3.28}$$

This is the *diffusion equation* for monoenergetic neutrons, or the *one-group diffusion equation*. It is remarkable that, despite the fact that the neutrons in a reactor are very far from being monoenergetic, it is possible, with a suitable choice of cross sections, to use the one-group diffusion equation to yield a reasonable approximation to the shape of the neutron flux in the reactor. Before pursuing this approximation further, however, it is worth indicating how the one-group approach may be expanded by dividing the complete energy range of the neutrons into a number of groups, each of which may be considered to satisfy an equation of the type of (3.28) above. This approach is known as the *multigroup method*.

In this method, the energy range between fission energies and thermal is divided into n groups, the first group containing the high-energy fission neutrons and the nth the thermal neutrons. The group boundaries are chosen so that the important parameters, such as the absorption cross sections of the constituents, vary sufficiently slowly over the energy region of the group that they may be adequately represented by an average value. In general, the boundaries are also chosen so that neutrons can be scattered into the energy range of a given group only from energies in the immediately adjacent higher-energy group, so the scattering into group i, for example, can be described by a term $\Sigma_s(i - 1 \rightarrow i)\phi_{i-1}$, where $\Sigma_s(i - 1 \rightarrow i)$ is a suitably averaged scattering cross section from one group to the other. The equation describing the neutron balance for any group i is then

$$D_i\nabla^2\phi_i - \Sigma_{ai}\phi_i - \Sigma_s(i \rightarrow i + 1)\phi_i + \Sigma_s(i - 1 \rightarrow i)\phi_{i-1} = 0 \tag{3.29}$$

Equations of this type will apply to all groups except those of highest and lowest energy. The neutron source for the first group is the fission neutrons produced by the thermal flux, ϕ_n. The total rate of capture of the thermal neutrons is $\Sigma_{an}\phi_n$ of which a fraction f are captured in the fuel. The number of fission neutrons produced per neutron captured in fuel is η, so that the neutron source term for the first group is $\eta f \Sigma_{an}\phi_n$. The balance equation for the first group is therefore

$$D_1\nabla^2\phi_1 - \Sigma_{a1}\phi_1 - \Sigma_s(1 \rightarrow 2)\phi_1 + \eta f \Sigma_{an}\phi_n = 0 \tag{3.30}$$

The equation for the thermal group is

$$D_n\nabla^2\phi_n - \Sigma_{an}\phi_n + \Sigma_s(n - 1 \to n)\phi_{n-1} = 0 \qquad (3.31)$$

We now have a set of n equations which may be solved to find the values for the fluxes in each of the groups. The solution of the multigroup equations in the general case requires complicated computer-based calculations, but, as mentioned earlier, it is possible to obtain reasonable results in many cases by the use of only one or two energy groups. The one-group and two-group approaches are summarized below.

3.2.1. One-Group Method

Here it is effectively assumed that the fission neutrons are produced at thermal energy, so that the only equation involved is that for the thermal neutrons. The source term for this single group is then $\eta f\Sigma_a\phi$, as for the first group of the multigroup case. The product ηf may be replaced by k_∞ (see Section 2.4), since both the fast fission factor and the resonance escape probability must obviously equal unity when the fission neutrons are produced at thermal energy. The equation for the one-group flux ϕ *for the critical reactor* is then

$$D\nabla^2\phi - \Sigma_a\phi + k_\infty\Sigma_a\phi = 0 \qquad (3.32)$$

or

$$\nabla^2\phi + \frac{\Sigma_a}{D}(k_\infty - 1)\phi = 0 \qquad (3.33)$$

At this point we introduce a quantity known as the *diffusion length* in the reactor, denoted by the symbol L. This is defined by the relation

$$L^2 = D/\Sigma_a \qquad (3.34)$$

so that equation (3.33) becomes

$$\nabla^2\phi + \frac{k_\infty - 1}{L^2}\phi = 0 \qquad (3.35)$$

It can be shown that the square of the diffusion length (*the diffusion area*) is equal to one sixth of the mean square distance from the point of origin of a

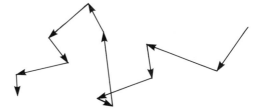

Fig. 3.5. Typical neutron path in a scattering medium.

thermal neutron to its point of capture. (The actual distance traversed by the neutron between these two points is, of course, considerably longer, since it consists of a sequence of random straight-line paths between successive collisions, as in Fig. 3.5.) The diffusion length is an important parameter in reactor design; among other things, the amount of neutron leakage from the reactor is found to depend directly on the value of the diffusion area. Values of the diffusion coefficient, diffusion length, and diffusion area for some common moderators are given in Table 3.3.

We can now illustrate the use of the one-group equation for the critical reactor by taking a particularly simple geometry. We assume that the reactor is in the form of a slab of thickness a, which is infinite in the other two dimensions. In this case the only spatial variable affecting the flux is the variable x (Fig. 3.6) and equation (3.35) reduces to

$$\frac{d^2\phi(x)}{dx^2} + \frac{k_\infty - 1}{L^2} = 0 \tag{3.36}$$

or

$$\frac{d^2\phi(x)}{dx^2} + B^2\phi = 0 \tag{3.37}$$

Table 3.3. Diffusion Parameters for the Common Moderators[a]

Moderator	Diffusion coefficient D (cm)	Diffusion length, L (cm)	Diffusion area, L^2 (cm^2)
Graphite	0.85	54	2916
H$_2$O	0.16	2.8	7.85
D$_2$O	0.84	147	21,600
Beryllium	0.50	21	441

[a] Based on *Reactor Physics Constants*, edited by L. J. Templin, ANL-5800, 2nd ed., Argonne National Laboratory (July 1963).

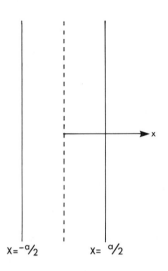

$$x = -a/2 \qquad\qquad x = a/2$$

Fig. 3.6. Coordinate system for slab reactor.

where

$$B^2 = \frac{k_\infty - 1}{L^2} \tag{3.38}$$

Equation (3.37) may be solved if a suitable boundary condition is imposed on the flux. It might seem that the appropriate condition would be that the flux go to zero at the edge of the reactor. In practice, however, the flux at the edge if not quite equal to zero, since neutrons are still streaming through the reactor surface into the space outside. A good approximation to the flux inside the reactor is obtained by assuming that the flux goes to zero at a certain distance, known as the *linear extrapolation distance*, outside the reactor boundary (see Fig. 3.7). It turns out that the linear extrapolation distance (d) is given approximately by the expression

$$d = \tfrac{2}{3}\lambda_s \tag{3.39}$$

where λ_s is the scattering mean free path referred to in connection with equation (3.26). The extrapolation distance is thus equal to approximately twice the diffusion coefficient, D (a more accurate relation, derived from transport theory, is that $d = 2.13D$).

In solving the diffusion equation, therefore, the actual boundary of the reactor is replaced by an artificial boundary at a distance d outside it, and the condition placed on the flux is that it vanish at this artificial boundary. The actual width of the reactor, a, is replaced by an *extrapolated width* $a' = a + 2d$, and the boundary condition on the flux is that it must vanish at $x = \pm a'/2$. The solution of equation (3.37) which is consistent with this requirement is

$$\phi(x) = A \cos Bx \tag{3.40}$$

and, on applying the boundary condition $\phi(a'/2) = 0$, we have

$$A \cos(Ba'/2) = 0 \tag{3.41}$$

Hence, B must satisfy the condition

$$\frac{Ba'}{2} = n\frac{\pi}{2} \qquad (n = 1, 3, 5 \dots)$$

or

$$B = \frac{n\pi}{a'} \tag{3.42}$$

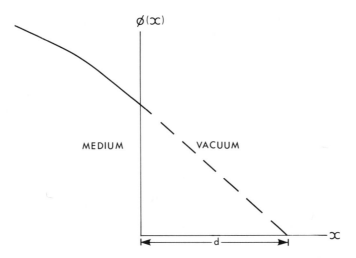

Fig. 3.7. Definition of linear extrapolation distance d. This is obtained by a linear extrapolation of the neutron flux into the region outside the reactor, using the flux gradient at the boundary.

It can be shown that, for the critical reactor, the only solution permissible for B is the fundamental solution, i.e., the case $n = 1$. Hence, for the critical reactor,

$$B = \frac{\pi}{a'} \tag{3.43}$$

and the flux distribution in the slab is given by

$$\phi(x) = A \cos \frac{\pi x}{a'} \tag{3.44}$$

This distribution is illustrated in Fig. 3.8. The arbitrary constant A is determined by the power level at which the reactor is being operated.

We can combine equations (3.38) and (3.43) to obtain the condition that must be satisfied if the reactor is in fact to be critical. Comparing these two equations, we have

$$\frac{k_\infty - 1}{L^2} = B^2 = \frac{\pi^2}{a'^2} \tag{3.45}$$

For a reactor of given composition, k_∞, which is determined by the proportions and nuclear cross sections of its constituents, is predetermined, and the condition that the reactor be critical is that the extrapolated width a' be given by

$$a' = \frac{\pi L}{(k_\infty - 1)^{1/2}} \tag{3.46}$$

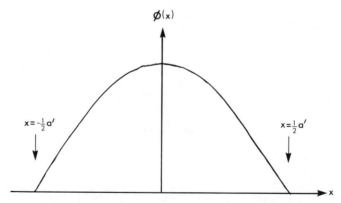

Fig. 3.8. Flux distribution $\phi(x) = A \cos(\pi x/a')$ for a slab reactor.

The quantity B^2 is known as the *buckling* for the critical reactor. It is related to the leakage, as may be seen by recalling equation (2.3), viz.,

$$k_{\text{eff}} = k_\infty \mathscr{L} \tag{2.3}$$

in which $\mathscr{L}$ is the *nonleakage probability*, i.e., the probability that the average neutron does not leak out of the reactor during its lifetime. Since, in the one-group treatment, we ignore any leakage which takes place while the neutrons are slowing down to thermal energies, the quantity we are actually dealing with is $\mathscr{L}_{\text{th}}$, the nonleakage probability for thermal neutrons only. Since, for the critical reactor, $k_{\text{eff}} = 1$, equation (2.3) becomes

$$k_\infty \mathscr{L}_{\text{th}} = 1 \tag{3.47}$$

Comparing with equation (3.38), we find

$$\mathscr{L}_{\text{th}} = \frac{1}{1 + L^2 B^2} \tag{3.48}$$

Equation (3.47), written in the form

$$\frac{k_\infty}{1 + L^2 B^2} = 1 \tag{3.49}$$

is known as the *one-group critical equation* for the reactor. As indicated above, the equation may be used to calculate B^2, and hence the size of the reactor needed for criticality, provided that its composition, and thus k_∞, is known. On the other hand, if the size of the reactor, and thus B^2, is known, the equation may be used to calculate the value of k_∞ required for criticality.

Example 3.3. An infinite slab reactor of extrapolated width 20 cm is made up of a homogeneous mixture of U^{235} and graphite. Calculate the ratio of carbon to U^{235} atoms required for criticality.

The infinite multiplication factor, k_∞, is given by the product $\eta \varepsilon p f$, where ε and p are both equal to unity in the one-group model. Hence

$$k_\infty = \eta f = \frac{\eta \Sigma_{a\text{U}}}{\Sigma_{a\text{U}} + \Sigma_{a\text{g}}} \tag{3.50}$$

where $\Sigma_{a\text{U}}$ and $\Sigma_{a\text{g}}$ are the thermal macroscopic absorption cross sections of U^{235} and graphite, respectively.

The diffusion area is

$$L^2 = \frac{D}{\Sigma_{aU} + \Sigma_{ag}} \tag{3.51}$$

so that equation (3.49) becomes

$$\frac{\eta \Sigma_{aU}}{\Sigma_{aU} + \Sigma_{ag}} = 1 + \frac{DB^2}{\Sigma_{aU} + \Sigma_{ag}}$$

which reduces to

$$\frac{\Sigma_{ag} + DB^2}{\Sigma_{aU}} = \eta - 1 \tag{3.52}$$

Taking the density of graphite to be 1.60 g cm^{-3}, we have

$$\Sigma_{ag} = \frac{\rho N_0}{A} \sigma_{ag} = \frac{1.60 \times 0.6}{12} \times 3.4 \times 10^{-3} = 2.72 \times 10^{-4} \text{ cm}^{-1}$$

The macroscopic absorption cross sections of the U^{235} and carbon are related by

$$\frac{\Sigma_{aU}}{\Sigma_{ag}} = \frac{\sigma_{aU}}{R\sigma_{ag}} = \frac{681}{R \times 3.4 \times 10^{-3}}$$

where R is the ratio of carbon to U^{235} atoms. Hence

$$\Sigma_{aU} = \frac{54.4}{R}$$

The buckling is given by

$$B^2 = \left(\frac{\pi}{a'}\right)^2 = \left(\frac{\pi}{20}\right)^2 = 2.467 \times 10^{-2} \text{ cm}^{-2}$$

The value of D for graphite is 0.85 cm, so that

$$DB^2 = 2.097 \times 10^{-2} \text{ cm}^{-1}$$

Substituting in equation (3.52), and taking $\eta = 2.07$, we find that $R = 2740$. This is the ratio of carbon to U^{235} atoms required for criticality.

□

The case chosen to illustrate the solution of the one-group diffusion equation was particularly simple in that the flux was a function of only one variable, x, because the reactor was infinite in the other two dimensions. For the finite reactor, the flux is expressed as the product of terms each of which is a function of a single spatial variable. For the cylindrical reactor of Fig. 3.9, for example, the flux at the point $P(r, z)$ is separable into radial and axial functions as below:

$$\phi(r, z) = X(r)Z(z) \tag{3.53}$$

In this case, the one-group diffusion equation (3.33) has to be solved by

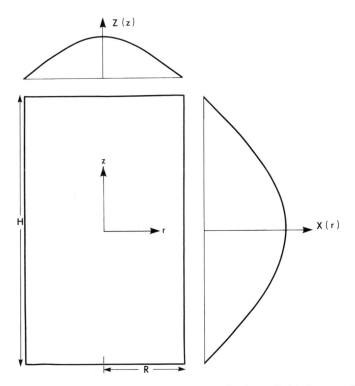

Fig. 3.9. Axial and radial dependence of thermal neutron flux in a cylindrical reactor. The core boundaries shown are extrapolated ones.

substituting the expression for ∇^2 in cylindrical polar coordinates. On doing so, it is found that the functions $X(r)$ and $Z(z)$ are of the form

$$X(r) = A_1 J_0 \frac{2.405r}{R} \tag{3.54}$$

and

$$Z(z) = A_2 \cos \frac{\pi z}{H} \tag{3.55}$$

where $J_0(2.405r/R)$ is the zero-order Bessel function of the first kind. The fluxes $X(r)$ and $Z(z)$ are shown in Fig. 3.9.

The expressions for the one-group fluxes for a number of geometries are listed in Table 3.4. Values of the bucklings for the critical reactor are also included in the table.

3.2.2. Two-Group Method

So far we have dealt with the simple case of a bare homogeneous reactor, where the solution is eased by the fact that the neutron spectrum has the same form throughout the whole system. This is no longer true when the reactor includes two or more regions of differing characteristics, as is commonly the case. It is advantageous, for example, to surround the multiplying region

Table 3.4. One-Group Fluxes and Buckling Values for Critical Bare Reactors

Reactor shape[a]	Flux	Buckling
Infinite slab (width a)	$\phi(x) = A \cos(\pi x/a)$	$(\pi/a)^2$
Sphere (radius R)	$\phi(r) = A \dfrac{\sin(\pi r/R)}{\pi r/R}$	$(\pi/R)^2$
Rectangular parallelepiped (sides a, b, c)	$\phi(x) = A_1 \cos(\pi x/a)$ $\phi(y) = A_2 \cos(\pi y/b)$ $\phi(z) = A_3 \cos(\pi z/c)$	$(\pi/a)^2 + (\pi/b)^2 + (\pi/c)^2$
Cylinder (radius R, height H)	$\phi(r) = A_1 J_0(2.405r/R)$ $\phi(z) = A_2 \cos(\pi z/H)$	$(2.405/R)^2 + (\pi/H)^2$

[a] Note: All dimensions quoted are extrapolated ones.

(*core*) with a region of moderating material, called the *reflector*. The presence of the reflector leads to a marked variation of the neutron spectrum across the core.

The simplest approach to the problem of a spatially varying spectrum is the use of two-group theory. The lower group here consists of the thermal neutrons and the other contains all those with energies above thermal. The fission neutrons originate at some energy E_0 and are assumed to diffuse at this energy until they have experienced the required average number of collisions to reduce their energy to E_{th}, the characteristic energy of thermal neutrons, at which point they are transferred to the other group.

For the core region, the two-group diffusion equations are then [cf. equations (3.30) and (3.31)] as follows:

Fast group (flux ϕ_{1c}),
$$D_{1c}\nabla^2\phi_{1c} - \Sigma_{sc}(1 \rightarrow 2)\phi_{1c} + \eta f\Sigma_{a2c}\phi_{2c} = 0 \qquad (3.56)$$

Thermal group (flux ϕ_{2c}),
$$D_{2c}\nabla^2\phi_{2c} - \Sigma_{a2c}\phi_{2c} + p\Sigma_{sc}(1 \rightarrow 2)\phi_{1c} = 0 \qquad (3.57)$$

where in this case the capture during the slowing-down process has been allowed for by including the resonance escape probability in the source term for the thermal group.

Since the reflector contains no fissile or fertile material, $\eta = 0$ and $p = 1$ and the equations become as follows:

Fast group (flux ϕ_{1r}),
$$D_{1r}\nabla^2\phi_{1r} - \Sigma_{sr}(1 \rightarrow 2)\phi_{1r} = 0 \qquad (3.58)$$

Thermal group (flux ϕ_{2r}),
$$D_{2r}\nabla^2\phi_{2r} - \Sigma_{a2r}\phi_{2r} + \Sigma_{sr}(1 \rightarrow 2)\phi_{1r} = 0 \qquad (3.59)$$

The solution methods for the two-group equations are outlined in most reactor physics texts. The solution will not be given here, but Fig. 3.10 shows typical flux variations obtained for a reflected cylindrical reactor using the two-group approach. By reflecting back into the core neutrons which would otherwise have escaped, the reflector reduces the mass of fuel required to make the reactor critical. The reflection of thermal neutrons produces a marked change in the thermal flux distribution in the core as compared with the equivalent bare reactor. The reason for the peaking of the thermal flux inside the reflector is that large numbers of thermal neutrons are being

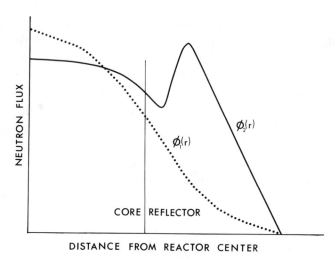

Fig. 3.10. Radial variation of fast flux $\phi_1(r)$ and thermal flux $\phi_2(r)$ in a cylindrical reactor.

produced there by the slowing down of neutrons entering from the core, while at the same time these neutrons are absorbed much less readily than in the core owing to the absence of fuel in the reflector.

As may be seen by comparing with Fig. 3.9, the thermal flux distribution in the core is much flatter than for the bare reactor. This, in turn, means that the *power density* (rate of power generation per unit volume) in the reflected reactor will be more uniform. The increase in uniformity of power density increases the power that can be extracted from the core for a given capacity of cooling system and also makes the burn-up rate of fuel more uniform throughout the core. The uniformity of power density may be further enhanced by differential fuel loading, i.e., by using fuel of higher enrichment in regions where the thermal flux is lower than average. This technique will be discussed in greater detail in Chapter 4.

3.3. The Heterogeneous Reactor

As has been mentioned already (see Section 2.4) the resonance capture by U^{238} in a homogeneous natural uranium reactor is too great to permit even an infinitely large system of this type to attain criticality; the maximum possible value of k_∞ for the reactor is in fact of the order of 0.74. It was realized at an early stage in the development of nuclear reactors that an appreciable increase in the resonance escape probability could be achieved, for a given ratio of fuel-to-moderator atoms, if the uranium were concentrated in lumps rather than being mixed uniformly with the moderator. A reactor of this kind is known as a *heterogeneous reactor.*

For a natural uranium reactor, the uranium is typically in the form of cylindrical rods which are located in a regular lattice array, usually in a square or hexagonal geometry, the space between the rods being occupied by moderator. The principal effects on the infinite multiplication factor as a result of adopting the heterogeneous arrangement are as follows.

i. The most important effect of lumping the fuel is to reduce the quantity of fuel which is exposed to neutrons in the *resonance* energy range. It will be recalled that the absorption cross section in the resonance region is characterized by numerous very high peaks. The peaks are high enough that a neutron which has acquired, during the slowing-down process, an energy corresponding to that of the resonance peak has a large probability of being captured if it encounters a U^{238} atom. If the fuel is in the form of fairly thick metal rods, for example, neutrons which enter the rod from the moderator with energies in the resonance range will mostly be absorbed before penetrating very far into the rod because of the high absorption. This means that the fuel in the interior of the rod is very effectively screened from neutrons in the resonance energy region, so that the resonance absorption is reduced by a considerable factor in comparison with the homogeneous reactor. Neutrons with energies which are not close to a resonance peak will be very likely to escape capture, since they lose energy only rather slowly by elastic collisions with the uranium atoms. They therefore have a good chance of passing through the rod back into the moderator, where the more rapid loss of energy may well result in their reaching thermal energies before they come in contact with another fuel rod.

ii. The flux of *thermal* neutrons in the fuel rod is also depressed below what it would be in a homogeneous reactor, again because of the high absorption cross section of the fuel at low neutron energies. This leads to a reduction in the thermal utilization, which tends to reduce the multiplication factor. Because the thermal cross section of the fuel is much less than the peak resonance cross sections, the flux depression effect on the thermal utilization is less important than the effect on the resonance escape probability, so that the net effect of the two is still to increase the value of k_∞.

iii. Since the fission neutrons are produced within the fuel rod, they will have a higher probability of causing fast fission before their energy is reduced by collision below the U^{238} fission threshold than would be the case in a homogeneous mixture of fuel and moderator. As a result, the fast fission factor is a few percent higher for the heterogeneous reactor.

Fuel lumping is, of course, only one of the sources of heterogeneity in a nuclear reactor. Coolant channels, pressure tubes (for the CANDU heavy water reactor) and control rods represent very large departures from homogeneity which produce *fine structure* variations in neutron flux and add to the problems of calculating the physics characteristics of the reactor. The degree to which these have to be taken account of in the calculations depends on the size of the heterogeneity compared to the neutron mean free path. In a fast reactor, where the average neutron energy is high enough that the mean free path may be of the order of tens of centimeters, fine structure effects are less marked, and may be treated by relatively cruder methods. For a light water reactor, on the other hand, the thermal neutron mean free path is of the same order as the fuel pin diameter, and the flux in the fuel will be considerably lower than in the moderator.

In order to take pronounced fine structure effects into account in calculations of the physics of a reactor core, the periodic array of the reactor lattice is divided into a number of identical cells and the fine structure of the flux distribution in each cell is calculated using transport theory. This can then be used to generate cell-averaged group constants which may be used in a multigroup calculation for the whole reactor core.

3.4. Reactor Kinetics and Control

3.4.1. Kinetics of the Reactor without Feedback

So far we have considered only the steady state critical reactor where the neutron flux does not vary with time. We now extend the discussion to cover the case where an initially critical reactor is rendered sub- or supercritical by some change such as the movement of a control rod. The dynamic response to changes in the multiplication factor is of obvious importance to the practical operation of the reactor and in particular to the assessment of its safety under operational conditions.

Initially, the discussion will be restricted to the reactor without feedback; that is, we assume that the changes in reactor power level consequent on the applied change in multiplication factor do not themselves produce further changes in multiplication. During the commissioning of a power reactor, for example, the reactor is often operated at very low powers (of the order of a kilowatt or less) in order that the physics characteristics may be measured and compared with theoretical predictions. Since the power generated at this stage is too low to produce significant heating effects in the core, there are no feedback effects to complicate the comparison with theory. If the multiplication factor of the critical reactor is changed to some value slightly greater or less than unity, causing the power to rise or fall, the multiplication factor can then be assumed to remain at the new value despite the change in power.

When the reactor is operating at its full power level, where it may be generating a total power of the order of 3000 MW, however, any change in that power level will produce changes in temperature, steam voidage, etc. which will alter the physical characteristics of the core and produce feedback effects on the multiplication factor. Under these conditions, the behavior of the reactor may be characterized in terms of operational parameters, such as the power coefficient, temperature coefficient, and voidage coefficient, which define how the multiplication factor changes with a change in the specified variable.

In discussing the dynamic behavior of a reactor, it is found convenient to introduce a new variable, the *reactivity*, which is related to the multiplication factor, k_{eff}, as shown below:

$$\rho = \frac{k_{eff} - 1}{k_{eff}} \tag{3.60}$$

The reactivity, like the multiplication factor, is dimensionless. It is frequently expressed as a percentage, a reactivity of 1% corresponding to a value of 0.01 for the ratio above. The critical reactor, of course, has zero reactivity, since k_{eff} is equal to unity in that case.

The basic problem in reactor dynamics is the prediction of the neutron flux $\phi(r, t)$ as a function of position and time following an imposed change in the reactivity. It is worth noting at this point that the flux solutions of the diffusion equation listed, for example, in Table 3.4, are the *fundamental mode* solutions of an equation which, as indicated by the expression (3.42), have higher harmonic solutions in addition. While the flux in the critical reactor may be taken as represented by the fundamental mode solution only, the imposition of any change in reactivity will induce transient higher harmonic terms which will change the flux shape from the fundamental distribution. For the purposes of this section, however, we shall ignore the existence of these spatial transient terms and assume that the time-dependent flux $\phi(r, t)$ may be separated into a constant shape function $R(r)$ and an amplitude function $T(t)$ which describes the time variation of the overall flux level, i.e., that

$$\phi(r, t) = R(r)T(t) \tag{3.61}$$

This simplified picture is known as the *point reactor model*.

As explained briefly in Chapter 2, the dynamic response of a reactor system is critically dependent on the fact that a small proportion of the secondary neutrons produced as a result of fission is emitted with a delay of up to several minutes after the fission has occurred. To illustrate the importance of the delayed neutrons, let us consider a very simplified argument leading to the rate of flux increase that would occur in a reactor if the neutrons produced by fission were all prompt neutrons, i.e., emitted within a time of the order of 10^{-14} s after fission.

Consider a reactor in which the multiplication factor is k_{eff}, where k_{eff} is greater than unity. The *excess multiplication factor* of the reactor is defined as

$$k_{ex} = k_{eff} - 1 \tag{3.62}$$

Let the total number of neutrons in the system at some instant be n. The multiplication factor is defined as the ratio of the number of neutrons in one generation to that in the previous generation. The time between generations is the *neutron lifetime*, l. If the number of neutrons in the system has increased to $n + dn$ after the time l, then

$$\frac{n + dn}{n} = k_{eff} \tag{3.63}$$

Hence

$$dn = n(k_{eff} - 1) = nk_{ex} \tag{3.64}$$

The rate of change of neutron number, dn/dt, is obtained by dividing by the neutron lifetime, i.e.,

$$\frac{dn}{dt} = \frac{nk_{ex}}{l} \tag{3.65}$$

or,

$$\frac{1}{n}\frac{dn}{dt} = \frac{k_{ex}}{l} \tag{3.66}$$

Integrating, we have

$$n = n_0\, e^{(k_{ex}/l)t} \tag{3.67}$$

where n_0 is the neutron number at $t = 0$.

Under these circumstances, the flux will rise exponentially with a period $T = l/k_{ex}$. Even for a graphite-moderated reactor, where the neutron lifetime may be as high as 10^{-3} s, the flux will rise at a very rapid rate if the reactivity is appreciably greater than zero. For $k_{eff} = 1.001$, for example, the period will be of the order of a second and in the absence of corrective action or rapid negative feedback the flux would increase by a factor of $e^{10} \sim 2 \times 10^4$ in 10 s. For an enriched-fuel H_2O system, with a correspondingly lower neutron lifetime of the order of 10^{-4} s, the period will be around 0.1 s and for a fast reactor it would be very much lower still. It is therefore clear that, but for the stabilizing effect of the delayed neutrons, the response time of a reactor would be so short as to make its control virtually impossible.

As explained in Chapter 2, the delayed neutrons are not emitted from the direct products of the fission, but from nuclei which are formed by subsequent β decay of these products. While many of the delayed neutron precursors have been identified, it is more convenient in practice to analyze the time behavior of the delayed neutrons by an empirical division into a number of groups, each characterized by a single decay constant, or half-life. It is found that the characteristics of the delayed neutrons from all the fissionable isotopes of interest can be adequately described by the use of six groups. The half-lives and yields of the delayed neutron groups for the fissile isotopes U^{233}, U^{235}, and Pu^{239}, and for the fertile isotope U^{238}, are summarized in Table 3.5.

Even though the neutrons which are delayed account for only a small fraction of the total, the much longer time between the initial fission and the appearance of the delayed neutrons means that the weighted average, or effective, lifetime is very much greater than that of the prompt neutrons alone. The mean lifetime is increased from l to the value $(1 - \beta)l + \sum \beta_i \tau_i$, where β_i is the fraction of the total fission neutron yield which appears in the ith group and τ_i is the mean delay time for the ith group (which is equal to the half-life quoted in Table 3.5 divided by ln 2, or 0.693). The quantity β is the total delayed neutron fraction, which is given by the sum of the fractional yields, i.e., $\beta = \sum \beta_i$. For U^{235}, the value of the quantity $(1 - \beta)l + \sum \beta_i \tau_i$ is approximately 0.1 s. If this value is inserted in equation (3.67), it is seen that the reactor period for 0.1 % excess reactivity ($k_{eff} = 1.001$) is around 100 s. On this argument, the reactor response is now so much slower that adequate

Table 3.5. Delayed Neutron Data

Isotope	Group No.	Decay constant, λ_i (s^{-1})	Half-life (s)	Yield (neutrons per fission)	Fractional yield (β_i)
		(i) Data on delayed neutron half-lives and yields in thermal fission[a]			
U^{235}	1	0.0124	55.72	0.00052	0.000215
	2	0.0305	22.72	0.00346	0.001424
	3	0.111	6.22	0.00310	0.001274
	4	0.301	2.30	0.00624	0.002568
	5	1.14	0.610	0.00182	0.000748
	6	3.01	0.230	0.00066	0.000273
	Total:			0.0158	0.0065
U^{233}	1	0.0126	55.00	0.00057	0.000224
	2	0.0337	20.57	0.00197	0.000777
	3	0.139	5.00	0.00166	0.000655
	4	0.325	2.13	0.00184	0.000723
	5	1.13	0.615	0.00034	0.000133
	6	2.50	0.277	0.00022	0.000088
	Total:			0.0066	0.0026
Pu^{239}	1	0.0128	54.28	0.00021	0.000073
	2	0.0301	23.04	0.00182	0.000626
	3	0.124	5.60	0.00129	0.000443
	4	0.325	2.13	0.00199	0.000685
	5	1.12	0.618	0.00052	0.000181
	6	2.69	0.257	0.00027	0.000092
	Total:			0.0061	0.0021
		(ii) Data on delayed neutron half-lives and yields from fast fission in U^{238}[a]			
U^{238}	1	0.0132	52.38	0.00054	0.00019
	2	0.0321	21.58	0.00564	0.00203
	3	0.139	5.00	0.00667	0.00240
	4	0.358	1.93	0.01599	0.00574
	5	1.41	0.490	0.00927	0.00333
	6	4.02	0.172	0.00309	0.00111
	Total:			0.0412	0.0148

[a] From *Physics of Nuclear Kinetics* by G. R. Keepin, Addison-Wesley, Reading, Massachusetts, pp. 86 and 90 (1965). Reprinted with permission.

time is available for taking any corrective action which may be necessary to control the reactor power level.

The type of argument given above provides only a very crude approximation to the time behavior of the reactor. A more detailed and rigorous approach is by the extension of the one-group diffusion equation (3.28) to the time-varying situation. The terms in this equation represent the contribution of leakage, absorption and neutron sources to the neutron balance in a unit volume of the reactor. In the steady state condition, the sum of these terms is equal to zero, since there is no net change in the neutron density with time. When a change in flux is occurring as a result of the reactivity having a nonzero value, however, the sum of the terms is equal to the rate of change of the neutron density, dn/dt, which may be written as $(1/v)(d\phi/dt)$, since $\phi = nv$. The equation which has to be solved to find the time-varying flux is therefore

$$\frac{1}{v}\frac{d\phi}{dt} = D\nabla^2\phi - \Sigma_a\phi + S \qquad (3.68)$$

In specifying the source term here, separate account has to be taken of the contributions of prompt and delayed neutrons. For accurate analysis of the time variance of the flux, the fractional yields and half-lives of the six delayed neutron groups have to be fed into the source term. The method of solution is given in many reactor physics texts, and will not be repeated here. To illustrate the time behavior predicted by the solution, we may consider a reactor where, starting from an initially critical condition, the reactivity is suddenly increased from zero to a small positive value, such as $\rho = 0.001$. Assuming six delayed neutron groups, the time dependence of the flux after the reactivity increase is given by an expression of the form

$$\phi(t) = \phi_0(A_0 e^{\omega_0 t} + A_1 e^{\omega_1 t} + \cdots + A_6 e^{\omega_6 t}) \qquad (3.69)$$

where ϕ_0 is the initial flux in the critical condition, and the A_i are numerical constants.

The seven coefficients occurring in this equation are the solutions of the equation

$$\rho = \frac{\omega l}{\omega l + 1} + \frac{1}{\omega l + 1}\sum_i \frac{\omega\beta_i}{\omega + \lambda_i} \qquad (3.70)$$

It turns out that, for a positive value of ρ, six of the seven roots of equation (3.70) are negative, while the other, ω_0, is positive. Immediately following the reactivity change, the flux will be represented as the sum of six terms which decrease with time (the *transients*) and one which increases. Once the transient terms have decayed away, however, the flux will diverge with an asymptotic period $= 1/\omega_0$, known as the *stable period*.

The relationship between the reactivity and the stable period for a U^{235}-fueled reactor is shown in Fig. 3.11 for a range of neutron lifetimes; for convenience the reactivity is expressed as a function of the total delayed neutron fraction, $\beta = \sum \beta_i$. This is equivalent to the introduction of a new unit, called the *dollar*, for measuring reactivity. One dollar is an amount of reactivity equal to the delayed neutron fraction; the reactivity in dollars is then obtained by dividing ρ, as defined by equation (3.60), by the delayed neutron fraction for the particular fuel in the reactor. The specification of reactivity values in terms of the delayed neutron fraction is convenient in that a given reactivity in dollars will produce essentially the same rate of flux rise for reactors containing U^{233}, U^{235}, or Pu^{239} fuel.

It will be seen from Fig. 3.11 that, for low values of reactivity, the stable reactor period does not depend significantly on the neutron lifetime, l, but that for positive values of ρ equal to or greater than the total delayed neutron fraction, β, the reactor period becomes very short and is also strongly dependent on the neutron lifetime. For $\rho > \beta$, the reactivity is such that the reactor is supercritical on the prompt neutrons alone, and delayed neutron effects are unimportant. This is equivalent to the situation described by

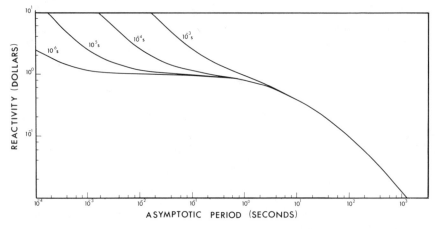

Fig. 3.11. Relation between reactivity and period for a U^{235}-fueled reactor for various values of the neutron lifetime l.

equation (3.67) and under these circumstances, the power will rise at a catastrophic rate. It is therefore of vital importance that the reactor be designed so as to eliminate as far as possible the chance of sudden positive reactivity changes of this order. The situation where $\rho = \beta$ is known as the *prompt critical* condition, and $\rho > \beta$ as *prompt supercritical*.

If the reactivity injected into the critical reactor is negative, i.e., if the multiplication factor is suddenly decreased, the neutron flux will again be described by the series of exponentials in equation (3.69), but in this case it turns out that all the roots of equation (3.70) are negative, so that the flux decays away as a sum of decreasing exponentials. Again, once the shorter-lived transients have decayed, the reactor flux will decrease exponentially with a negative stable period. As the magnitude of the imposed reactivity change increases, i.e., for $|\rho| \gg \beta$, the stable period approaches more and more closely to the decay period of the longest-lived of the delayed neutron groups; this means that, even for a very large negative reactivity injection, the flux cannot fall faster than with a "half-life" of about 55 s (see Table 3.5).

The fact that, after the transients have decayed, the reactor flux rises or falls with an asymptotic period which is a function of the imposed reactivity change provides a useful method of measuring the reactivity values associated with the control system of the reactor. For example, a standard method of calibrating a control rod, in terms of the reactivity change produced by a given movement of the rod, is to raise the rod out of the critical reactor by the amount required to set the reactor flux diverging with a conveniently measurable period. The determination of the period then leads immediately to a knowledge of the reactivity value associated with the section of rod removed from the reactor.

For the measurement of large negative insertions of reactivity such as are produced, for example, by dropping a whole control rod rapidly into the critical reactor, it is possible to make use of the flux behavior immediately following the reactivity change. This is illustrated in Fig. 3.12. If a positive reactivity, ρ, is injected into the reactor, the flux rises almost instantaneously by the ratio $(\beta/(\beta - \rho))$; this rapid change in flux level is known as the *prompt jump*. If the reactivity added is negative, the ratio of the flux soon after the change to the flux before is also given by the ratio $\beta/(\beta - \rho)$, which in this case will be less than unity. Measurement of the flux change when a control rod is dropped into the reactor, using rapid-response instrumentation, therefore allows a value to be obtained for the reactivity worth of the rod.

So far we have assumed that the appropriate β value to use in the kinetic equation is simply the total fraction of neutrons that is delayed. In fact, because the energies of the delayed neutrons are much lower than those of the

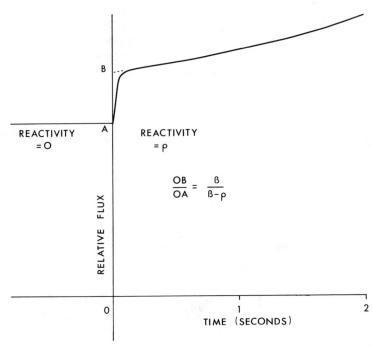

Fig. 3.12. Flux variation following step change in reactivity, showing "prompt jump" by ratio $\beta/(\beta - \rho)$.

prompt neutrons, fewer of the former will leak out of the reactor before becoming thermal, so that the fraction of delayed neutrons available for causing fission may be appreciably greater than β. The additional effectiveness of the delayed neutrons due to reduced leakage during slowing down may be allowed for by defining an *effective delayed neutron fraction*, β^*, which is the fraction of neutrons reaching thermal energy which originated as delayed neutrons.

The ratio β^*/β will obviously depend on the degree of fast leakage from the reactor. The greatest effect will be observed for a light water reactor, where there is a high leakage during slowing down. In this case the ratio β^*/β may exceed 1.20. For a large natural uranium graphite-moderated system, on the other hand, the differential leakage effect will be small, but the effective delayed neutron fraction will be increased by the addition of the delayed neutrons from fast fission of U^{238}, which has a much greater value of β (0.0148) than does U^{235} (0.0065). For reactors where the fuel contains a mixture of isotopes, an appropriate weighting of the contributions from each has to be made in calculating the effective delayed neutron fraction.

3.4.2. Reactivity Effects Arising from Power Operation

3.4.2.1. Temperature Coefficient of Reactivity

When bringing a power reactor up to its normal operating condition from the cold shutdown state, the reactor is first brought to criticality at low power and the power is then gradually increased until the plant has reached the full power conditions of temperature and pressure. Since, for reasons to be discussed below, the change in temperature will produce a corresponding change in reactivity, it is necessary to design the control system so that it will be possible to compensate for this reactivity change by altering the amount of control rod absorption in the core. For this reason, it is important that one should be able to predict the change in reactivity produced by changes in reactor temperature. In addition, the stability of the reactor during normal operation will depend on the sign and magnitude of the change in reactivity following the increase in temperature caused by an increase in power level. In this case, a rapid negative feedback is valuable in providing an automatic limitation to transient power surges owing to an unexpected increase in reactivity.

The response of the reactor to changes in temperature may be described by the *temperature coefficient* α, which is the fractional change in k_{eff} per unit change in temperature. We have

$$\alpha = \frac{1}{k_{\text{eff}}} \frac{dk_{\text{eff}}}{dT} \tag{3.71}$$

Since

$$k_{\text{eff}} = k_\infty \mathscr{L}_{\text{th}} \mathscr{L}_f \tag{3.72}$$

where $\mathscr{L}_{\text{th}}$ and $\mathscr{L}_f$ are the thermal and fast nonleakage probabilities, we can write

$$\frac{1}{k_{\text{eff}}} \frac{dk_{\text{eff}}}{dF} = \frac{1}{k_\infty} \frac{\partial k_\infty}{\partial T} + \frac{1}{\mathscr{L}_{\text{th}}} \frac{\partial \mathscr{L}_{\text{th}}}{\partial t} + \frac{1}{\mathscr{L}_f} \frac{\partial \mathscr{L}_f}{\partial t} \tag{3.73}$$

Since $k_\infty = \eta \varepsilon p f$, the first term on the right-hand side may be written

$$\frac{1}{k_\infty} \frac{dk_\infty}{dT} = \frac{1}{\eta} \frac{\partial \eta}{\partial T} + \frac{1}{\varepsilon} \frac{\partial \varepsilon}{\partial T} + \frac{1}{p} \frac{\partial p}{\partial T} + \frac{1}{f} \frac{\partial f}{\partial T} \tag{3.74}$$

The significance of the temperature-induced changes in k_∞, $\mathscr{L}_{th}$, and $\mathscr{L}_f$ may be illustrated first by considering the simple case of a uniform temperature change occurring in a large homogeneous reactor. This does represent a considerable simplification, since in practice there will usually be marked differences in temperature between one part of the reactor and another. In addition, for the heterogeneous reactor, where the fuel and moderator are separated, the temperature coefficients of the two have to be considered separately, since in a power transient the fuel will heat up much more rapidly than the moderator.

Looking first at the temperature variation of k_∞, the second term in equation (3.74) is negligible, since events in the high-energy region are not affected by changes in the thermal neutron spectrum. The third term is also small for the homogeneous reactor, although it is important, as will be discussed later, for the heterogeneous case.

The parameter η is sensitive to temperature because it involves the ratio of two cross sections which are functions of the neutron velocity, i.e.,

$$\eta = \frac{\sigma_f(v)}{\sigma_a(v)} \, v \tag{3.75}$$

where $\sigma_f(v)$ and $\sigma_a(v)$ are the microscopic fission and absorption cross sections of the fissile isotope concerned, and v, the number of neutrons per fission, is effectively constant at thermal energies.

Since an increase in the temperature of the *moderator* leads to an increase in the mean thermal energy of the moderator nuclei and hence to an upward shift in the peak of the Maxwellian distribution of neutron energies, any variation in the ratio of $\sigma_f(v)$ to $\sigma_a(v)$ with neutron velocity will give rise to a temperature coefficient of η. For natural uranium, the variation in η arises not only from the change in σ_f/σ_a for U^{235}, but also from a change in the ratio of capture in U^{235} to capture in U^{238} with neutron temperature. For the Calder Hall natural uranium reactor, for example, the value of $(1/\eta)(\partial\eta/\partial T)$ is found to be of the order of -2×10^{-5} per °C, for a freshly loaded core. The value of the temperature coefficient of η becomes more negative as irradiation proceeds, owing to the buildup of Pu^{239}, which exhibits a stronger decrease of η with temperature than does U^{235}; the magnitude of $(1/\eta)(\partial\eta/\partial T)$ has changed to the order of -6×10^{-5} per °C halfway through the life of the fuel charge. For the fissile isotope U^{233}, on the other hand, the temperature coefficient of η is positive at thermal energies.

The temperature coefficient of thermal utilization would be zero if all the thermal cross sections showed the same $1/v$ dependence. This is approximately the case for a U^{235} or natural uranium reactor but, when appreciable amounts of Pu^{239} are present, a strong positive temperature coefficient of f arises from a shift of the thermal spectrum peak towards the region of the large Pu^{239} resonance at 0.3 eV (see Section 2.6). For the Calder Hall natural-uranium graphite-moderated reactor, for example, the buildup of Pu^{239} causes an increase in $(1/f)(\partial f/\partial T)$ from the start-of-life value of $+0.5 \times 10^{-5}$ per °C to about $+12 \times 10^{-5}$ per °C halfway through the life of the fuel charge. For a light water reactor, the absorption by the moderator is reduced because of its expansion out of the core as the temperature increases, leading to a positive contribution to the thermal utilization coefficient.

Turning now to the changes in neutron leakage induced by changes in reactor temperature, we note that the thermal nonleakage probability is given by equation (3.48) as

$$\mathscr{L}_{\text{th}} = \frac{1}{1 + L^2 B^2} \tag{3.48}$$

The diffusion length is defined by equation (3.34) as

$$L^2 = \frac{D}{\Sigma_a} = \frac{1}{3\Sigma_s \Sigma_a} \tag{3.34}$$

using equation (3.26). Both Σ_s and Σ_a are proportional to the atom density (atoms per cm^3) of the moderator. Since the moderator expands with increasing temperature, the atom density will decrease, and hence L^2 will increase. The decrease of Σ_a will be enhanced by the fact that the nuclear cross section, σ_a, of most of the reactor materials will vary as $1/v$, so that the increase in neutron velocity produced by increase of moderator temperature will also reduce the absorption.

Since the buckling B^2 varies inversely with the size of the reactor [e.g., $B^2 = \pi^2/(a')^2$ for an infinite slab reactor, as shown in Section 3.2], the expansion of the reactor due to temperature increase will produce a reduction in B^2 which will affect the thermal leakage in the opposite sense to the change in L^2. Overall, however, the effect of the change in diffusion length is the predominating term, so that the thermal leakage coefficient will be negative. Similar considerations will apply to the fast leakage, which can also be shown to have a negative temperature coefficient.

For a heterogeneous reactor, the changes in the resonance escape probability with temperature become significant. In a water-moderated reactor, for example, an increase in reactor temperature will cause expansion of the coolant out of the core, with a resulting decrease in the ratio of moderator-to-fuel atoms. It can be shown that the resonance escape probability has an exponential dependence on this ratio, so that an increase in overall core temperature will cause a reduction in p and thus in the reactivity. An effect in the opposite direction will be caused by the increase in f due to the reduction in water density. The light water reactor is "undermoderated" in the sense that the addition of hydrogen atoms to the core would produce a positive effect on the resonance escape probability which would be larger than the negative effect due to the decrease in the thermal utilization. Thus, in general, the net effect on reactivity due to moderator expansion will be negative, and this will be enhanced by the increase in thermal and fast leakage caused by the reduction in moderator density. If, however, the moderator has been heavily poisoned with a chemical "shim," such as boron, for reactivity control purposes, it is possible that the increase in f due to moderator expansion may be sufficiently great that the net effect becomes positive.

A second factor affecting the resonance escape probability is the *nuclear Doppler broadening* effect associated with the heating of the fuel itself. At low temperatures, the absorption cross section of a material such as U^{238} exhibits sharp peaks as shown in Fig. 1.16. The variation of cross section with neutron energy shown is based on the assumption that the U^{238} atom is at rest. In fact, it is the relative velocity of the neutron and the nucleus that determines the probability of absorption occurring. If the uranium is heated to a higher temperature, the atoms will vibrate more vigorously, with the result that a neutron with an energy which originally lay outside the peak may now encounter a U^{238} atom which is moving at a speed such that their relative velocity does correspond to the peak region. The result is a *broadening* of the peak to include neutrons which were previously outside the energy band of the resonance (see Fig. 3.13).

Despite the change in the peak shape, the area under the curve remains effectively constant, so that it might appear that the resonance absorption would be unchanged, since the decreased cross section is compensated for by the larger spread of neutron energies at which absorption can take place. For a dilute homogeneous reactor, where the U^{238} atoms are essentially isolated from one another, this would be true. For the lumped fuel element, however, as described in the previous section, most of the U^{238} is shielded from the flux of resonance neutrons because the peak cross section is so high that they are absorbed near the surface of the fuel rod. The reduction in peak height caused by the Doppler effect reduces the resonance flux depression in the fuel, so

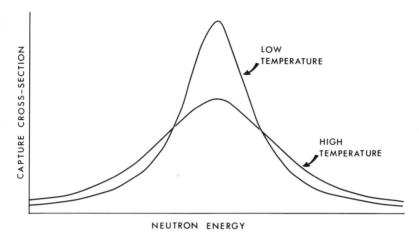

Fig. 3.13. Nuclear Doppler effect.

that there is less self-screening when the fuel rod is hot. The result is a rise in the absorption of neutrons in the U^{238} resonances.

In a thermal reactor, where the great majority of the fissions are taking place at neutron energies which are well below the resonance range, the sign of the Doppler reactivity effect is always negative. For a fast reactor, on the other hand, where most of the fission occurs as a result of neutron absorption in the resonance region, there is a compensating positive contribution to the Doppler coefficient due to an increased fission rate. Since, in a transient, the fuel heats up more rapidly than the other components of the reactor, it is important for stability reasons that the fast reactor be designed so that the overall Doppler coefficient is negative. This point will be discussed in more detail in Chapter 11.

It is clear that the overall temperature coefficient of the reactor is the resultant of a combination of effects, some of which will be difficult to calculate accurately. In addition, an increase in power level will cause the various components of the reactor to change in temperature by different amounts; the temperature rise of the fuel, for example, will in general be greater than that of the moderator when the reactor changes from one steady state to another. For the prediction of the reactivity response of the reactor to a sudden power change, the time constants of the heating of the different components have to be taken into account. In practice, calculation of the temperature coefficients are usually supplemented by an analysis of measurements carried out on a prototype reactor during its commissioning stage.

3.4.2.2. Void Coefficient of Reactivity

One effect of an increase in temperature which is important enough to merit discussion as a separate topic is the reactivity change due to formation of voids in the coolant, either as part of the normal operating condition, as in the boiling water reactor, or under accident conditions in reactors such as the pressurized water reactor or the fast breeder reactor. The effects of void formation are similar to those of reduction in coolant density already considered, e.g., reduced moderation, giving increased leakage and resonance capture, and reduced absorption, leading to an increase in the thermal utilization, but the effects can be more dramatic on account of the more rapid possible variation of coolant density.

The void coefficient, defined typically as the reactivity change per % voidage, is strongly negative for the pressurized water reactor (PWR), since the core is highly undermoderated. For the boiling water (BWR) system, where in-core boiling is permitted, it is also desirable to have a negative void coefficient to provide an automatic stabilizing mechanism for normal operation. On the other hand, too large a negative void coefficient in a BWR can lead to instability due to the coupling between the reactivity and the voidage. Also, it implies a large decrease of reactivity in going from the cold to the normal operating temperature, and hence the need for a large investment of reactivity in the control system. The void coefficient can be used as a standard method of controlling the power level, since by increasing the speed of the coolant circulators one can obtain a lower voidage, and thus an increase of reactivity.

For the CANDU type of heavy-water-moderated reactor, the degree of moderation required for operation with natural uranium fuel is such that the void coefficient is always positive. The magnitude of the coefficient is small for the standard CANDU, which is both moderated and cooled by heavy water, and the degree of subdivision of the cooling pipework characteristic of the pressure tube design ensures that the power transient which might be caused by an accident leading to a loss of coolant can easily be terminated by the normal reactivity control mechanism. The highest positive value of the void coefficient arises in the CANDU–BLW design which is moderated by heavy water, but cooled by light water. In this case, fuel pins of larger diameter than usual are used, in order to increase the time constant governing the transfer of heat from the fuel to the coolant and thus allow time for the control system to reduce the reactor power before significant voidage formation has taken place.

The void coefficient is also of great importance for the sodium-cooled fast breeder reactor. Voidage induced in the liquid sodium coolant produces two main effects. The first is an increase in the mean neutron energy in the core due to a reduction in scattering; this results in a positive reactivity contribution due to an improvement in η, coupled with increased fission in the elements, such as U^{238}, which have a high fission threshold. An effect in the opposite direction is produced by increased leakage due to the reduction in coolant density. The sign of the overall coefficient depends, among other things, on the geometry of the core. It is possible to ensure that the sodium void coefficient is always negative by a suitable choice of the reactor geometry (e.g., by encouraging high leakage), but it is possible also to counteract any positive void coefficient by designing the reactor to have a strong negative Doppler coefficient. This is discussed in greater detail in Chapter 11.

3.4.3. Fission Product Poisoning

Each fission of a fuel nucleus in the reactor leads to the production of two nuclei of intermediate mass, known as the fission products. Not only is there a large number of isotopes produced directly by the fission process, but radioactive decay of these products leads to the formation of further species. Some of the fission products or their daughters have large cross sections for the capture of thermal neutrons, and consequently their presence reduces the reactivity of the core. The control system has to be able not only to compensate for the long-term effect of the buildup of stable isotopes but also to deal with major fluctuations of the concentrations of the radioactive ones, of which the most important is xenon-135. The fission products of large cross section are listed in Table 3.6, along with their half-lives and relative fission yields. The latter are total cumulative yields, including not only direct production in fission, but also subsequent formation as a result of decay of a radioactive parent.

Table 3.6. Characteristics of Most Important Fission Products

Fission product	Half-life	Decay constant (s^{-1})	Effective yield (atoms per fission)			Thermal absorption cross section (b)
			U^{233}	U^{235}	Pu^{239}	
I^{135}	6.7 h	2.87×10^{-5}	0.051	0.064	0.055	—
Xe^{135}	9.2 h	2.09×10^{-5}	—	0.003	—	2.65×10^6
Pm^{149}	53 h	3.57×10^{-6}	0.0066	0.0113	0.019	1.7×10^3

The fission product which has an important effect on the short-term behavior of the reactor is Xe^{135}. The direct fractional yield of this isotope from fission is only some 0.3%, and most of it is in fact derived from the radioactive chain shown below:

$$Te^{135} \xrightarrow[11\,s]{\beta-} I^{135} \xrightarrow[6.7\,h]{\beta-} Xe^{135} \xrightarrow[9.2\,h]{\beta-} Cs^{135} \xrightarrow[2.3 \times 10^6\,yr]{\beta-} Ba^{135} \text{ (stable)}$$

The half-life of Te^{135} is so short that for practical purposes we may assume that iodine-135 is formed as a direct fission product, with the high yield of 6.4% for U^{235} fission.

The absorption cross section of Xe^{135} is so much greater than that of any other member of the chain that the reactivity of the reactor following a major change in power level is governed by the buildup and decay of this isotope alone. We now consider the variation of Xe^{135} concentration as a function of time following such a change.

Let the concentration of I^{135} be I nuclei per cm^3. If the flux is ϕ n cm^{-2} s^{-1} and the macroscopic fission cross section is Σ_f cm^{-1}, the rate at which I^{135} is being formed is $\gamma_I \Sigma_f \phi$, where γ_I is the effective fractional yield of the iodine isotope in fission. Two factors operate to cause a reduction in the I^{135} concentration. The first is its own decay, which proceeds at a rate of $\lambda_I I$ per cm^3 per second, where λ_I is the decay constant of I^{135}. The second effect is its conversion to I^{136} by neutron capture, the rate of which is equal to $\sigma_I \phi I$, where σ_I is the thermal capture cross section. The net rate of change of iodine concentration is therefore given by

$$\frac{dI}{dt} = \gamma_I \Sigma_f \phi - \lambda_I I - \sigma_I \phi I \tag{3.76}$$

The capture cross section of I^{135} is only about 7 b, and even for the very high flux value of 10^{15} n cm^{-2} s^{-1} the capture term in the above equation will be less than $10^{-8} I$. The decay constant λ_I, on the other hand, is approximately 3×10^{-5} s^{-1}, so that the third term can be neglected in comparison with the second, and we can write

$$\frac{dI}{dt} = \gamma_I \Sigma_f \phi - \lambda_I I \tag{3.77}$$

The time variation of Xe^{135} will include terms of the same nature as those in equation (3.76), since the same processes are involved, but there will be an additional term allowing for its formation by decay of I^{135} The equation for the Xe^{135} concentration, X, is therefore

$$\frac{dX}{dt} = \gamma_X \Sigma_f \phi + \lambda_I I - \lambda_X X - \sigma_X \phi X \tag{3.78}$$

where the symbols have the same significance as in equation (3.76).

The most interesting feature of the behavior of Xe^{135} is its variation following a shutdown of the reactor after it has been running for a long period at steady power. If this period is long compared to the half-lives of the iodine and xenon, these will have reached saturation values, I_0 and X_0, respectively. Taking $dI/dt = 0$ in equation (3.77) yields for the equilibrium iodine concentration

$$I_0 = \frac{\gamma_I \Sigma_f \phi}{\lambda_I} \tag{3.79}$$

From equation (3.78), with $dX/dt = 0$, and using the above expression for the iodine concentration, the equilibrium xenon is

$$X_0 = \frac{(\gamma_I + \gamma_X)\Sigma_f \phi}{\lambda_X + \sigma_X \phi} \tag{3.80}$$

Now assume that at $t = 0$ the reactor, which has been running in the steady state, is shut down. Over a period which is short compared to the half-lives of the fission products involved, the neutron flux drops to zero, and equation (3.77) for the iodine variation becomes

$$\frac{dI}{dt} = -\lambda_I I \tag{3.81}$$

or

$$I = I_0 e^{-\lambda_I t} \tag{3.82}$$

Equation (3.78) for the xenon concentration is now

$$\frac{dX}{dt} = \lambda_I I - \lambda_X X \tag{3.83}$$

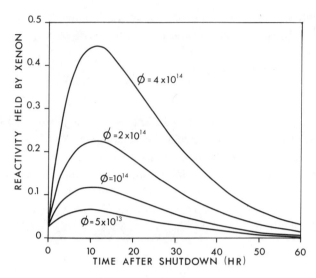

Fig. 3.14. Reactivity held by Xe^{135} as a function of time after shutdown, for various preshutdown flux levels.

Solving, and using equation (3.79) for the iodine concentration at $t = 0$, we have for the xenon concentration as a function of time after shutdown,

$$X = \Sigma_f \phi \frac{\gamma_I}{\lambda_X - \lambda_I} (e^{-\lambda_I t} - e^{-\lambda_X t}) + \frac{\gamma_I + \gamma_X}{\gamma_X + \sigma_X \phi} e^{-\lambda_X t} \qquad (3.84)$$

Typical variations of the reactivity of xenon with time after shutdown, for selected values of the pre-shutdown neutron flux ϕ, are shown in Fig. 3.14. It will be seen that, for high pre-shutdown fluxes, the xenon poisoning rises to a maximum value some 10 h after the shutdown has occurred. The reason for the rise is that xenon is still being built up from decay of I^{135} but, in the absence of a neutron flux, its rate of destruction is limited to its natural decay rate, rather than being a combination of this effect plus neutron capture. Since the decay constant of xenon is less than that of iodine, the result is a temporary peaking followed by a gradual fall-off as both the iodine and xenon concentrations are reduced by radioactive decay. Since the ratio of the equilibrium inventories of iodine and xenon is proportional to the quantity $\lambda_X + \sigma_X \phi$, as indicated by equations (3.79) and (3.80), the ratio of the maximum to the steady state poisoning increases with the flux level. As shown in Fig. 3.14 the maximum reactivity for a flux of 2×10^{14} n cm^{-2} s^{-1}, which is typical of a high-flux reactor, can reach a value as high as 25%.

A period of up to three days must elapse before the reactivity returns to the value it had before the shutdown. If it is a requirement that it be possible to start the reactor up again at any time during this period, a high percentage of excess reactivity has to be built into it in order to overcome the xenon transient. This is known as *xenon override capacity*. In normal operation this built-in reactivity excess has to be held down by a large equivalent negative reactivity in the form of control rods or some other mechanism.

For a natural uranium reactor, where the possible excess reactivity is limited by the low k_∞ value, it may not be possible to overcome the peak in the transient, and it will then be necessary to wait for some time after the peak is reached before the reactor can be started up again. One way of avoiding such a situation, employed in the CANDU heavy water reactor, is the use of *booster rods* of enriched uranium which can be inserted into the core to provide a temporary increase in reactivity. For any reactor with a limited capacity for xenon override, it is desirable to restart after an unscheduled reactor trip as soon as possible, before the xenon transient has had a chance to build up. The re-establishment of the xenon burn-out due to restoring the operational flux level then allows the equilibrium xenon concentration to be regained without any large change in the reactivity having taken place.

In addition to its overall effect on the reactivity of the reactor, the presence of xenon can lead to a form of spatial instability, known as *xenon oscillations*, in large high-flux reactors. For a reactor whose dimensions are large compared with the diffusion length, e.g., a large light-water-moderated reactor, the effect of the movement of a control rod tends to be relatively localized. A small withdrawal of a rod located, say, at the right side of the core, will produce a local increase in power, which will in turn cause the burn-out rate of xenon in the neighborhood to increase, raising the local reactivity still further. If the reactor is maintained critical simply by some overall reactivity control mechanism, the result will be to produce a flux tilt, the increased flux in the region of the initial disturbance being balanced by a corresponding flux reduction on the left side of the core.

After some hours, the increased buildup of I^{135}, and its decay to Xe^{135}, will reduce the flux on the right side of the core, while the lower flux on the left side will have led to reduced I^{135} and Xe^{135} concentrations, causing a gradual flux rise on that side. The result is that the flux across the reactor will eventually come to tilt in the opposite direction. By this mechanism, an oscillation in the power profile, with a period of about 24 h, can be set up. This effect will not be detectable as a change in the overall reactivity of the reactor, since this will remain approximately constant, the increased xenon on one side being balanced by the reduced xenon on the other. The oscillation is potentially damaging, however, in that it may cause local overheating in the regions where the flux is temporarily above normal.

To avoid localized power increases, it is necessary to use *sector control*, where the core is divided into sections, each having separately operable control rods and flux sensing instrumentation. This method is employed, for example, on the large graphite-moderated magnox reactors of the British generating system. The problem does not arise for the boiling water reactor, since a localized increase in reactivity due to a flux rise causing local xenon burn-out immediately leads to increased boiling in that region, the resulting increase in voidage restoring the reactivity to its initial value.

Another important isotope of high thermal cross section is Sm^{149}, which shows some similarity in behavior to Xe^{135}, although the detailed variation after shutdown differs because, unlike Xe^{135}, Sm^{149} is a stable nuclide. It is the daughter of 53-h Pm^{149}, which may essentially be regarded as being

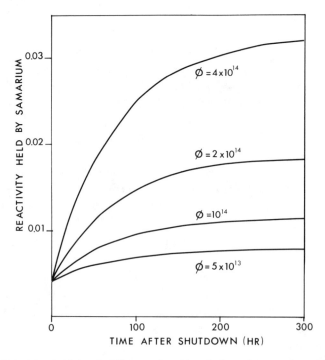

Fig. 3.15. Reactivity held by Sm^{149} as a function of time after shutdown, for various pre-shutdown flux level.

produced directly in U^{235} fission with a yield of 1.13%. The effective cross section of Sm^{149} in a Maxwellian spectrum is approximately 6.0×10^4 b. The equilibrium concentration of Pm^{149} is given by an expression of the same form as equation (3.79) for the I^{135}, while the Sm^{149} concentration is given by a modification of (3.80), with $\gamma_S = 0$ since the direct production of this isotope is negligible, and $\lambda_S = 0$ since Sm^{149} is stable. The equilibrium Sm^{149} is therefore

$$S_0 = \frac{\gamma_P \Sigma_f}{\sigma_S} \tag{3.85}$$

For a U^{235}-fueled reactor, the reactivity associated with the equilibrium Sm^{149} poisoning is about 0.45%, independent of the flux level of the reactor. After a shutdown, the samarium concentration will increase, since none of it is being burned out, while it is still being produced by the decay of promethium. The variation of samarium reactivity as a function of time after shutdown is shown in Fig. 3.15 for several values of the pre-shutdown flux. For a flux of 2×10^{14} n cm^{-2} s^{-1}, the reactivity held by Sm^{149} rises asymptotically over a period of about 20 days to a maximum value of approximately 2%. In contrast to the Xe^{135} situation, where the reactivity decreases again due to xenon decay, the Sm^{149} will remain at this asymptotic value until the reactor is restarted, when the renewed burn-out of the isotope eventually reduces its concentration back to the equilibrium value.

Problems

(The data required for the solution of the problems are given in Tables A.1–A.6).

3.1. Assuming that a thermal neutron spectrum is given by equation (3.15), show that the *average* neutron velocity is equal to $(8kT/\pi m)^{1/2}$.

3.2. Calculate the most probable speed for thermal neutrons in a moderator at a temperature of 400°C.

3.3. Calculate the average number of collisions required to slow a neutron down from an energy of 10 MeV to 0.0253 eV if the moderator is (a) hydrogen, (b) deuterium, (c) beryllium.

3.4. Given that the average logarithmic energy decrement of oxygen is 0.121, calculate the value of ξ for (a) H_2O and (b) D_2O.

3.5. If neutrons are being produced in a graphite moderator at a rate of 10^{13} neutrons per second, calculate the value of the neutron flux per eV at an energy of 10 eV.

3.6. A large reactor consisting of a dilute homogeneous mixture of Pu^{239} and graphite runs at a power of 2 MW. If the total mass of Pu^{239} is 10 kg, and the reactor temperature is 300°C, calculate the average 2200-ms^{-1} neutron flux in the reactor. Take the recoverable energy release in the fission of Pu^{239} as 210 MeV.

3.7. A homogeneous reactor is in the form of a cylinder of height 2 m and radius 1 m. If the thermal flux at the center is 10^{13} n cm^{-2} s^{-1}, calculate the flux at a point which is at a distance of 0.5 m from the axis of the cylinder, and 0.5 m above its central plane.

3.8. A homogeneous spherical reactor consists of a uniform mixture of U^{235}, boron, and graphite, with the atomic ratios $C/U^{235} = 10,000$ and $B/U^{235} = 0.5$. Calculate (a) the diffusion length in the mixture and (b) the buckling.

3.9. For the reactor of the previous problem, calculate the radius of the sphere required for criticality, using one-group theory.

3.10. Using one-group theory, calculate the radius of a critical reactor in the form of a sphere made up of a uniform mixture of U^{235} and graphite, in the atomic ratio $C/U^{235} = 20,000$.

3.11. If the reactor of the previous problem is running at a power of 1 kW, calculate the value of the thermal flux at its center.

3.12. Calculate the concentration (in g cm^{-3}) of Pu^{239} required to make an infinite homogeneous mixture of water and Pu^{239} critical.

3.13. An infinite slab reactor of (extrapolated) thickness 1.5 m is composed of a uniform mixture of U^{235} and graphite. Calculate (a) the ratio of carbon to U^{235} atoms required for criticality and (b) the thermal nonleakage probability.

3.14. If the thermal flux value at the central plane of the reactor in the previous question is 10^{12} n cm^{-2} s^{-1}, calculate the average power density (in W cm^{-3}) in the slab.

3.15. Calculate the equilibrium concentrations (in atoms per cm^3) of I^{135} and Xe^{135} in a U^{235}-fueled reactor after it has been running for a long period at a power of 10 MW.

3.16. A reactor in the form of a cube is composed of a uniform homogeneous mixture of U^{233} and graphite in the atomic ratio $C/U^{233} = 15,000$. If the reactor is at a temperature of 20°, calculate the dimensions required for achieving criticality.

4

Nuclear Fuels and Their Characteristics

4.1. Nuclear Properties of Fissile Materials

The isotopes which are *thermally fissile*, that is, which can undergo fission as a result of the capture of a low energy neutron, are U^{235}, U^{233}, Pu^{239}, and Pu^{241}. Of these, only the first is naturally occurring, since the radioactive half-lives of the others are sufficiently short that any quantities which were formed in the original nucleosynthesis event which gave rise to the material of the solar system have long since decayed. U^{233} is formed by neutron capture in the *fertile* isotope Th^{232}, which is the only stable isotope of the element thorium, while Pu^{239} is formed similarly from fertile U^{238}. The higher plutonium isotope, Pu^{241}, is formed following two successive neutron captures in Pu^{239}, and is present as a relatively small proportion in currently operating reactor systems.

Although the relative proportions of the isotopes U^{235} and U^{238} in natural uranium must have been rather similar at the time of nucleosynthesis, the shorter radioactive half-life of the former (0.71×10^9 yr as compared with 4.5×10^9 yr) has resulted in its forming only a very small percentage of present-day uranium. The composition of natural uranium is shown in Table 4.1. The minute amount of U^{234} may be ignored for the present purposes.

As has been mentioned in Chapter 2, the relatively abundant isotopes U^{238} and Th^{232} will undergo fission provided that the energy of the interacting neutron is above a "fission threshold" of the order of 1 MeV. The term "fissionable" is applied to isotopes which will undergo fission when bombarded with neutrons of suitable energy. The only isotopes where the neutron energy required is low enough for the process to be of practical interest are those having atomic numbers of 90 or greater. The contribution of U^{238} and Th^{232} to direct fission neutron production in a thermal reactor is small because elastic or inelastic scattering rapidly reduces the energies of the neutrons below the fission threshold, and their primary role in the neutron

Table 4.1. Isotopic Composition of Natural Uranium

Isotope	Relative abundance (%)	Half-life (yr)
U^{234}	0.0057	2.5×10^5
U^{235}	0.720	7.1×10^8
U^{238}	99.275	4.51×10^9

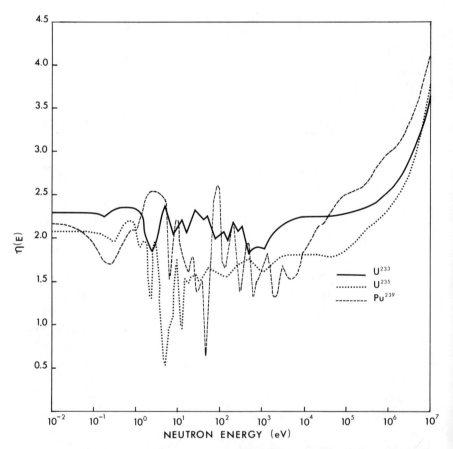

Fig. 4.1. Variation of η with neutron energy for U^{233}, U^{235}, and Pu^{239} [from J. J. Duderstadt and L. J. Hamilton, *Nuclear Reactor Analysis*, John Wiley and Sons (1976). Reprinted with permission].

economy of the reactor is the production of higher fissile isotopes by neutron capture.

The parameter which determines the value of a fissile material is η, the average number of secondary fission neutrons produced *per neutron absorbed in the fuel*. As shown in Chapter 2, η is given by

$$\eta = v\frac{\Sigma_f}{\Sigma_a} \tag{4.1}$$

where v is the average number of neutrons produced *per fission*. Here Σ_a, the macroscopic absorption cross section, is equal to the sum of the fission and capture cross sections, so that

$$\eta = v\frac{\Sigma_f}{\Sigma_f + \Sigma_c} = \frac{v}{1 + \alpha} \tag{4.2}$$

where α is the ratio of the capture to the fission cross section.

For all fissile isotopes, v is approximately constant with incident neutron energy up to about 100 keV, and increases slowly thereafter. The behavior of η is more complicated, however, owing to the variation of the capture-to-fission ratio, α, at the lower energies. The variation of η with energy for the fissile isotopes U^{233}, U^{235}, and Pu^{239} is shown in Fig. 4.1, and the η, v, and α values and cross sections at thermal energies are given in Table 4.2. It will be noted that, although the value of v for Pu^{239} at thermal energy is some 19% greater than for U^{235}, the η values are about the same, owing to the higher capture-to-fission ratio of the former isotope.

Figure 4.1 shows that, over the thermal and intermediate energy region, U^{233} has the highest η value, while Pu^{239} becomes the most efficient fuel at high neutron energies. The minimum requirement for *breeding*, i.e., the generation of more fissile material by neutron capture in fertile isotopes than

Table 4.2. Values of v, η, and α and Fission and Capture Cross Sections for the Fissile Isotopes at Thermal Energy[a]

Fissile isotope	v	η	σ_f (b)	σ_c (b)	α
U^{233}	2.49	2.285	531	48	0.090
U^{235}	2.42	2.07	582	99	0.170
Pu^{239}	2.87	2.11	742.5	269	0.362
Pu^{241}	2.93	2.145	1009	368	0.365

[a] Based on S. F. Mughabghab and D. I. Garber, *Neutron Cross Sections*, BNL-325, 3rd ed., Vol. 1, Brookhaven National Laboratory, Upton, New York (1973).

is consumed in fission, is that η be greater than 2. In practice, however, because of the neutrons lost by leakage and capture in materials other than fuel, the value of η has to be appreciably greater than 2 before breeding can occur. For a thermal reactor, the possibility of breeding is effectively restricted to the U^{233}–Th^{232} cycle, but at high neutron energies breeding is possible for all four fuels.

The efficiency of the process of production of replacement fuel is measured by the *conversion* (or *breeding*) *ratio*, which is defined as

$$C = \frac{\text{rate of production of new fissile material}}{\text{rate of destruction of fissile material}} \tag{4.3}$$

The processes taking place in the U^{233}–Th^{232} and the U^{238}–Pu^{239} cycles are illustrated in Fig. 4.2. Each of the isotopes in the chains is subject to two competing processes, radioactive decay and neutron capture, the relative magnitude of the two effects depending on the flux level to which the fuel is exposed. The horizontal arrows represent neutron capture, while the radioactive decay is indicated by vertical arrows, with the appropriate half-life given alongside. The neutron captures may be characterized by an effective "half-life" equivalent to that of the radioactive decay process, and related in the same way to the rate of change of concentration. If the number of atoms of a given isotope per cm^3 in a reactor is N, then the rate at which these are being destroyed by neutron capture is

$$\frac{dN}{dt} = -N\sigma_a\phi \tag{4.4}$$

where ϕ is the neutron flux and σ_a the nuclear absorption cross section. If this equation is compared with that for radioactive decay, where

$$\frac{dN}{dt} = -\lambda N$$

it is seen that the quantity $\sigma_a\phi$ is equivalent to the decay constant, λ. By analogy with the radioactive half-life ($T_{1/2} = 0.693/\lambda$), the neutron capture half-life for a given flux level is $0.693/\sigma_a\phi$. In the diagrams the flux has been taken at the typical value of 10^{14} n cm^{-2} s^{-1}. The radioactive half-lives of Th^{232}, U^{233}, U^{238}, and Pu^{239} are so long that their decay may be assumed to be negligible over the time of residence in the reactor.

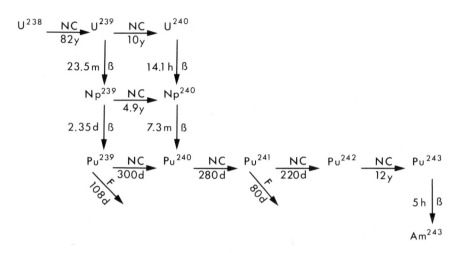

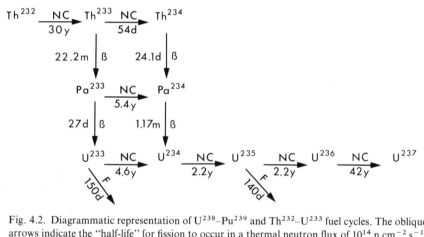

Fig. 4.2. Diagrammatic representation of U^{238}–Pu^{239} and Th^{232}–U^{233} fuel cycles. The oblique arrows indicate the "half-life" for fission to occur in a thermal neutron flux of 10^{14} n cm^{-2} s^{-1}.

4.2. Buildup of Plutonium in a Natural Uranium Reactor

In deriving an expression for the conversion ratio, account has to be taken of both thermal and epithermal capture of neutrons by the fertile material. Considering a reactor where the fissile material is U^{235} and the fertile U^{238}, the rate of thermal neutron capture per unit volume in U^{238} is simply $\Sigma_{28}^a \phi$, where Σ_{28}^a is the thermal macroscopic absorption cross section of U^{238} and ϕ the thermal neutron flux. (In the present chapter, we use the standard convention of specifying isotope cross sections by Σ_{mn}, where m is the last figure of the atomic number of the isotope and n the last figure of the mass number). For the resonance capture, we first note that the production rate of fast neutrons is $\eta \Sigma_{25}^a \phi$, which is increased to $\eta \varepsilon \Sigma_{25}^a \phi$ if significant fast fission is present. Of these fast neutrons, a fraction $\mathscr{L}_f$ survive to reach the resonance region, where $\mathscr{L}_f$ is the fast nonleakage probability. The rate of resonance capture is therefore $\eta \varepsilon \Sigma_{25}^a \phi \mathscr{L}_f (1 - p)$, where p is the resonance escape probability. Hence the total rate of formation of Pu^{239} by neutron capture in U^{238} is

$$\Sigma_{28}^a \phi + \eta \varepsilon \Sigma_{25}^a \phi \mathscr{L}_f (1 - p)$$

and the conversion ratio is obtained by dividing this quantity by the rate of destruction of U^{235}, i.e.,

$$C = \frac{\Sigma_{28}^a \phi + \eta \varepsilon \Sigma_{25}^a \phi \mathscr{L}_f (1 - p)}{\Sigma_{25}^a \phi} \tag{4.5}$$

or

$$C = \frac{N_{28}}{N_{25}} \frac{\sigma_{28}^a}{\sigma_{25}^a} + \eta \varepsilon \mathscr{L}_f (1 - p) \tag{4.6}$$

where N_{28} and N_{25} are the number densities of U^{238} and U^{235}, respectively, and σ_{28}^a and σ_{25}^a the corresponding thermal cross sections.

Equation (4.5) may be put in another form by introducing the neutron balance condition for the critical reactor. The production rate of *thermal neutrons*, which is given by the product $\eta \varepsilon \Sigma_{25}^a \phi \mathscr{L}_f p$, is equal to the thermal neutron loss rate due to absorption in U^{235}, U^{238}, moderator, and fission product poisons, and to leakage. Hence,

$$\eta \varepsilon \Sigma_{25}^a \phi \mathscr{L}_f p = (\Sigma_{25}^a + \Sigma_{28}^a + \Sigma_m + \Sigma_p) \phi - D \nabla^2 \phi$$

where Σ_m and Σ_p are the macroscopic absorption cross sections for moderator and poisons. From equations (3.35) and (3.38), the leakage term $-D \nabla^2 \phi$ may be written as $D B^2 \phi$, which is equal to $B^2 L_m^2 \Sigma_m \phi$, where B^2 is the buckling and L_m the diffusion length in pure moderator. Hence,

$$\eta \varepsilon \mathscr{L}_f p = 1 + \frac{\Sigma_{28}^a}{\Sigma_{25}^a} + \frac{\Sigma_m (1 + L_m^2 B^2)}{\Sigma_{25}^a} + \frac{\Sigma_p}{\Sigma_{25}^a}$$

Substituting in equation (4.5) yields

$$C = \eta \varepsilon \mathscr{L}_f - 1 - \frac{\Sigma_m (1 + L_m^2 B^2)}{\Sigma_{25}^a} - \frac{\Sigma_p}{\Sigma_{25}^a} \tag{4.7}$$

It is readily apparent from equation (4.7) why the best conversion ratio is obtained for a fast reactor system. Apart from the more favorable value of η at high energies (see Fig. 4.1), the fast fission factor can rise to a value as high as 1.3 in a fast reactor spectrum. In addition, the third factor on the right-hand side is reduced owing to the absence of moderator. Consequently, values of C greater than unity can be readily achieved for a fast reactor.

It is of interest to consider the way in which Pu^{239} builds up in a thermal reactor originally fueled with U^{235}, with U^{238} as fertile material. We assume that the initial number densities of U^{235} and U^{238} are $N_{25}(0)$ and zero, respectively. N_{28}, the number density of U^{238}, is assumed to remain constant, since its destruction over the life of the fuel charge is small in comparison with the total quantity present. To simplify the calculation, we ignore the capture of neutrons in the intermediate products U^{239} and Np^{239} (see Fig. 4.2) and assume that Pu^{239} is produced directly from U^{238}, an assumption which is reasonable in view of the relatively short half-lives of its precursors.

The equation for the rate of change of U^{235} concentration is then

$$\frac{dN_{25}}{dt} = -N_{25} \sigma_{25}^a \phi \tag{4.8}$$

which has the solution

$$N_{25} = N_{25}(0)\, e^{-\sigma_{25}^a \phi t} \tag{4.9}$$

The rate of formation of Pu^{239} from capture of thermal neutrons in U^{238} is $N_{28} \sigma_{28}^a \phi$. The resonance absorption is given by an expression similar to the second term in the numerator of equation (4.5), with a corresponding term having to be added to include the fast neutron source from Pu^{239}, once this has built up enough to be a significant factor. If we assume that ε is the same for U^{235} and Pu^{239}, the resonance production of Pu^{239} is then

$$(\eta_{25} N_{25} \sigma_{25}^a + \eta_{49} N_{49} \sigma_{49}^a)\varepsilon L_f (1 - p)\phi$$

The rate of destruction of Pu^{239} is $N_{49}\sigma_{49}^a\phi$, so that the net rate of change is

$$\frac{dN_{49}}{dt} = N_{28}\sigma_{28}^a\phi + (\eta_{25}N_{25}\sigma_{25}^a + \eta_{49}N_{49}\sigma_{49}^a)\varepsilon L_f(1-p)\phi - N_{49}\sigma_{49}^a\phi$$

(4.10)

Now let

$$\sigma_{49}' = \sigma_{49}^a[1 - \eta_{49}\varepsilon L_f(1-p)]$$

(4.11)

and

$$\sigma_{25}' = \eta_{25}\sigma_{25}^a\varepsilon L_f(1-p)$$

(4.12)

Using these relations and equation (4.9), we have

$$\frac{dN_{49}}{dt} = N_{28}\sigma_{28}^a\phi + N_{25}(0)\sigma_{25}' \, e^{-\sigma_{25}^a\phi t} \, \phi - N_{49}\sigma_{49}'\phi$$

(4.13)

The solution for the case where $N_{49}(0) = 0$ is

$$N_{49} = N_{28}\frac{\sigma_{28}^a}{\sigma_{49}'}\left(1 - e^{-\sigma_{49}'\phi t}\right) + N_{25}(0)\frac{\sigma_{25}'}{\sigma_{49}' - \sigma_{25}^a}\left(e^{-\sigma_{25}^a\sigma t} - e^{-\sigma_{49}'\sigma t}\right)$$

(4.14)

Similar solutions for the buildup of the higher plutonium isotopes, Pu^{240}, Pu^{241}, and Pu^{242}, may be obtained by solving differential equations of the form (for Pu^{240}),

$$\frac{dN_{40}}{dt} = \gamma_{49}N_{49}\sigma_{49}^a\phi - N_{40}\sigma_{40}^a\phi$$

(4.15)

where γ_{49} is the fraction of thermal neutron captures in Pu^{239} which leads to the (n, γ) rather than the fission reaction ($\gamma = \alpha/(1 + \alpha)$, where α is the capture-to-fission ratio). The decay of Pu^{240} has been neglected since the effective capture half-life is so much shorter than its decay half-life (6540 yr).

It will be seen from equations (4.9) and (4.14) that the concentrations of U^{235} and Pu^{239} are functions of the product ϕt (and indeed in the early stages the Pu^{239} concentration is approximately proportional to this product). The time-integrated flux ϕt, known as the *fluence*, is thus a convenient measure of the degree of burn-up of the fuel; its units are n cm^{-2}.

An alternative measure of burn-up is the integrated amount of energy which has been generated by unit mass of the fuel. The amount of energy released from a mass M of fuel, exposed to a fluence ϕt, is

$$E = \frac{MN_0}{A} \sigma_f \chi \phi t \qquad (4.16)$$

where N_0 is Avogadro's number and χ is the average energy released per fission (200 MeV). The energy is commonly expressed in megawatt-days (MW d) and the mass in tons. In these units, the expression for the burn-up becomes (for a reactor fueled with pure U^{235})

$$E/M \text{ (MW d}/t) = 5.5 \times 10^{-16}(\phi t) \qquad (4.17)$$

It should be noted that the term *megawatt* as used above refers to the rate of production of *heat* in the reactor. In practice, only about a third of this is converted into electrical output in a power reactor. The efficiency of a reactor is simply the ratio of the electrical power (in megawatts) to the *thermal* power, or total number of megawatts of heat generated.

Typical curves for the decay of U^{235} and the buildup of plutonium isotopes in a natural uranium reactor are shown in Fig. 4.3. It will be seen that, beyond an exposure of approximately 7500 MW d/tonne *of natural uranium*, the concentration of Pu^{239} exceeds that of U^{235}. Because the fission cross section of Pu^{239} is higher than that of U^{235}, the majority of fissions in the reactor will be taking place in plutonium rather than uranium even before the 7500-MW d/tonne exposure is reached.

The changes in concentration of the uranium and plutonium isotopes naturally lead to a change in the reactivity of the reactor as burn-up proceeds. In the four-factor formula for the infinite multiplication factor, $k_\infty = \eta \varepsilon p f$, the parameters ε and p do not change significantly with burn-up for a natural uranium reactor, since both are functions of the concentration of U^{238}, which remains approximately constant. The reactivity change therefore depends on the variation of the product ηf. The overall reactivity for a natural uranium system as a function of irradiation, and the main components giving rise to the total, are illustrated in Fig. 4.4.

Although the rate of replacement of U^{235} by Pu^{239} atoms in this example is less than unity, the higher absorption cross section of the latter more than compensates for the reduction in the number density of fissile

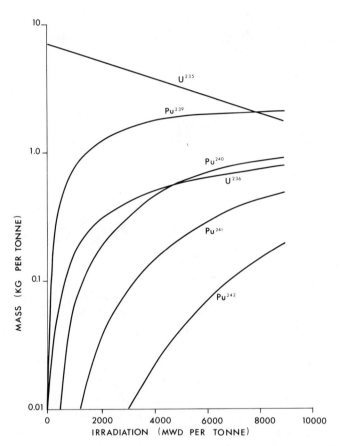

Fig. 4.3. Burn-up of U^{235} and buildup of plutonium isotopes in a natural uranium reactor [from H. Rose and J. J. Syrett, *J. Brit. Nucl. Energy Conf.*, 179 (July 1959). (Courtesy of the British Nuclear Society.)]

atoms, with the result that there is an increase in the thermal utilization, leading to an increase in the overall reactivity. The positive reactivity change is also aided slightly by the marginally higher value of η for the Pu^{239}. At higher irradiations, the rate of buildup of Pu^{239} levels off, as shown by equation (4.13), and, above about 3000 MW d/tonne, the net change in fissile material concentration produces a reduction in reactivity. The steady growth of Pu^{240}, with its large absorption resonance at 1 eV, provides an increasingly negative contribution to the reactivity. Coupled with the increasing absorption due to the buildup of fission products, the result is that the reactivity returns to zero at an irradiation of about 3000 MW d/tonne, and thereafter steadily decreases with irradiation. The decrease would be considerably more rapid were it not for the positive contribution due to the buildup of fissile Pu^{241} by neutron capture in Pu^{240}.

The reactivity rise due to plutonium buildup is most marked for the natural uranium reactor, where there is a relatively high conversion ratio. For a reactor of higher fuel enrichment, with a correspondingly lower conversion ratio, the rise would be less and the irradiation time before return to the initial reactivity would be shorter.

A light water reactor produces about 250 g of plutonium per year for each megawatt of electrical output. While the most efficient method for the long-term utilization of plutonium is in the fast breeder reactor (see Chapter 11), the rate at which plutonium stocks are being built up in the current thermal reactor programs is such that, in some countries at least, there is considerable interest in extracting the plutonium from the discharged uranium and recycling this as a thermal reactor fuel. Plutonium recycle in thermal reactors is an attractive alternative to the use of enriched uranium for countries that do not possess an enrichment plant, and wish to be

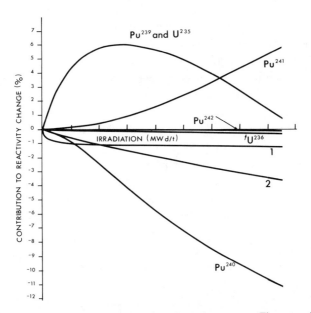

Fig. 4.4. Long-term reactivity variation in a natural uranium reactor. The curves labeled 1 and 2 are the reactivity contributions due to high cross-section fission products (excluding Xe^{135}) and low cross-section fission products, respectively [from H. Rose and J. J. Syrett, *J. Brit. Nucl. Energy Conf.*, 179 (July 1959). (Courtesy of the British Nuclear Energy Society.)]

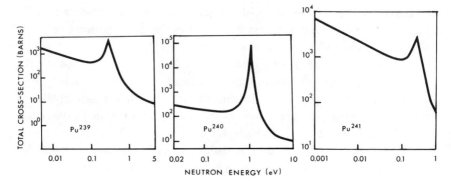

Fig. 4.5. Total cross sections of Pu^{239}, Pu^{240}, and Pu^{241} at low neutron energies.

independent of foreign suppliers of enriched fuel. The strategy for the use of accumulated plutonium depends on the estimated date at which a full-scale program of fast breeder construction can be initiated. An early start to such a program could imply that plutonium should currently be stockpiled to ensure that the fuel demand for commissioning breeders would not exceed the supply. Delay in the fast reactor program, however, could make it economically advantageous to use some of the plutonium stocks for thermal recycle; the extent of the benefit will depend on the cost of constructing complex reprocessing plants for the production of plutonium fuel elements.

The fuel loading for a reactor operating on plutonium recycle will in general contain a mixture of standard enriched uranium oxide fuel elements and elements containing a mixture of plutonium and uranium oxides. When the fuel is recycled in the same reactor in which it was produced, the amount available would mean that some one third of the core would be composed of plutonium elements. On the other hand, it may be advantageous to load a whole core with plutonium elements only, since this would permit the lattice to be redesigned to take advantage of the nuclear characteristics of plutonium. The more important distinctions between uranium and pluto-nium fuel are summarized briefly below:

i. The recoverable energy *per thermal fission* is some 4% greater for Pu^{239} than for U^{235}. On the other hand, because the thermal capture-to-fission ratio for Pu^{239} is 0.36 compared with 0.17 for U^{235}, the energy released *per unit mass of fuel destroyed* is actually 11% less for Pu^{239}.

ii. As mentioned earlier, the capture and fission cross sections of Pu^{239}, and also the cross sections of the higher isotopes Pu^{240} and Pu^{241}, vary much more markedly in the thermal region than do those of U^{235} (see Fig. 4.5).

This behavior is associated with the large resonances at approximately 0.3 eV in Pu^{239} and Pu^{241} and at 1.0 eV in Pu^{240}. The capture-to-fission ratio also varies considerably; for example, the value of η ($= \nu/1 + \alpha$) in the 0.3-eV resonance of Pu^{239} is as low as 1.67. As a result of the resonance structure, the sensitivity of a plutonium-fueled reactor to changes in the thermal flux spectrum shape is much more pronounced than for a uranium system, with a corresponding complication in the prediction of temperature coefficients. For a light water reactor, the temperature coefficient with a plutonium core tends to be more negative than for uranium, owing to the shift of the spectrum into the low-η Pu^{239} resonance and to increased absorption in the Pu^{240} resonance. The Doppler effect in the resonances can also give a significantly negative contribution to the overall temperature coefficient.

iii. The Pu^{239} and Pu^{241} absorption cross sections are considerably greater than for U^{235} (see Table 4.2). Consequently, plutonium competes more successfully for thermal neutrons with absorbers present in the core than does U^{235}. One effect is a lower reactivity worth for control rods in the plutonium core, leading to a requirement for a greater number of rods. For a boiling water reactor with fuel assemblies (see Fig. 9.7) containing a mixture of plutonium and uranium pins, the latter may have to be grouped as far as possible from the control rod blades to minimize the reduction of control rod worth due to the localized high plutonium absorption.

iv. Pu^{240} plays a crucial role in the neutron economy of a plutonium-fueled core. Because of the high α value of Pu^{239}, the production of Pu^{240} per neutron absorbed in fuel is more than twice as great as the production of U^{236} from capture in U^{235}. While U^{236} is a weak parasitic absorber, giving a gradual fall in reactivity, the Pu^{240} has an absorption cross section of 290 b to yield Pu^{241}, which has a high value of ν and a fission cross section considerably greater than that of Pu^{239}. By loading fuel of high Pu^{240} concentration (i.e., fuel which has undergone a long irradiation) into the reactor, the life of the fuel charge can be extended considerably. The Pu^{240} is much more effective in this respect than is, for example, U^{238}, since the much higher absorption cross section of the Pu^{240} means that its burn-out rate, and consequently the reactivity rise, is much more rapid. The fuel element geometry constitutes an additional variable in the choice of plutonium isotope ratios to achieve long-life fuels, since all the isotopes have prominent resonances which give rise to self-screening effects.

v. As discussed in Chapter 3, the kinetic behavior of the reactor depends on the value of the delayed neutron fraction, which is only 0.0021 for Pu^{239}, compared with 0.0065 for U^{235}. The overall delayed neutron fraction for a plutonium core is increased by the presence of Pu^{241}, which has a value of

0.0053. The lower margin to prompt critical leads to more stringent limitations on the possible reactivity changes that might occur under fault conditions.

The desirability of using plutonium to conserve uranium stocks has to be balanced against the fact that the cost of its recovery from irradiated fuel is inevitably high, partly because of the radiation levels of the discharged fuel and partly owing to the highly toxic nature of plutonium itself. The specific toxicity of Pu^{239} is more than a factor of 10^4 higher than that of U^{235}, owing to the greater α activity, while the permissible body burden is more than a factor of 10^5 lower, owing to the tendency of the plutonium to concentrate in the bone. Very strict precautions against leakage are therefore necessary. Considerable shielding is also required against the γ activity of highly irradiated plutonium, the chief contributors being Pu^{238}, Pu^{239}, Pu^{241}, and Am^{241}; in addition, one has a significant hazard from neutron emission due to spontaneous fission of Pu^{240}. A further complication, of course, is the necessity for avoiding the buildup, at any stage of the reprocessing and fabrication process, of a critical mass of plutonium. The consequent restriction on batch size increases the production costs.

4.3. The Th^{232}–U^{233} Fuel Cycle

The Th^{232}–U^{233} cycle is attractive because it is based on the conversion of the relatively abundant element thorium into an isotope of uranium which has nuclear characteristics superior to those of U^{235} in a thermal reactor. Since, unlike uranium, thorium in the natural state does not incorporate a thermally fissile isotope, the Th^{232}–U^{233} cycle has lagged behind that of U^{238}–Pu^{239}, which was based on the automatic inclusion of the U^{238} contained in natural uranium. In addition, the radiological problems involved in the handling of highly irradiated thorium are somewhat more severe than those encountered in the plutonium cycle.

Experience of the Th^{232}–U^{233} cycle was first obtained in the Indian Point boiling water reactor, where the first core, loaded in 1962, contained pellets of urania–thoria mixture. The main interest, however, has centered on its use in the high-temperature gas-cooled reactor (HTGR), and thorium has been employed as fertile material both in the prismatic fuel elements of the Dragon reactor in the United Kingdom and the Peach Bottom reactor in the United States, and in the spherical elements of the pebble-bed AVR in West Germany. There is also a possibility of adopting the thorium cycle in the

Canadian CANDU heavy water reactor, and in a circulating fuel reactor designed to achieve thermal breeding, the molten salt breeder reactor (MSBR).

The fuel management strategy of the Th^{232}–U^{233} cycle (see Fig. 4.2) is strongly influenced by the need to minimize radiation levels in the reprocessing stages. One of the main problems in reprocessing the irradiated thorium to obtain U^{233} is the high activity associated with the decay chain starting with the isotope U^{232}. This is formed following the (n, 2n) reaction of fast neutrons with Th^{232}. This and subsequent processes lead, as shown below, to the formation of U^{232}, an α emitter of 72-yr half-life:

$$Th^{232} + n \rightarrow Th^{231} + 2n$$
$$Th^{231} \rightarrow Pa^{231} + \beta^-$$
$$Pa^{231} + n \rightarrow Pa^{232}$$
$$Pa^{232} \rightarrow U^{232} + \beta^-$$

The U^{232} decays by α emission to Th^{228}, which is a strong γ emitter, and the first member of a decay chain which contains several sources of γ rays. Since all the other members have half-lives much less than that of Th^{228} (1.9 yr), the overall rate of decay of the whole chain is governed by the Th^{228}.

On chemical separation of the uranium from the irradiated thorium, the U^{232} is, of course, removed along with the U^{233}. It does not itself represent a serious radiation hazard and, since it has a long half-life, the buildup of its daughter Th^{228} in the extracted uranium is relatively slow. Consequently, it is possible to avoid the need for heavy shielding and remote handling of the recycle U^{233} provided that the fabrication of the fresh fuel elements is done within a short time of the separation of the uranium from the irradiated fertile material.

On account of the long half-life of U^{232}, the problem of the activities of its daughter products becomes more severe each time the fuel is recycled through the reactor. One way in which the buildup of U^{232} can be reduced is by the physical separation of the fertile thorium from the fissile fuel in the reactor. The separation of the two by a significant thickness of moderator reduces the kinetic energies of most of the fission neutrons below the threshold of the (n, 2n) reaction in Th^{232} before they have the opportunity of reaching the fertile material. It is also necessary to limit the exposure of the fertile material, since the buildup of U^{233} will gradually supply a new source of fission neutrons in contact with the thorium.

The separation of the fertile from the fissile material is also desirable on the grounds of neutron economy and conversion ratio. The reason for this is the relatively long half-life (27.4 d) of Pa^{233}, the precursor of U^{233}, compared, for example, with the 2.3-d half-life of the equivalent member of the plutonium cycle, Np^{239}. The capture of neutrons in Pa^{233} is doubly disadvantageous in that one loses not only a neutron but also a potential nucleus of fissile U^{233}. The loss may be reduced by segregating the fertile material into a region of lower neutron flux, such as a blanket surrounding a fissile core. The opportunity for continuous removal of Pa^{233} is one of the advantages of the molten salt breeder reactor.

A further consequence of the long half-life of Pa^{233} is that, in order to obtain a high overall yield of U^{233}, the discharged elements should be left in storage for several months to allow for Pa^{233} decay. It may in practice be more economic to accept a lower yield of U^{233} rather than suffer the disadvantage of holdup of large thorium inventories.

The thorium removed from the irradiated fertile material can also be recycled, but here again the Th^{228} activity presents a handling problem. It may be necessary to store the thorium for at least five half-lives of Th^{228} (about 10 yr) to allow the activity to decay to a manageable level.

Because of the small reactivity margin available for breeding in a thermal reactor, the use of the thorium cycle has mainly been associated with reactors with very good neutron economy based on low parasitic absorption, such as the high-temperature gas-cooled reactor, where graphite is used in place of metal for the fuel cladding, or heavy water reactors, with very low moderator absorption. A special case is the molten salt breeder reactor, where circulation of the fissile and fertile materials allows continuous removal not only of Pa^{233} but also of fission products.

For the high-temperature reactor, the U^{235}–Th^{232}–U^{233} cycle is in competition with the "low-enrichment" uranium cycle, where the fuel is uranium enriched to about 10% (U^{238} being the only fertile material). While the former cycle is probably more economically attractive provided that adequate supplies of U^{235} are available, the latter has the advantage of eliminating the cost of the highly enriched uranium and may also fit better in countries where a program of fast breeder reactors, with reprocessing facilities for plutonium fuel, is also planned. The economics of the U^{233}–Th^{232} system are very dependent on the conversion ratio achieved since, the smaller this is, the greater the demand for "makeup" U^{235} (which cannot be recycled because of the buildup of U^{236} poison). In the HTGR design, where the fuel is in the form of small particles coated with impervious layers of carbon or silicon carbide (see Chapter 5), the U^{235} particles are made a different size from the fertile in order to facilitate separation of the U^{235} from the bred U^{233} in the discharged fuel.

4.4. Aspects of Fuel Management in Thermal Reactors

We may start by considering the way in which the reactivity of a reactor changes between the initial condition where it is cold and loaded with fresh fuel and the final state where it runs out of reactivity and has to be refueled. In order to allow for reactivity loss in going from cold to operating temperature, and to provide a sufficient margin of reactivity to compensate for fuel burn-up over a period of about a year before refueling, the enrichment of the initial fuel loading must be high enough to give the core a large potential reactivity at start of life. Initially, therefore, this reactivity must be held down either by control rods or by some sort of distributed poison, such as a boron salt dissolved in the moderator.

As the reactor is brought up to the operating temperature, with a negative temperature coefficient, control rods have to be moved out of the core to maintain criticality, or the boron concentration reduced. Build-up of Xe^{135} causes a further reduction of reactivity. Finally, over the life of the core, the gradual burn-up of the fuel and the build-up of long-lived fission products causes a decrease of reactivity until the reactor has to be refueled.

The calculation of the variation of the reactivity over the life of the core is a very complicated exercise. Allowance has to be made for (i) decrease in the overall fuel concentration, (ii) net buildup of fresh fissile material from conversion, and (iii) changes in the spatial distribution of fuel and fission products as burn-up proceeds. For example, the flux will generally be highest at the core center, and consequently the fuel will burn up and fission products accumulate more rapidly there, causing in turn a change in the flux distribution in the reactor.

In order to predict the behavior of the reactor over its core life, a series of multigroup calculations has to be done for selected time intervals. The first calculation gives the multigroup fluxes at the start of core life, and it is assumed that these fluxes prevail throughout the whole of the first interval. The change in the material concentrations over the first interval is then calculated from the fluxes. The multigroup fluxes at the end of the interval are then derived using the new material concentrations, and so on. At each stage, the control rod configuration or the boron concentration required to maintain criticality is included in the calculation. The computations are continued until the stage when all the control reactivity has been removed, at which point it will be necessary to refuel the reactor.

A number of methods has been proposed for extending the useful life of the fuel charge. The use of a large number of control rods as a means of permitting a high initial reactivity of the core is undesirable because of the waste of neutrons by absorption in the rods and the local flux perturbations which the rods produce when inserted. The latter effect may be removed in the case of a liquid-moderated reactor by the addition of soluble poison. The maximum concentration of poison which can be used in this so-called *chemical shim* technique is limited by the need to maintain a negative temperature coefficient (see Section 3.4).

An alternative to poisoning the moderator is the inclusion of a burnable poison in or adjacent to the fuel elements themselves. The poison is chosen to have an absorption cross section and a concentration which cause it to burn out at a rate which will match the rate of loss of reactivity of the fuel. The elements which have been used as burnable poisons include boron and gadolinium (in the form of gadolinia, Gd_2O_3).

In a boiling water reactor, it is possible to make use of the reactivity controlled by voidage to "stretch out" the life of the fuel charge. As the 'all rods out" reactivity approaches zero, the feedwater temperature is reduced, causing a reduction in the amount of coolant boiling and thus of the core voidage fraction. Since the voidage coefficient is negative, the result is an increase in the core reactivity. The thermal power output of the reactor remains unchanged, but the electrical output decreases progressively due to a gradual decrease in steam flow. The technique is only suitable for relatively short extensions of the core lifetime, of the order of a few weeks, but it gives a measure of flexibility which permits firm forward planning of major refueling.

It is worth noting an ingenious suggestion, due to Edlund, for reactivity control in a pressurized water reactor. In this method, known as spectral shift control, the moderator consists of a mixture of light and heavy water, and the reactivity is adjusted by varying the relative proportions of the two over the core life. The lattice is sufficiently closely packed that an increase in the H_2O to D_2O ratio produces a corresponding rise in reactivity, since the reduction in leakage and resonance absorption produced by the increased slowing down power overcomes the effect of the greater neutron absorption. The main contribution to reactivity control, in fact, arises from the change in the U^{238} resonance capture; the method is efficient in that, during the early life of the core, neutrons are being soaked up productively by U^{238}, rather than nonproductively in a control absorber. Disadvantages of spectral shift control include the cost of the heavy water and the complexity involved in adjustment of the D_2O/H_2O ratio, and the system has not in fact been utilized in a commercial reactor.

For natural uranium reactors, such as magnox or CANDU, the attainable excess reactivity at start of life is considerably smaller than can be achieved in an enriched reactor, such as the BWR or PWR. Thus, the interval between refueling shutdowns would be inconveniently small. The natural uranium reactors, therefore, employ an *on-load refueling system*, in which the fuel elements may be removed while the reactor is running at full power. The advantages of this method include: avoidance of the need for periodic shutdowns for refueling (typically about 30 days outage for a light water reactor); easy discharge of any failed fuel element, leading to a cleaner primary circuit; better neutron economy, since the flux to which a given element is exposed can be varied to achieve optimum irradiation by changing the position of the element; limitation of the effects of possible reactivity transients, since the excess reactivity of the core can be kept to a modest value. The disadvantage of on-load refueling is the cost and complexity of the associated equipment; virtually all continuous refuelling systems have given trouble in the early commissioning stages.

An aspect of fuel management which is closely related to obtaining the maximum lifetime for the core is the control of the shape of the thermal flux in the reactor. The maximum power at which the reactor can be operated is set in general by the attainable heat removal rate at the hottest fuel position. The limit may be set by the need to avoid either excessive fuel cladding temperature or the onset of "burn-out" conditions due to the breakdown of the heat transfer mechanism at the fuel surface (see Chapter 6). The important parameter is therefore the ratio of the maximum to average power production rate in the core, known as the *form factor*. Ideally, for a core with uniform heat removal capacity, the form factor should be unity. In practice, it is affected by factors such as the axial and radial flux variations in the core, local flux depressions due to control rods, or local flux peaking due, for example, to water gaps in a light water reactor.

Variations in power to heat removal ratio may be minimized either by making the power distribution more uniform (*flux flattening*) or by varying the local heat removal rate to match the existing power distribution. The latter may be done, for example, by having variable gaps at the entrance to the coolant channels, as is done in the advanced gas-cooled reactor (AGR). The power distribution may be made more uniform by employing control rods preferentially in the central region of the core or by varying the enrichment of the fuel as a function of core position. With a multibatch scheme, the core may be divided into radial zones, with the lowest enrichment in the central zone and the highest in the outermost. At reloading, partially depleted fuel from the outer zones is moved inwards and replaced by fresh fuel. The use of a differential enrichment scheme of this kind can improve the radial form factor by up to 30%, and provide a more uniform irradiation of the fuel elements than would be possible with a single-batch loading. The enrichments of the individual fuel pins comprising an element may also be varied in order to equalize the power distribution within the element itself.

Problems

(The data required for the solution of these problems are given in Tables A.1–A.6).

4.1. A homogeneous graphite-moderated reactor has U^{235} as fuel and Th^{232} as fertile material. The atomic ratios are carbon/U^{235} = 10,000, Th^{232}/U^{235} = 40. The 2200-ms^{-1} flux is $\phi = 10^{13}$ n cm^{-2} s^{-1}. If $\mathscr{L}_f = 0.77$, $p = 0.89$, and $\varepsilon = 1.02$, calculate the conversion ratio and the rate of formation of U^{233} per cm^3.

4.2. A homogeneous graphite-moderated research reactor is fueled with uranium enriched to 2% in U^{235}. The carbon/U^{235} atomic ratio is 12,000 and the total U^{235} content of the core is 10 kg. The reactor operates at a thermal power of 12 MW, at a temperature of 400°C. If $\mathscr{L}_f = 0.75$, $p = 0.91$, and $\varepsilon = 1.00$, calculate the average 2200-ms^{-1} flux in the core, the burn-up rate of U^{235}, and the rate of formation of Pu^{239}, per cm^3 at the start of core life.

4.3. For the reactor in the previous question, calculate the concentration of Pu^{239} per cm^3 after the reactor has been operating for 1 yr at a power of 12 MW.

4.4. A light water reactor has a conversion ratio of 0.40. Calculate the mass of Pu^{239} produced per year per megawatt of electrical output, if 3 MW of thermal output is required to produce 1 MW of electrical power.

4.5. A reactor fueled with natural uranium operates at a total thermal power of 600 MW. If the conversion ratio is 0.89, calculate the total rate of production of Pu^{239} in kg/yr.

5

Materials Problems for Nuclear Reactors

5.1. Effects of Irradiation on Reactor Materials

The materials in a conventional power plant have to be capable of operating reliably under conditions of high temperature and pressure and of withstanding chemical corrosion for long periods of time. These problems are well understood on the basis of extensive experience. The nuclear power plant, however, introduces a range of new problems associated with the use of materials in a high radiation environment.

In the first place, the choice of structural materials for the core of the reactor is greatly restricted by the need to minimize neutron absorption for reasons of reactivity. This factor eliminates many common elements altogether, and imposes rigorous limitations on the levels of high absorption impurities which can be tolerated in the elements which are acceptable.

Materials exposed in or near the reactor core experience a marked change in properties due mainly to bombardment with fast neutrons, which cause displacements of the constituent atoms from their regular positions in the crystal lattice. In a typical graphite reactor, for example, a large fraction of the atoms will suffer displacement during the life of the reactor. The buildup of displacements gives rise to dimensional changes in the graphite and also to a storage of energy which will be released when the atoms fall back to their regular positions. In structural materials such as steel, fast neutron bombardment causes changes in strength and loss of ductility. The phenomenon of radiation damage is complicated by a marked dependence on the temperature of the irradiated material, due to the tendency for displacement damage to be removed by annealing.

A further consequence of irradiation is the change in composition arising from neutron interactions with the nuclei of the material. This is predominantly a slow neutron effect, but the formation of helium gas as a result of the fast neutron-induced (n, α) reaction in boron has an important effect on the structural stability of steel in which the boron may be present even at a very low impurity level.

Chemical interaction between fuel and cladding, or between coolant and cladding or structural materials, must be limited to a level giving an acceptably low probability of fuel element or plant component failure. The normal problems of chemical compatibility are increased by the effect of irradiation in enhancing the rate of chemical attack. The radiolytic oxidation of graphite in a CO_2-cooled reactor, for example, raises problems of graphite erosion and mass transport to cooler regions of the primary circuit.

It is on the fuel itself that the potential effects of irradiation are most severe. In addition to the changes induced by neutron bombardment, the fuel is subject to intense localized damage produced by the recoiling fission fragments. The buildup of gaseous fission products, such as xenon and krypton, leads to the formation of gas bubbles causing swelling of uranium metal fuel rods. In addition, pronounced dimensional changes are produced by the irradiation of nonisotropic uranium metal by fast neutrons; the avoidance of this effect requires preirradiation treatment of the uranium to give a material with random grain orientation.

The limitations on possible linear rating and burn-up levels imposed by the characteristics of metallic uranium have led to the development of ceramic forms of fuel such as uranium oxide (UO_2) and uranium carbide (UC) pellets. While fuel performance is considerably enhanced by the use of the ceramic form, problems arise due to densification of the fuel and fuel-cladding interaction in a radiation environment, leading to the formation of interpellet gaps and clad flattening.

A more advanced development in nuclear fuel technology is the coated-particle concept, originally designed for the high-temperature gas-cooled reactor. The use of small particles of fuel and fertile material, individually coated with impermeable surface layers, and dispersed in a refractory medium such as graphite, has permitted the operation of reactor cores at high temperatures with minimum release of activity into the coolant gas of an HTGR.

5.2. Uranium—Production and Enrichment

Uranium is a widely distributed element in the oxide form, but there are relatively few areas where it occurs in economically recoverable concentrations. Among the major potential sources are Australia, Canada, South Africa, and the Western United States. Recent estimates of the main

Table 5.1. Estimated Uranium Resources in Kilotonnes, for Non-Communist Countries[a]

Source	$26/kg RAR	$26/kg EAR	$26–$39/kg	Total
United States	300	500	1200	2000
Canada	190	190	340	720
Australia[b]	150	75	50	275
South Africa	200	8	88	296
Sweden	—	—	330	330
France	37	23	45	105
Western and Central Africa	49	28	38	115

[a] From *Nuclear Engineering International*, February 1975.
[b] Extent of Australian resources uncertain owing to rapid rate of discovery.

resources of uranium in the non-communist world are given in Table 5.1. The estimates are broken down into two price categories, those recoverable (in oxide form) at a cost of around $26 per kilogram uranium and those in the range from $26 to $39 per kilogram. The former group is divided into two parts, known as *reasonably assured resources* (*RAR*) and *estimated additional resources* (*EAR*). The first category implies that the extent of the deposit is known and samples have been analyzed, while the second comprises potential sources which are likely to exist either as extensions of known fields or in places where uranium is known to occur but has not been worked before.

The higher-grade ore may contain up to 4% of uranium, although mining is carried out down to a concentration of about 0.1%. In addition to large amounts of uranium in the earth's crust, some 4×10^9 tonnes are estimated to be present in seawater. The concentration of the latter, however, is very low, of the order of 3 parts per billion (ppb), so that recovery will not be economic until a large proportion of the ore reserves has been exhausted.

To extract the uranium from the ore, the latter is first concentrated by standard metallurgical techniques and then leached in strong acid or alkali. The dissolved uranium is then recovered from the leach solution by solvent extraction or ion exchange. The product, known as 'yellow cake", contains in the range of 70%–90% uranium oxide, as a mixture of UO_2 and U_3O_8.

The production of natural or enriched uranium fuel elements, in either metallic or oxide form, requires as an intermediate stage the production of pure uranium dioxide, UO_2. By reacting the mixed-oxide concentrate given by the previous process with hydrogen in a fluidized-bed reactor, the U_3O_8 is converted to UO_2:

$$U_3O_8 + 2H_2 \rightarrow 3UO_2 + 2H_2O$$

Alternatively, the uranium concentrate may be treated with nitric acid and fed to an extraction column, the eventual product being uranyl nitrate hexahydrate. This is then decomposed by heating to form uranium trioxide, UO_3 ("organe oxide"). By reaction with hydrogen in a fluidized-bed reactor, the UO_3 is reduced to UO_2, which is of sufficient purity to be used directly in the fabrication of natural uranium oxide fuel elements.

For conversion to uranium metal, or as a preliminary to enrichment, the UO_2 is converted to uranium tetrafluoride, UF_4, by reaction with anhydrous hydrogen fluoride:

$$UO_2 + 4HF \rightarrow UF_4 + 2H_2O$$

Pure metallic uranium may be formed by reaction of the tetrafluoride with calcium or magnesium, e.g.,

$$UF_4 + 2Mg \rightarrow 2MgF_2 + U$$

The difference in density between the liquid metal and the molten slag leads to effective separation of the two.

If the uranium is to be enriched in the U^{235} isotope, it must be converted to the gaseous form, as uranium hexafluoride, UF_6, which is gaseous at around 60°C. This is achieved by reaction with fluorine gas:

$$UF_4 + F_2 \rightarrow UF_6$$

After enrichment, the hexafluoride is converted back to UF_4 by reaction with hydrogen:

$$UF_6 + H_2 \rightarrow UF_4 + 2HF$$

The UF_4 may be treated, as above, with magnesium or calcium to yield enriched uranium metal or reacted with water and a hydroxide salt to give UO_3, from which enriched UO_2 suitable for fuel element manufacture may be obtained by reaction with hydrogen.

The enrichment of uranium has to be accomplished by some physical process which depends on differential behavior of the two isotopes of uranium. The great majority of separative work has to date been carried out by the *gaseous diffusion process*, where the distinguishing feature between the isotopes is the rate at which the molecules of different mass diffuse through a porous barrier under an applied pressure differential. In a sample of uranium hexafluoride gas, the mean kinetic energy of the lighter molecules is equal to that of the heavier, i.e.,

$$\tfrac{1}{2}m_5 v_5^2 = \tfrac{1}{2}m_8 v_8^2$$

where m_5 and m_8 are the masses of the hexafluoride molecules associated with the U^{235} and U^{238} isotopes. Hence the average speed of the molecules of the lighter isotope is higher than that of the heavier, and their initial rate of diffusion through a porous barrier containing a large number of very fine channels will be more rapid. The flow of a given isotope is governed by Knudsen's law and is consequently inversely proportional to the square root of the molecular weight. The gas passing through the barrier will therefore ideally be enriched in U^{235} by a factor (α) proportional to the square root of the ratio of the molecular masses, i.e., with uranium hexafluoride gas

$$\alpha = \left(\frac{m_8}{m_5}\right)^{1/2} = \left(\frac{352}{349}\right)^{1/2} = 1.0043$$

With such a small fractional mass difference between the isotopes, the separation obtained by diffusion through a single barrier is very small, and a practical gaseous diffusion plant must consist of a very large number of barrier stages arranged in a *diffusion cascade*. The cascade principle is illustrated diagrammatically in Fig. 5.1.

The basic unit of the cascade is a chamber divided into two sections by a thin and finely porous barrier. Gas is pumped into the first section under pressure and allowed to diffuse through the barrier into the other section; the more rapid diffusion of the molecules of the lighter isotope causes a slight enrichment of the gas flowing through the barrier. The pressure drops in the circuit are arranged so that about one half of the gas entering the high-pressure section of the chamber is allowed to diffuse into the low-pressure section, the remainder being fed back to be recycled through the previous stage. The enriched gas from the low-pressure section is pumped into the high-pressure section of the next enrichment stage. Hence the successive stages above the feed level in Fig. 5.1 (*enrichment section*) lead to a

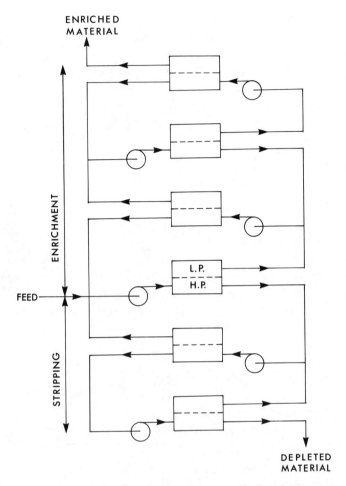

Fig. 5.1. Schematic diagram of gaseous diffusion plant.

progressive enrichment of the hexafluoride in U^{235} and the stages below the
feed level (*stripping section*) to a progressive depletion.

The point at which the product is withdrawn from the cascade is
determined by the degree of enrichment required. Typically, the number of
stages required for a high-enrichment product (over 90% U^{235}) may be of the
order of several thousand.

One of the principal problems of the gaseous diffusion technique is the
design of the porous barriers used for isotope separation. The UF_6 gas is

highly chemically reactive, forming fluorides with most metals, and reacting with any water vapor which may be present to form HF and UO_2F_2. The choice of metals for the barrier is therefore restricted to a very few, such as aluminum or nickel, which form stable fluoride layers and consequently are protected from extensive corrosion by the gas. A very high standard of leak tightness is necessary to avoid inleakage of air, which would cause decomposition of the UF_6 into particulate UO_2F_2, resulting in blockage of the very fine pores in the barriers.

The problems of barrier design for the gaseous diffusion process, coupled with the high capital cost of the plant, have led to a search for more economical means of enrichment. One alternative technique is the centrifugal process, which depends on the separation of molecules of different mass in a high-speed centrifuge. Although the difficulty of producing satisfactory high-speed centrifuges resulted in the development of the centrifugal method lagging behind the diffusion technique, the potential for reduced costs owing to the much lower electric power demand for the centrifugal process has stimulated research effort in recent years, particularly among the countries of Western Europe, where plans for the construction of major centrifugal enrichment facilities are at an advanced stage.

The capacity of a uranium enrichment plant is specified in terms of a quantity known as the *separative work*. This in turn is defined in terms of a quantity called the *separation potential*, which can be used to specify the value of a unit mass of material at a given concentration of U^{235}. The separation potential ϕ_c at concentration c is given by

$$\phi_c = (2c - 1)\ln\frac{c}{1-c} \tag{5.1}$$

The function ϕ_c is arbitrary in the sense that it is chosen so that the separation potential is zero when the concentration of the U^{235} is $c = 0.5$. The value of ϕ_c increases as c becomes either less than or greater than 0.5, so that if we start with material which has equal concentrations of U^{235} and U^{238}, the passage of this material through the enrichment plant will produce an enriched and a depleted fraction, both of which will have a positive value of the separation potential.

Suppose that a mass M of material at U^{235} concentration c is used as the *feed*, and that the plant produces a mass M_p of enriched *product* at concentration c_p, together with a residue (the *tails*) of mass M_w at concentration c_w. Then the separative work involved is defined as the amount by which the product (mass × separation potential) has been increased in the process, i.e.,

$$S = M_p\phi_p + M_w\phi_w - M\phi \qquad (5.2)$$

Since ϕ is dimensionless, separative work has the units of mass. Conservation of total mass and mass of U^{235} impose the following constraints on the masses and concentrations involved:

$$M_p + M_w = M \qquad (5.3)$$

and

$$M_p c_p + M_w c_w = Mc \qquad (5.4)$$

The usefulness of the function chosen for ϕ lies in the fact that the separative work which a plant can produce is independent of the concentrations of the particular feed, product, and waste concentrations involved. This is illustrated in the example below.

Example 5.1. An enrichment plant fed with natural uranium produces 400 tonnes per year of enriched uranium at a concentration of 2 wt % (percent by weight). The tails concentration is 0.25 wt %. What is the separative work of the plant in tonnes per year? How much uranium at 3 wt % could the same plant produce per year, assuming that the tails enrichment remains at 0.25 wt %? What would be the output of 3 wt % uranium if the tails enrichment were raised to 0.35 wt %?

Using equation (5.1), and noting that natural uranium contains 0.00711 wt % of U^{235}, we have for the separation potentials of product, waste and feed the following values:

$$\phi_p = 3.736$$
$$\phi_w = 5.959$$
$$\phi = 4.869$$

If the mass of waste per year is M_w tonnes, the total mass of feed is $(M_w + 400)$ tonnes and, using equation (5.4), we have

$$(400 \times 0.02) + (M_w \times 0.0025) = (M_w + 400) \times 0.00711$$

which gives $M_w = 1118$ tonne/yr and the total mass of feed, M, as 1518 tonne/yr. Hence, using equation (5.2), the separative work is

$$S = (400 \times 3.736) + (1118 \times 5.959) - (1518 \times 4.869) = 765 \text{ tonne/yr}$$

If the output concentration is changed to 3 %, we now have $\phi_p = 3.2675$, and we substitute in equations (5.2) and (5.4) to give

$$(M_p \times 3.2675) + (M_w \times 5.959) - (M_p + M_w) \times 4.869 = 765$$

and

$$(M_p \times 0.03) + (M_w \times 0.0025) = (M_p + M_w) \times 0.00711$$

Solving, we find

$$M_p = 201 \text{ tonne/yr}$$

i.e., the amount of 3 wt % uranium which the plant can produce per year is half the output of 2 wt %.

For a tails enrichment of 0.35 %, we have $\phi_w = 5.612$. Substituting this value in equation (5.2), and taking $c_w = 0.0035$ in equation (5.4) yields for the output of 3 wt % uranium the value

$$M_p = 246 \text{ tonne/yr}$$

In this case, the increase in production rate from 201 to 246 tonne/yr is achieved at the expense of a higher concentration of U^{235} in the waste, and a corresponding increase in the feed requirement.

5.3. Properties of Uranium Metal

For physics reasons, uranium in the form of metal rods was extensively employed as fuel for the first generation of nuclear reactors. The requirement for metallic fuel for a natural uranium graphite-moderated reactor is based on the need for a high fuel density and a fuel rod of sufficiently large diameter to reduce the resonance capture to a level where criticality may be achieved. Only uranium in metallic form has sufficiently high thermal conductivity to permit adequate heat removal from rods of the required diameter.

Table 5.2. Phases of Uranium Metal

Phase	Crystal structure	Temperature range (°C)
α	Orthorhombic	Below 666
β	Tetragonal	666–771
γ	Body-centered cubic	771–1130

Metallic uranium has one of three possible crystal structures in the solid state, depending on temperature. The structures and corresponding temperature ranges are listed in Table 5.2.

For pure uranium, repeated temperature cycling through the transition temperatures, with resultant transformation from one phase to another, leads to serious distortion which can only be prevented by strong external restraint in the form of heavy canning. An alternative method of avoiding *transformation damage* is the employment of uranium alloys which retain the γ phase throughout the complete working range of temperature. Suitable alloying materials include molybdenum, niobium, titanium, and zirconium.

For natural uranium reactors, the α phase is the only one of importance, since the use of thick cans or extensive addition of alloying material is not permissible on reactivity grounds. Even when the temperature range of operation is wholly within the stability range of the α phase, temperature cycling can lead to severe distortion owing to the highly anisotropic behavior of the orthorhombic lattice. The coefficients of expansion parallel to the a and c directions are strongly positive, while that in the b direction is negative, as shown in Table 5.3.

Distortion under temperature cycling is particularly marked when the structure contains grains with a preferred orientation in the [010] direction, as is typical of cold-worked uranium. On heating above about 350°C, stress relaxation takes place mainly by grain boundary flow. On cooling, the lattice is too rigid to allow the process to be reversed, and the stress relaxation which

Table 5.3. Coefficients of Thermal Expansion of α-Uranium

Direction	Coefficient of linear expansion in range 25–650°C (10^{-6} per °C)
a[100]	36.7
b[010]	−9.3
c[001]	34.2

does take place is mainly by twinning within the grains. In the presence of preferred grain orientation, the net result of temperature cycling is a growth in the original working direction, with a corresponding shrinkage in a direction at right angles to this. This mechanism is termed *thermal rachetting*.

The preferential growth may be eliminated by pretreatment of the rod to produce a random grain orientation, for example, by heating into the β region, followed by quenching back to the α phase. With randomly oriented grains, the differential growth of the grains manifests itself in the form of surface roughening, which can be minimized by the use of fine-grained material.

Neutron irradiation of uranium produces two types of dimensional change. The first is a *growth* effect similar to that produced by thermal cycling and the second is an overall *swelling* due to the buildup of gaseous fission products. The magnitude of the growth effect depends on the temperature and the degree of preferred grain orientation in the metal, but can be very pronounced at low temperatures in material which has not been treated to produce random grains. The typical effect is a marked growth in the [010] direction, with a corresponding reduction in the [100] and little change in the [001] direction. The total volume, and therefore the density, remains effectively unchanged. The growth effect decreases markedly with temperature, becoming zero at about 500°C.

The mechanism responsible for growth in uranium appears to be generated by the damage caused by recoiling fission fragments. The passage of the relatively massive ions of high kinetic energy through the lattice produces intense damage in the form of *vacancies* and *interstitials*, in a cylindrical volume surrounding the path of the fragment, which typically travels several micrometers before coming to rest. The lattice vacancies are concentrated towards the center of the affected region, the displaced atoms appearing as interstitials towards the outside. It is believed that the interstitials tend to diffuse in the [010] direction and the vacancies in the [100] or [001] directions, producing growth in the former direction and shrinkage in the others. As in the case of thermal rachetting, the effects of irradiation can be minimized by treating the uranium to produce a fine-grained random orientation.

5.4. Ceramic Uranium Fuel

Metallic uranium has been largely superseded by ceramics, in particular uranium dioxide, as fuel for power reactors. Among the advantages of UO_2

are suitability for operation at high temperature (melting point 2750°C as compared to 1130°C for pure uranium metal), absence of phase changes, greater resistance to neutron irradiation, leading to much higher permissible burn-up, and chemical compatibility with common coolants, particularly water, in the event of coolant leakage. Among the disadvantages are the lower density of the oxide compared with the metal (10.5 g cm^{-3} compared with 19.0 g cm^{-3}) and the lower thermal conductivity, which necessitates a highly divided fuel element to avoid melting at the center of the fuel column.

The production of UO_2 powder is normally done by precipitation of uranium from a uranyl nitrate solution with either ammonia or hydrogen peroxide, to yield a precipitate of uranium diuranate or uranium peroxide, which is then calcined to UO_3 by heating. This is subsequently reduced by heating in hydrogen to give UO_2 powder. The fuel is mostly used in the form of pellets, which are normally produced by cold pressing and sintering. It is possible to obtain densities of more than 95% of the theoretical value of 10.97 g cm^{-3}, the maximum density being desirable both for maximizing the uranium content of the element and to provide a good fission product retention capability.

Columns of the pellets are enclosed in a thin-walled can of suitable cladding material, such as zircaloy or stainless steel, and a number of these fuel pins are then assembled to form a fuel bundle (Fig. 5.2). The pins are filled with helium to provide a thermal bond between pellets and cladding and so enhance the heat transfer from the fuel to the coolant. The pins are held together by grids spaced at intervals along the element to provide a structure which allows a certain flexibility to accommodate dimensional

Fig. 5.2. Uranium oxide fuel bundle for CANDU reactor (photo by courtesy of Atomic Energy of Canada Limited).

changes, while allowing the coolant to circulate freely around the pins. The high cost of fabrication of the fuel elements, due to the close tolerances and quality control required, makes an important contribution to the overall generating costs of the reactor, and efforts have been made to simplify the process. For example, the need for pelleting can be eliminated by the use of so-called vibrocompacted fuel, in which loose UO_2 powder is allowed to settle in the cladding while the latter is continuously vibrated.

The thermal conductivity of the oxide fuel varies markedly with the degree of porosity and also shows a pronounced decrease with increasing temperature. The low conductivity results in a high temperature gradient across the fuel pellet, giving rise to thermal stress cracking and eventually to radial movement of fuel components and fission products. The thermal stress cracking and the creation of voidage due to lattice defects and fission gas production result in a gradual decrease in the thermal conductivity, which can drop by more than an order of magnitude as a result of prolonged irradiation.

The effect of prolonged irradiation of a fuel pellet at high power is illustrated by the cross section shown in Fig. 5.3. Three distinct zones may be identified. In the outermost zone the microstructure of the fuel appears similar to that of the unirradiated pellet since the temperature is too low to cause restructuring. The intermediate zone (known as the *equiaxed zone*) is characterized by a larger grain size due to grain growth, while the central (*columnar*) region contains columnar grains produced by the dragging of grain boundaries by pores migrating up the temperature gradient. If the power density is high enough, the effect of pore migration into the center can cause the formation of a central void. Under these conditions there will be a large release of gaseous fission products at the surface of the central void; highly rated fuel pins, such as those of the fast breeder reactor, require a considerable plenum space above the fuel column to accommodate the released fission gas and avoid overpressurizing the pin.

The swelling of the fuel due to fission products can lead to stress on the cladding. While the effect of fuel swelling in producing cladding stress is most marked for high burn-up FBR fuel, localized stress can be generated even in low burn-up LWR fuel by displaced pieces of cracked pellets bearing on the clad. As mentioned earlier, a circumferential helium-filled gap is left between pellet and cladding to allow for expansion. Further provision for accommodation of fission products can be made by using annular pellets or by increasing the porosity of the manufactured fuel. The pellets are usually "dished" at the ends to allow for differential swelling due to the high temperature gradient. As a further insurance, the fuel rods for a modern

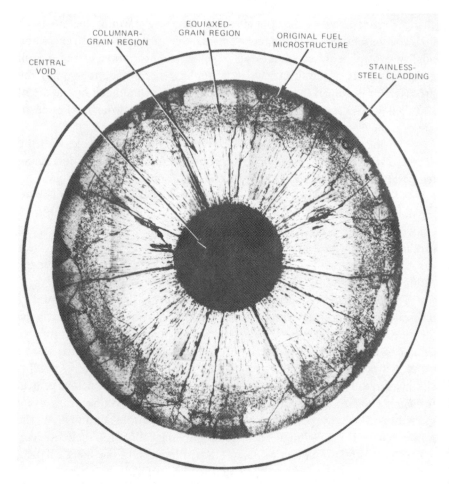

Fig. 5.3. Cross section of irradiated fuel pellet [from D. R. O'Boyle, F. L. Brown, and J. E. Sanecki, *J. Nucl. Mater.* **29**, 27 (1969). Courtesy of Argonne National Laboratory and North-Holland Publishing Company].

PWR are prepressurized to reduce the pressure differential between the coolant and the interior of the rod, and thus oppose the tendency of the zircaloy cladding to creep down into contact with the fuel.

Assuming that pellet–clad interaction should happen to occur with high burn-up fuel, there is a possibility of the cladding being damaged due to the stresses that would be created by a too rapid increase in the reactor power. This effect was in fact observed in a small number of cases in the mid-1970s and, as a result the rate of power increase after a refueling shutdown in PWRs is now subject to a limitation of about 3% increase per hour. Based on experience accumulated over the past few years, this restriction would appear to have solved the problem of cladding failure induced by pellet–clad interaction.

While fuel pellet swelling due to fission products may be minimized by increasing the porosity of the fuel, the provision of excessive porosity can lead to an effect known as *fuel densification*, which caused problems in a few of the earlier PWRs. The effect was first observed on the Ginna PWR in 1972. For fuel of rather low initial pellet density, operating within a particular temperature range, the swelling due to fission gas generation may be more than compensated for by a sintering process which gives rise to shrinkage of the fuel pellets causing an appreciable decrease in the axial length of the pellet stack. If there is contact between the pellet stack and the cladding part way up the pin, it is possible for the pellets below this point to creep down, leaving a gap in the stack. The cladding may then be compressed by the coolant pressure into the gap, causing the pin to collapse at that position. This is no longer a problem for PWRs, since experience has shown that it can be avoided by starting off with a higher pressure of helium in the pin and also having a higher initial fuel density (more complete sintering in manufacture).

Another ceramic fuel of interest is uranium carbide, which has a higher thermal conductivity than uranium oxide and can be produced at a somewhat higher density. Carbide fuel is ruled out for use in the LWR because it reacts strongly with water in the event of a fuel pin leak. It has a possible future as a fast reactor fuel, however, since its high conductivity would permit the use of larger fuel pins. In addition, it has a higher fission gas retention than does oxide, thus reducing the required cladding thickness.

To summarize, the suitability of UO_2 as a reactor fuel for both thermal and fast reactors has been confirmed by a wealth of experience, and the failure rate of fuel pins can be made acceptably low (down to about 1 in 10^4) by careful quality control in manufacturing. Particular problems which have arisen, such as fuel densification and damage due to pellet–clad interaction, have been solved by adjustment of the initial fuel density and pin gas pressure, and by limiting the rate at which large power increases are allowed to occur. Alternatives to the oxide fuel, such as uranium carbide and uranium nitride, have been investigated less extensively, but hold promise of superior performance in fast reactor systems.

5.5. Dispersion-Type Fuel Elements

Dispersion-type elements are those in which discrete particles of fissile material are dispersed in a metallic or ceramic matrix. The matrix effectively acts as a structural element, which will retain its strength over the irradiation life of the fuel provided that the fuel particles are sufficiently widely spaced that their fission fragment damage zones do not overlap. The existence of a continuous region of undamaged matrix can result in good fission product retention and resistance to fission gas swelling. A further advantage is that by choosing a material of good thermal conductivity as the matrix one can achieve lower fuel operating temperatures.

The dispersion-type fuel concept can be applied to both metal and ceramic fuel and matrix materials. When the fuel is in the form of a ceramic compound dispersed in a metal matrix, the combination is known as a *cermet*. The fuel materials used include uranium, uranium dioxide, uranium carbide, and uranium nitride, while the commonest matrix elements are steel, aluminum, zircaloy, and graphite. The fuel and matrix are mixed in powder form, cold- or hot-pressed after blending, and sintered. Typical particle sizes for fuel and matrix are in the range 30–300 µm. The dispersed fuel can be made up into clad plate-type elements by hot pressing slabs of fuel-bearing material between sheets of the pure metal. A typical element of this type is that used in the materials testing reactor (MTR), where the fuel material is highly enriched UAl_4 in a matrix of Al and UAl_4, surrounded by aluminum clad. This type of fuel element has proved highly reliable, although, for highly-rated fuel, UO_2 in stainless steel is more suitable.

A dispersed fuel of particular interest is the *coated-particle* type adopted for the high-temperature gas-cooled reactor (HTGR). In order to achieve the high operating temperature and power density needed in the HTGR, it was necessary to eliminate the metal cladding used in earlier gas-cooled reactors. This was done by having the fuel in the form of small spheres (a few hundred micrometers in diameter), each sphere being coated individually with multiple layers of material impervious to fission products. The spheres are then dispersed through a graphite matrix to form the fuel element. In addition to preventing the escape of fission products, the coatings preserve the fuel from chemical attack by impurities, such as water vapor, in the helium coolant and also protect the graphite matrix from fission fragment recoil damage.

The structure of a typical coated particle is shown in Fig. 5.4. The fuel "kernel" is a sphere of uranium carbide with a diameter in the range of 100–400 µm. The innermost coating consists of a layer of pyrolytic carbon laid down by deposition from hydrocarbon gases in a high-temperature fluidized bed. This inner layer absorbs fission fragment recoils and is made relatively porous to provide voidage for the accommodation of fission gases. The next layer is made up of silicon carbide, which is particularly effective in the

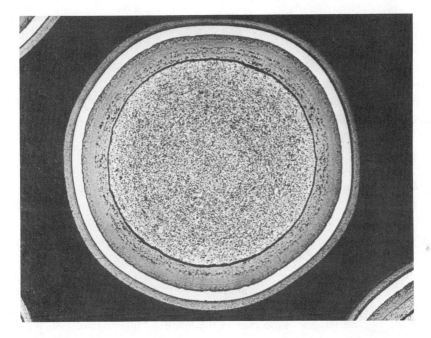

Fig. 5.4. Coated-particle fuel. The "kernel" of uranium carbide is surrounded by successive layers of pyrolytic carbon, silicon carbide, and high-density pyrolytic carbon (photo by courtesy of the United Kingdom Atomic Energy Authority).

retention of certain alkali earth and rare earth metallic fission products which diffuse rather readily through pyrolytic carbon. The final layer is of high-density pyrolytic carbon, which is impervious to gaseous fission products.

The original design of HTGR fuel element incorporated a continuous purge of gas in order to flush away fission products escaping from the fuel particles. The newer multiple-layer coated particles, however, have proved to have such an excellent retention capacity that the purge system is no longer necessary. Fuel testing in reactors such as Dragon and Peach Bottom has shown extremely low levels of activity in the coolant even after prolonged operation at fuel temperatures in excess of 1200°C. Even in the unlikely event of a major ingress of air or water vapor into the reactor, leading to a rapid attack on the core graphite, the triple coating would prevent a high fission product release. A further safety advantage is that the particles can withstand a marked rise of temperature for several days before permanent damage occurs due to chemical interaction of the kernels with the coating.

The fuel particles in the Fort St. Vrain HTGR are designed for an average burn-up of around 100,000 MW d/tonne, corresponding to about 10%–20% fissions per initial fissile atom. The fuel residence time is six years, one sixth of the core being replaced at each loading. Since all the moderator of the HTGR is incorporated in the fuel elements themselves, their regular replacement eliminates the buildup of irradiation damage that would be experienced in a fixed moderator operating at the HTGR flux levels.

5.6. Graphite

Graphite has been in common use as a moderator since the early days of the application of nuclear energy and, in particular, for the gas-cooled reactor, where the great majority of the currently operating units are of the graphite-moderated type. In addition to its good moderating ratio, graphite has the advantages of ready availability, high thermal conductivity, good mechanical properties, and the ability to operate at high temperatures. Among the limitations of graphite is the tendency to react with oxygen or carbon dioxide at high temperature, although it is possible to control the latter reaction in a CO_2-cooled reactor by the addition of small amounts of a gas such as methane to the coolant. A further disadvantage is that of dimensional change under irradiation, requiring the use of special production techniques to yield a highly isotropic graphite.

Graphite is a form of carbon which is produced commercially by graphitization of petroleum coke, a by-product of petroleum refining. The process involves the mixing of ground coke with a pitch "blender," extrusion of the mixture, and subsequent baking and final graphitization at temperatures up to 3000°C by electrical resistance heating. The latter process is essential for the production of high-purity graphite, since the impurity content is greatly reduced by volatization.

The extrusion part of this process is responsible for the marked anisotropy of physical properties shown by graphite. The graphite crystal has a layered structure as shown in Fig. 5.5, the atoms in the sheets being arranged in a hexagonal array held together by covalent bonding. The binding between the sheets is due to relatively weak Van der Waals forces, with the result that the tensile strength is considerably greater in the direction at right angles to this. A similar anisotropy is shown by the other physical properties, such as electrical and thermal conductivity. Graphite produced by the process outlined above displays a bulk asymmetry, since extrusion results in an

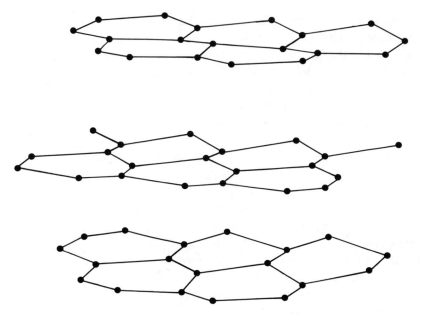

Fig. 5.5. Structure of graphite crystal. The atoms are arranged in hexagonal arrays with strong binding in the planes and weaker binding between planes.

alignment of the coke particles with their long axes parallel to the direction of extrusion.

Neutron irradiation of graphite at low temperature (below 300°C) produces an elongation perpendicular to the direction of extrusion, due to displaced atoms coming to rest in the space between the atomic planes. The growth in this direction is accompanied by a shrinkage in the direction parallel to extrusion. At higher temperatures, self-annealing reduces the number of interplane defects, and as the temperature goes above 300°C, the general effect of irradiation is a shrinkage in both directions. At still higher temperatures, in the range 500–550°C, there is a fairly rapid growth rate, probably owing to the development of additional porosity or to clustering of vacancies.

Excessive change in dimensions is undesirable in a large graphite-moderated reactor, such as the British magnox system, where the graphite is an important structural element of the core. In addition to inducing a slackness in the reactor structure, misalignment due to shrinkage could lead to difficulties in fuel loading and control rod operation. Methods of limiting structural changes have included the use of keying on the graphite bricks and the use of zirconium pins to maintain the lattice pitch. For the later advanced gas-cooled reactor system (AGR), the higher neutron irradiation dose has necessitated the use of graphite produced by a pressed mold technique, which gives a greater degree of isotropy than can be obtained by the extrusion process. Despite this, it is necessary, as discussed in Section 8.3, to provide a reentrant flow of cooler gas to maintain the graphite sleeve surrounding the fuel element at a temperature which minimizes dimensional changes.

Another effect of atomic displacement due to irradiation, which was the cause of some concern in the early days of nuclear power, is the so-called *Wigner energy*. After irradiation at a relatively low temperature, the accumulation of displaced atoms in interlattice positions results in a storage of energy which can be released by raising the graphite temperature to a level where the increased thermal agitation allows the atoms to drop back to fill the lattice vacancies. In early graphite reactors, it was the practice to carry out a controlled release of the stored energy at intervals by allowing the graphite temperature to rise in this way. Close control of the rate of temperature rise had to be exercised, since in the absence of such control the process is self-accelerating because of positive feedback. It was, in fact, an uncontrolled Wigner release, due to inadequate instrumentation, which led in 1957 to the well-known incident in the Windscale reactor in the United Kingdom, which eventually resulted in the destruction of the reactor core and the dispersion of quantities of radioactive material to the environment. The temperature of irradiation in a modern gas-cooled graphite-moderated reactor is fortunately sufficiently high that self-annealing prevents large-scale storage of Wigner energy in the graphite.

The oxidation of graphite in air has led to the use of carbon dioxide or helium as coolant in power reactors. The use of carbon dioxide is limited by the onset of the reaction

$$C + CO_2 \rightleftharpoons 2CO$$

which results in erosion of the graphite and mass transport of carbon which is then deposited in the cooler parts of the circuit. In the AGR, it has been found possible to limit the weight loss of the moderator to acceptable limits by injection of methane gas, CH_4. For operation at still higher temperature, as in the high-temperature gas-cooled reactor (HTGR), it is necessary to use helium as coolant.

During the baking process in the production of high-grade graphite, the evolution of gas from the pitch binder gives the material a porous structure, and the density of the final product is in the range 1.6–$1.8\,\mathrm{g\,cm^{-3}}$, considerably lower than the theoretical density of $2.26\,\mathrm{g\,cm^{-3}}$, which is predicted from the dimensions of the graphite lattice. With the advent of the HTGR, there has been an interest in the development of a graphite with a density approaching the theoretical value, which could be used as an impermeable coating for particle fuels. This material, the pyrolytic carbon mentioned in the previous section, is produced by deposition on a heated surface as a result of the thermal decomposition of a gaseous hydrocarbon, such as methane. The pyrolytic carbon is highly anisotropic, the atomic sheets being aligned parallel to the surface on which the carbon is deposited. The impermeability of the material is such that even a layer 50-μm thick is essentially impermeable to gaseous fission products from a fuel particle.

5.7. Stainless Steel

Many of the early light water reactors used stainless steel as the fuel-cladding material, in the expectation that, by a combination of corrosion-resistant alloys and high water purity, corrosion problems could be eliminated. In practice, stress corrosion cracking proved to be much more serious than anticipated, especially where conditions were particularly unfavorable, for example, when previously irradiated fuel was exposed to an increase in rating, leading to an increase in clad stress due to expansion simultaneously with an enhanced release of fission products, in particular iodine. The susceptibility of stainless steel to stress corrosion cracking has led to its abandonment in favor of zirconium alloys for cladding in water reactors.

The main interest in the use of stainless steel cladding is now for the sodium-cooled fast reactor, where the operating temperatures are high enough to result in seriously reduced strength and corrosion resistance for most of the alternative cladding materials. An interesting effect found in fast reactor cladding material is fast-neutron-induced swelling. Since an appreciable number of the neutrons in the fast reactor have energies in excess of the (n, p) or (n, α) threshold for the steel constituents, hydrogen and helium are produced in the cladding. At the typical operating temperature, the hydrogen diffuses fairly readily out of the material, but the helium, on account of its low solubility, precipitates in the form of small bubbles, principally at dislocation sites and at grain boundaries. It is found, however, that the total volume of bubbles produced in the cladding is much greater than can be explained by the helium produced in this way, and that the bubbles are largely empty voids.

The explanation appears to lie in the accumulation of lattice vacancies at the nucleation sites of the helium bubbles. The average energy of a neutron in the fast reactor is above 100 keV, greatly in excess of the 25 eV or so which is required to displace an atom from a lattice site. Neutron collisions, followed by multiple cascade processes, therefore lead to a high density of vacancies and interstitials. The interstitials have a higher mobility than the vacancies and tend to be more rapidly absorbed at grain boundaries and dislocations, where they lose their identity. The surplus vacancies are then available for the formation of voids at the nucleation centers of the helium produced by the (n, α) reactions.

The target burn-up for a fast breeder reactor is around 100,000 MW d/tonne (10% heavy atom burn-up), which corresponds to a fast neutron fluence of the order of 5×10^{23} n cm^{-2}. The fast-neutron-induced voidage for this fluence can result in substantial dimensional changes in the cladding size. This has led to changes in fast reactor design, including increased initial clearances between fuel pins to allow for swelling, fuel assemblies with provision for expansion (such as the "leaning post" design of the U.K. Prototype Fast Reactor described in Section 11.4), and the adoption of periodic rotation of fuel elements in regions of high flux gradient to reduce differential swelling.

Since the requirement for a high start-of-life clearance between the fuel pins results in an increase in the coolant-to-fuel volume ratio and thus a degradation of the neutron spectrum with a consequent reduction in breeding ratio, there is an obvious incentive to develop cladding materials with minimum voidage production. Also, in addition to the voidage effect described, the helium production leads to embrittlement and loss of ductility. It is found that cold working of stainless steels reduces the fast neutron swelling by producing a greater dislocation density and thus more sinks for the mutual annihilation of vacancies and interstitials. The hardening caused by accumulation of helium at grain boundaries can be reduced by the precipitation of particles of a second phase within the grains; these particles act as alternative nucleation sites for the helium bubbles, thereby reducing the amount of helium at the grain boundaries.

It should be noted that irradiation of steels in thermal reactors also produces embrittlement, though in this case the main contributor of helium is the (n, α) reaction of slow neutrons with boron impurity in the steel. Because of the high thermal absorption cross section of the boron, however, this burns out at fairly moderate exposures, leading to saturation of the effect.

This is in contrast to the fast neutron case, where the reaction cross sections are much lower and the helium content increases steadily with increasing fluence.

5.8. Zirconium

Metals intended for use inside the reactor core—for fuel cladding, for example, or for pressure tubes in a CANDU reactor—must have low thermal neutron absorption cross sections. The four potentially useful low-absorption metals are aluminum, beryllium, magnesium, and zirconium; their thermal neutron cross sections and melting points are listed in Table 5.4. Magnesium and aluminum have low melting points which restrict their usefulness to modest temperatures, as in the magnox reactor. Beryllium is brittle, difficult to fabricate, and expensive. Of the four, therefore, zirconium is the most attractive, being reasonably abundant and having good corrosion resistance and a high melting point. In refining zirconium for reactor use, great care has to be taken to remove the hafnium with which it is always associated, on account of the high neutron absorption of the latter.

In practice, the strength and corrosion resistance of zirconium can be greatly improved by minor alloying additions. The principal alloys are zircaloy-2 and zircaloy-4. The former has 1.2–1.7 wt % tin, with minor additions of iron, chromium, and nickel. Zircaloy-4 is similar, but without the nickel. Another important alloy is $Zr–2\frac{1}{2}\%Nb$, which offers a further improvement in strength and thus permits the use of thinner tubes to give better neutron economy.

The zirconium alloys are strongly resistant to corrosion in hot water, which is a highly corrosive substance at a typical operating temperature of

Table 5.4. Thermal Neutron Absorption Cross Sections and Melting Points of Some Low-Absorption Metals

Metal	Thermal absorption cross section (mb)	Melting point (°C)
Aluminum	230	660
Beryllium	9.2	1280
Magnesium	63	651
Zirconium	185	1845

around 300°C. On exposure of pure zirconium to water in the range of 300–360°C, the initial corrosion rate is slowed down by the formation of a thick adherent film of zirconium oxide, according to the reaction

$$Zr + 2H_2O \rightarrow ZrO_2 + 2H_2$$

Further corrosion then depends on the rate of diffusion of metal and oxygen ions through the oxide film. When the oxide film reaches a thickness known as the *critical thickness*, breakaway of the film occurs, with a sharp increase in corrosion rate. One of the principal factors in determining the onset of breakaway of the film is the takeup of the hydrogen produced in the reaction above. Initially, the hydrogen goes into solution in the metal, but when the solubility limit is reached, precipitation of zirconium hydride occurs, causing a loss of coherency between the metal and the oxide layer. The process of breakaway is also accelerated by impurities, particularly carbon and nitrogen, present in the zirconium. The effects of these impurities are countered by the presence of the alloying additions used to produce the zircaloy series. Alloys such as zircaloy-2 and zircaloy-4 are greatly superior to even pure zirconium as regards corrosion resistance.

Although zircaloy-2 has an excellent corrosion resistance, it may eventually suffer embrittlement as a result of pickup of hydrogen. While the hydrogen is in solution in the metal, it has little or no effect on the mechanical properties but, when the solubility limit is exceeded, the formation of platelets of zirconium hydride can cause a considerable decrease in impact strength and ductility. The magnitude of the effect depends critically on the orientation of the platelets relative to the applied stress. The ductility of zircaloy pressure tubes, for example, is virtually unaffected by the formation of platelets in the circumferential direction, while it is severely reduced by platelets in the radial direction. One of the advantages of cold-drawn pressure tubes is that they give rise to circumferential plating.

Exposure to fast neutron fluxes increases the corrosion rate of zirconium-based alloys. Even in the highly oxidizing environment of a BWR coolant, however, the corrosion rate of the principal zirconium alloys is low enough that cladding corrosion is not a limiting factor on the life of the fuel element.

An important consideration for zircaloy pressure tubes is the enhancement of the creep rate by exposure to neutron irradiation. This is illustrated in Fig. 5.6, which shows measurements of the transverse strain in a zircaloy-2 pressure tube which had been machined to different thicknesses along its length, so that it was possible to investigate the effects of stress variation as

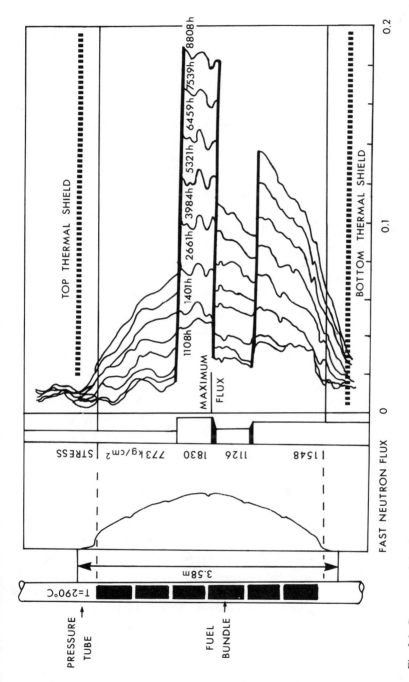

Fig. 5.6. Creep rate of zircaloy under neutron irradiation. The figure shows measurements of the transverse strain in a zircaloy-2 pressure tube machined to different thicknesses along its length, as a function of fast neutron flux and time of exposure (from E. C. W. Perryman, AECL-5125).

well as that of the fast neutron flux. The results show that the in-reactor creep is proportional to the stress, the fast neutron flux, and the time of exposure. The creep rates of the zircaloy pressure tubes in the early Pickering units turned out to be much higher than expected, creating the unfortunate necessity for replacing these at some stage during the plant life.

In the later reactors at Pickering and Bruce, zirconium-$2\frac{1}{2}\%$ niobium was used for the pressure tubes, partly in order to improve the neutron economy, but also because it was hoped that the creep rate under irradiation would be lower than for the zircaloys. In practice, the creep rate of the newer alloy turned out to be similar to that of zircaloy. An additional problem which emerged first in Pickering unit 3 was the development of cracks in certain of the zircaloy-$2\frac{1}{2}\%$ niobium tubes. This was found to be due to an incorrect rolling technique having been used in sealing the pressure tube to its end fitting, with the result that the end of the tube was left in a highly stressed condition. Precipitation of zirconium hydride occurred preferentially in the high-stress region, leading eventually to cracking and leakage of heavy water from the pressure tubes. Measures taken to deal with the problem included the replacement of defective tubes in Pickering units 3 and 4 and the annealing of over-rolled tubes in the Bruce reactor.

5.9. Light and Heavy Water

Heavy water (deuterium oxide) differs from the vastly more abundant light water, H_2O, in having the higher isotope of hydrogen, $_1H^2$, in place of $_1H^1$. Typical concentrations of deuterium in natural sources are in the range 130–160 atoms of deuterium per million atoms of hydrogen. The ratio varies with meteorological factors, such as seasonal temperature, and with proximity to the ocean. On account of the rarity of the isotope, the naturally occurring deuterium is almost invariably in the form of the monodeuterated molecule, HDO.

Since heavy water is chemically identical to light water, it has to be separated through some process whose rate depends on the different masses of the molecules involved. The commonest method is the so-called GS hydrogen sulfide process. This depends on the exchange of deuterium between streams of water and hydrogen sulfide gas, flowing in opposite directions through a large tower (see Fig. 5.7). The water flows down through a series of perforated trays, with the gas flowing upwards through the perforations. The process depends on the reversible reaction

$$HDO + H_2S \rightleftharpoons H_2O + HDS$$

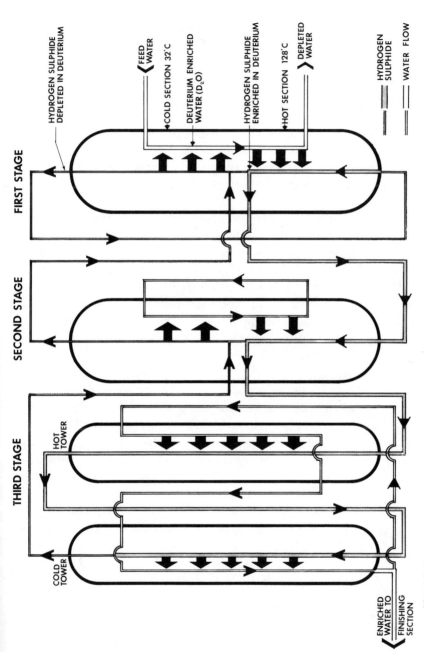

Fig. 5.7. Production of heavy water by GS hydrogen sulfide process (courtesy of Atomic Energy of Canada Limited).

At ordinary temperatures, the equilibrium favors the concentration of deuterium in the water, but at a temperature of around 130°C the equilibrium favors the concentration of deuterium in the hydrogen sulfide. The tower is therefore divided into two sections, the upper or cold section increasing the concentration of deuterium in the water, which is then used as feed for the lower or hot section, where the exchange leads to a further enrichment, this time in the hydrogen sulfide stream. The enriched gas from this section is then led to the second stage for further concentration. In a final stage, deuterium from the enriched gas is transferred to water, which is then fed to a vacuum distillation system for final enrichment to almost pure D_2O (99.75%).

The large quantity of water which has to be processed results in the first-stage towers being very large structures. At the Bruce plant in Canada, for example, the first stage consists of three towers in parallel, each some 9 m in diameter and 90 m high. The power requirements for the process are large, some 300 MW of heat being required to produce an output of 400 tonnes of D_2O per year. Development is proceeding on alternative processes, using deuterium exchange between hydrogen and ammonia, or hydrogen and aminomethane, which offer the advantages of smaller plant and reduce power requirements and also avoid the corrosion problems associated with the hydrogen sulfide method. Exchange between water and hydrogen gas is also a possibility, provided that a suitable catalyst can be developed to provide a sufficiently rapid exchange rate.

The leakage of heavy water from hot, pressurized reactor circuits presents problems both on account of the economic penalty and the associated escape of radioactive tritium which is formed by neutron capture in the deuterium. Early experience with the Douglas Point CANDU reactor, for example, indicated the need for sealing of building volumes around the coolant circuit, and the extraction and recovery of the D_2O vapor which had leaked into them.

A feature common to both light water and heavy water reactors is the radiolysis of the coolant as a result of exposure to ionizing radiation. On exposure to radiation, water breaks up into hydrogen atoms and free hydroxyl radicals:

$$H_2O \rightarrow H + OH$$

The two products may either recombine:

$$H + OH \rightarrow H_2O$$

or combine in pairs:

$$H + H \rightarrow H_2$$
$$OH + OH \rightarrow H_2O_2$$

The rate at which the pair combination reactions occur is greater when the specific ionization of the radiation is high, due to the greater density of the decomposition products. Subsequent decomposition of the hydrogen peroxide leads to the evolution of appreciable amounts of hydrogen and oxygen gas, and arrangements have to be made in water reactors for the removal and recombination of the hydrogen (or deuterium) and oxygen. Since the oxygen represents a corrosion hazard, reactors such as the pressurized water reactor frequently employ hydrogen injection into the coolant to maintain the oxygen level at a low value.

Problems

5.1. An enrichment plant produces an annual output of 150 tonnes of uranium enriched to 2.8% in U^{235}. The feed is natural uranium and the tails enrichment is 0.25% by weight. Calculate the natural uranium feed rate and the separative work per year.

5.2. An enrichment plant has a capacity of 1000 tonnes of separative work per year. How much 3% enriched uranium can it produce from natural uranium feed if the tails enrichment is 0.2% by weight?

5.3. In the previous question, what would be the change in the annual output if the tails enrichment were altered to 0.25% by weight?

6

Heat Generation and Removal in Nuclear Reactors

6.1. Introduction

In a power reactor, the heat generated in the fuel as a result of fission is removed by passing a *coolant* through the reactor core. The coolants most commonly used are water, heavy water, carbon dioxide, and liquid sodium. In only one type of reactor, the boiling water reactor, is the coolant allowed to boil during its passage through the core; in other reactors, steam production takes place in boilers, or steam generators, where the heat from the primary coolant is transferred to water in a *secondary circuit*. The steam produced in this secondary circuit is then fed to a *turbine* coupled to an *electrical generator*. The limit to the power at which the reactor can be operated is set by the heat transfer capacity of the coolant, so that a knowledge of the heat removal process is necessary both for efficient and safe operation of the plant.

In the present chapter, we shall first consider the details of the heat generation process within the reactor core. The transfer of heat from the fuel elements to the coolant will then be covered for both the nonboiling and boiling cases. Finally, the overall design of the nuclear steam supply system (NSSS) will be discussed.

6.2. Maximum Heat Generation Rate in Fuel

Even for a reactor in which all fuel elements are identical, the local heat generation rate will vary because of the variation of thermal neutron flux with position in the core. We illustrate by calculating the maximum local rate of heat generation as a function of total core power for the case of a bare reactor with vertical fuel elements, all of which contain fuel of the same enrichment.

175

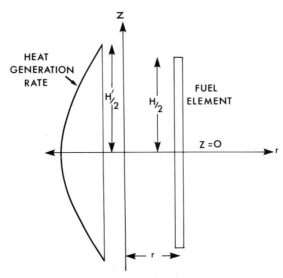

Fig. 6.1. Heat generation in fuel element. The axial variation of the heat generation rate is given by a cosine distribution as shown on the left of the diagram.

The fueled length of the element is H, while the extrapolated height of the core is H' (see Fig. 6.1). The cross-sectional area of the element is A_s.

If the width of the element is small compared with the core radius, we may ignore the radial variation of neutron flux across it. It is then only necessary to consider the axial variation, which depends on the function $\cos(\pi z/H')$, as shown in Table 3.4. The central plane of the reactor is taken as $z = 0$.

Using the standard heat transfer notation, the heat generation rate per unit volume, at height z, for the element at the radial position r, is $q_r'''(z)$. If the fuel is uniformly distributed in the element, the heat generation rate per unit volume of fuel is proportional to the flux and, hence,

$$q_r'''(z) = q_{0r}''' \cos(\pi z/H')$$

where q_{0r}''' is the heat generation rate per unit volume in the fuel at the central plane of the core ($z = 0$).

The heat generation rate in a small slice of the element, of height dz, is $q_r'''(z)A_s\, dz$ and the total heat generation rate for the whole element at the radial position r is therefore

$$q_{tr} = q_{0r}''' A_s \int_{-H/2}^{+H/2} \cos\frac{\pi z}{H'}\, dz$$

which gives

$$q_{tr} = \frac{2}{\pi} q_{0r}''' A_s H' \sin \frac{\pi H}{2H'} \tag{6.1}$$

If the extrapolation length is small compared with the core height, this reduces to

$$q_{tr} = \frac{2}{\pi} q_{0r}''' A_s H \tag{6.2}$$

For a pin-type fuel element of radius a, $A_s = \pi a^2$ and

$$q_{tr} = 2a^2 H q_{0r}''' \tag{6.3}$$

We now extend the argument to a full core of such elements. For simplicity, both the axial and radial extrapolation distances are ignored, and we consider a core of height H and radius R (see Fig. 6.2). The expression for the neutron flux in a finite cylindrical core is (see Table 3.4)

$$\phi(r, z) = \phi_0 \cos \frac{\pi z}{H} J_0 \frac{2.405r}{R} \tag{6.4}$$

where ϕ_0 is the flux at the core center.

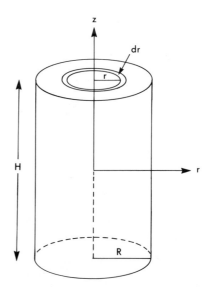

Fig. 6.2. Geometry of reactor core.

For the element at radius r, the heat generation rate for the fuel is

$$q_{tr} = \frac{2}{\pi} q_{0r}''' A_s H \tag{6.2}$$

If the total number of elements in the reactor is n, the number of elements for a unit area of the core cross-section is $n/\pi R^2$. The heat generation rate per unit cross-sectional area at radius r, due to the heat generated in the fuel, is therefore

$$q_r = \frac{2n}{\pi^2 R^2} q_{0r}''' A_s H$$

Now,

$$q_{0r}''' = q_0''' J_0 \frac{2.405r}{R}$$

where q_0''' is the heat generation rate per unit volume at the core center. Hence,

$$q_r = \frac{2n}{\pi^2 R^2} A_s H q_0''' J_0 \frac{2.405r}{R} \tag{6.5}$$

Taking a volume element as a cylindrical shell of thickness dr at radius r, the heat generation rate from the fuel in this element is $q_r 2\pi r \, dr$, so that the corresponding heat generation rate over the whole core, Q_t, is given by the integral

$$Q_t = \int_0^R q_r 2\pi r \, dr$$

$$= \frac{4n}{\pi R^2} A_s H q_0''' \int_0^R r J_0 \frac{2.405r}{R} \, dr \tag{6.6}$$

From the properties of Bessel functions, the integral in equation (6.6) is known to be equal to $0.216R^2$. Hence,

$$Q_t = 0.275 n A_s H q_0''' \tag{6.7}$$

For circular elements of radius a, $A_s = \pi a^2$ and

$$Q_t = 0.864 n a^2 H q_0''' \qquad (6.8)$$

The maximum heat generation rate in the core is then

$$q_0''' = 1.157 \frac{Q_t}{na^2 H} \qquad (6.9)$$

In this expression, Q_t is the total heat generated in the fuel itself. As discussed in Section 6.6, for each 200 MeV of recoverable energy, some 184 MeV, or 92%, is generated in the actual fuel, the remainder being deposited in moderator, coolant, and structural materials. Thus, Q_t is approximately equal to $0.9P$, where P is the total heat generation rate in the reactor. In terms of P, then, the maximum fuel heat generation rate in the core is

$$q_0''' = 1.04 \frac{P}{na^2 H} \qquad (6.10)$$

(It should be noted that the total heat available for power generation will in practice be somewhat less than P, since some of the heat will be generated in components not cooled by the primary coolant.)

Example 6.1. A typical pressurized water reactor (PWR) has a thermal power of 3560 MW. The core contains a total of 193 elements, each of which has 204 fuel pins. The height of the core is 13 ft and the fuel diameter is 0.36 in. Calculate the maximum heat generation rate per unit volume.

Using the conversion factor in Table A.2, the total core power is 3560 $\times 10^6 \times 3.412 = 1.215 \times 10^{10}$ Btu/h. The radius of the pin is 0.015 ft. Hence, from equation (6.10),

$$q_0''' = \frac{1.04 \times 1.215 \times 10^{10}}{193 \times 204 \times (0.015)^2 \times 13} = 1.097 \times 10^8 \text{ Btu/h ft}^3$$

The corresponding value in the SI system, using the conversion factor from Btu/h ft^3 to W m^{-3}, is 1.135×10^9 W m^{-3}. $\square$

In practice, equation (6.10) tends to overestimate the maximum heat generation rate because the peak-to-average flux is usually less than the value predicted for the bare homogeneous reactor because of flux flattening effects due to the presence of reflectors and nonuniform fuel loading. For most reactors, the value of q_0''' will only be some 60–70 % of that given by equation (6.10).

6.3. Heat Conduction in Fuel Elements

6.3.1. The Poisson Equation

Heat may be transferred from the fuel elements by *conduction, convection*, and *radiation*. For all but high-temperature gas-cooled reactors, the last of these is of little significance under normal operating conditions. *Convection* is the process by which the heat from the surface of the fuel element is carried into the coolant and subsequently removed from the reactor. Before this happens, however, the heat generated in the fuel must travel by *conduction* from the interior of the element to its surface. In the present section, we consider the application of the standard heat conduction equations to determine the temperature distribution within fuel elements of various geometries.

We start by taking a small cubic volume, of sides Δx, Δy, Δz, which is within a region of fuel where heat generation and conduction is taking place (Fig. 6.3).

Conduction takes place as a result of a temperature gradient in the medium. The rate of flow of heat into the cube in the x direction is related to the *heat flux*, q_x'', defined as the quantity of heat transferred per unit time

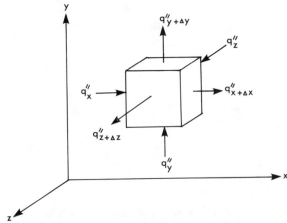

Fig. 6.3. Net conduction out of volume element in a medium with a temperature gradient.

across a unit area perpendicular to the x direction. This is given in terms of the temperature gradient in the x direction, $\partial T/\partial x$, by the Fourier equation

$$q_x'' = -k_f \frac{\partial T}{\partial x} \tag{6.11}$$

where k_f is the thermal conductivity of the fuel (which is assumed to be independent of temperature).

The rate of heat conduction out of the volume element in the x direction is related to the heat flux $q_{x+\Delta x}''$, which is given by

$$q_{x+\Delta x}'' = q_x'' + \frac{\partial q_x''}{\partial x} \Delta x$$

so that from equation (6.11)

$$q_{x+\Delta x}'' = -k_f \frac{\partial T}{\partial x} - k_f \frac{\partial^2 T}{\partial x^2} \Delta x \tag{6.12}$$

The net rate of conduction of heat into the volume element in the x direction is

$$(q_x'' - q_{x+\Delta x}'') \Delta y \, \Delta z$$

which, from equations (6.11) and (6.12) is equal to

$$k_f \Delta x \, \Delta y \, \Delta z \frac{\partial^2 T}{\partial x^2}$$

Similar expressions may be found for the net conduction rates in the y and z directions. The total net conduction rate into the volume is then equal to

$$k_f \Delta x \, \Delta y \, \Delta z \, \nabla^2 T$$

where $\nabla^2 T = \partial^2 T/\partial x^2 + \partial^2 T/\partial y^2 + \partial^2 T/\partial z^2$ is the Laplacian operator in Cartesian coordinates.

If the fuel has attained a state of equilibrium, i.e., for the steady state temperature distribution, the sum of the heat generation rate in the volume element and the net conduction rate into the element must be equal to zero, i.e.,

$$q''' \, \Delta x \, \Delta y \, \Delta z + k_f \, \Delta x \, \Delta y \, \Delta z \, \nabla^2 T = 0$$

or

$$\nabla^2 T + \frac{q'''}{k_f} = 0 \tag{6.13}$$

where q''' is the heat generation rate per unit volume. Equation (6.13) is known as the *Poisson equation*.

For the special case where there is no heat generation, the equation becomes

$$\nabla^2 T = 0 \tag{6.14}$$

which is known as the *Laplace equation*.

6.3.2. Solutions of the Poisson Equation

6.3.2.1. The Plate-Type Fuel Element

We consider the element to be a thin uniform slab of fuel-bearing material of thickness $2t$ (see Fig. 6.4). The heat generation rate per unit volume (q''') is taken as uniform throughout the fuel, a situation which would be approximately realized in an element whose length was much less than the total height of the core, or in a short section of a full-length element. The length and width of the slab are long in comparison with its thickness, so that heat loss through the ends and edges is small enough to be neglected. The problem is therefore reduced to a one-dimensional one, with heat flow in the x direction only. The origin is taken, as shown in Fig. 6.4, at the center of the plate. The Poisson equation reduces to

$$\frac{d^2 T(x)}{dx^2} + \frac{q'''}{k_f} = 0 \tag{6.15}$$

Integrating, we have

$$\frac{dT(x)}{dx} = -\frac{q'''}{k_f} x + C_1 \tag{6.16}$$

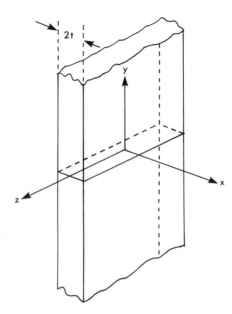

Fig. 6.4. Plate-type fuel element.

and, following a second integration,

$$T(x) = -\frac{q'''}{2k_f} x^2 + C_1 x + C_2 \qquad (6.17)$$

where C_1 and C_2 are numerical constants.

By symmetry, the temperature gradient at $x = 0$, where the temperature has its maximum value T_m, is zero. Hence

$$\frac{dT(x)}{dx} = 0 \quad \text{at } x = 0$$

which, from equation (6.16), give $C_1 = 0$. Substituting $T(x) = T_m$ at $x = 0$ in equation (6.17) gives

$$C_2 = T_m$$

Hence equation (6.17) becomes

$$T(x) = T_m - \frac{q'''}{2k_f} x^2 \qquad (6.18)$$

The temperature at the surface of the plate ($x = \pm t$) is then

$$T_s = T_m - \frac{q'''}{2k_f} t2 \tag{6.19}$$

The total rate of heat generation in the element is $2q'''At$, where A is the area of one of the large surfaces of the plate. The heat conduction rate out of each surface (q) is one half of this, so that

$$q = q'''At \tag{6.20}$$

Combining equations (6.19) and (6.20) gives the alternative expression

$$q = 2k_f A \frac{T_m - T_s}{t} \tag{6.21}$$

where T_s is the surface temperature of the plate.

In practice, the fuel region will be separated from the coolant by cladding of thickness c (Fig. 6.5). Since there is no heat generation in the cladding, equation (6.15) reduces to the Laplace form

$$\frac{d^2 T(x)}{dx^2} = 0 \tag{6.22}$$

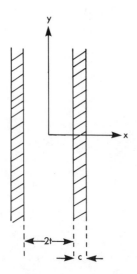

Fig. 6.5. Fuel element with cladding.

By integration, we obtain for the temperature within the cladding $(t < x < t + c)$

$$T(x) = C_1 x + C_2$$

where C_1 and C_2 are the constants of integration.

Using the boundary conditions that $T(x) = T_s$ at $x = t$ and $T(x) = T_c$ at $x = t + c$, we obtain

$$T(x) = T_s - \frac{x - t}{c}(T_s - T_c) \qquad (6.23)$$

The temperature therefore decreases linearly through the cladding.

The heat conducted per unit time through the cladding on one surface is given by the Fourier equation as $k_c A (T_s - T_c)/c$ where k_c is the thermal conductivity of the cladding. Equating this to the rate of heat conduction into the cladding, as given by equation (6.20), we obtain the relation

$$T_c = T_s - \frac{ct}{k_c} q''' \qquad (6.24)$$

Combining equations (6.19) and (6.24) gives

$$T_m - T_c = \left(\frac{t}{2k_f} + \frac{c}{k_c} \right) q''' t$$

The rate of heat conduction out of each surface may then be written, using equation (6.20), as

$$q = \frac{T_m - T_c}{(t/2k_f A) + (c/k_c A)} \qquad (6.25)$$

This equation has an obvious similarity to the Ohm's law formula for the current i which flows through a resistance R when a potential difference V is applied across it, i.e.,

$$i = \frac{V}{R} \qquad (6.26)$$

Here the quantity corresponding to current is the heat conduction rate q, while the temperature difference corresponds to the potential difference in the electrical analog. The resistance to heat flow consists of two resistances in series, one being the *thermal resistance* of the fuel region and the other of the cladding. The rate of heat conduction out of the plate-type element may then be written as

$$q = \frac{T_m - T_c}{R_{mc}} \tag{6.27}$$

where R_{mc} is the total thermal resistance of fuel and cladding. If there are other thermal resistances, such as that due to bonding material or a gas layer between fuel and cladding, these have to be added to the denominator of the expression (6.25).

6.3.2.2. The Solid Cylindrical Element

Consider a long cylindrical fuel rod of height H and radius R, surrounded by closely fitting cladding of thickness c (see Fig. 6.6). The heat production rate q''' in the rod is assumed to be uniform so that the temperature variation is essentially a radial one, and the Poisson equation reduces to the one-dimensional form in cylindrical coordinates

$$\frac{d^2 T(r)}{dr^2} + \frac{1}{r}\frac{dT(r)}{dr} + \frac{q'''}{k_f} = 0 \tag{6.28}$$

The solution of this equation is

$$T(r) = -\frac{q'''}{4k_f} r^2 + C_1 \ln r + C_2 \tag{6.29}$$

Inserting the boundary conditions, that $dT(r)/dr = 0$ at $r = 0$ and $T = T_m$ at $r = 0$, we have

$$T(r) = T_m - \frac{q'''}{4k_f} r^2 \tag{6.30}$$

The temperature at the surface of the fuel, at $r = R$, is

$$T_s = T_m - \frac{q'''}{4k_f} R^2 \tag{6.31}$$

Since the total heat production rate in the fuel is $\pi R^2 H q'''$ and since this is

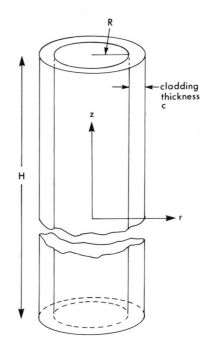

Fig. 6.6. Cylindrical fuel element with cladding.

equal to q, the rate of heat conduction out of the fuel, we find by substitution in (6.31) that

$$q = 4\pi k_f H (T_m - T_s) \tag{6.32}$$

The surface area of the fuel $A = 2\pi RH$ so that

$$q = 2k_f A \frac{T_m - T_s}{R} \tag{6.33}$$

The Poisson equation in the cladding is

$$\frac{d^2 T(r)}{dr^2} + \frac{1}{r}\frac{dT(r)}{dr} = 0 \tag{6.34}$$

the solution to which is

$$T(r) = \frac{T_s \ln(R + c) - T_c \ln R - (T_s - T_c)\ln r}{\ln(1 + c/R)} \tag{6.35}$$

The heat conduction rate through the outer surface (A_c) of the cladding is given by Fourier's equation as

$$q = -k_c A_c \frac{dT(r)}{dr} \qquad (6.36)$$

where the derivative is taken at $r = R + c$.

Using equation (6.35) to evaluate the derivative, and equating the heat conduction rate to the heat generation rate in the fuel, we find for the cladding surface temperature

$$T_c = T_s - \frac{R^2 \ln(1 + c/R)q'''}{2k_c} \qquad (6.37)$$

Combining equations (6.31) and (6.37) with the expression $\pi R^2 H q'''$ for the heat production rate in the fuel, which is equal to q, the heat conduction rate through the cladding, we have

$$q = \frac{T_m - T_c}{1/4\pi k_f H + \ln(1 + c/R)/2\pi k_c H} \qquad (6.38)$$

Here the terms in the denominator are, respectively, the thermal resistances of the fuel rod and the cladding.

Again, the equation may be written in the same form as equation (6.27) for the plate-type element, with the total thermal resistance for the cylindrical fuel element being

$$R_{mc} = \frac{1}{4\pi k_f H} + \frac{\ln(1 + c/R)}{2\pi k_c H} \qquad (6.39)$$

As will be seen in the following section, this equation can be extended to incorporate the heat transfer from the clad to the bulk coolant.

6.4. Temperature Distributions along a Coolant Channel

6.4.1. Heat Transfer by Convection

So far, we have not dealt with the mechanism by which the heat is transferred from the surface of the fuel element to the coolant. For

PLATE   COOLANT

Fig. 6.7. Convection flow past element. The lengths of the flow vectors
give the rates of flow at different distances from the surface.

illustration, we may consider the coolant being pumped past the surface of a
hot plate-type element (Fig. 6.7), for which the outer cladding temperature is
equal to T_c. This situation is known as *forced convection*. If no external
pumping agency were present, a movement of the coolant would be induced
by the density gradient near the plate; this is known as *natural convection*.

Because of viscosity effects, the velocity of the fluid is zero at the fuel
surface. The velocity vectors of the fluid layers increase as one moves further
away from the surface, as indicated in Fig. 6.7. Since the fluid layer in contact
with the fuel is at rest, heat actually flows by *conduction* from the element to
this layer, and the rate of heat flow will depend on the temperature gradient
of the fluid at the surface. This in turn, however, depends on the rate at which
heat is being carried away by the moving layers of the coolant, and so is a
function of the coolant flow rate. The equation which gives the heat flux from
the fuel plate to the coolant is

$$q'' = h(T_c - T_f) \tag{6.40}$$

where T_f is the temperature at a reference point in the fluid at some distance
from the heated surface. The quantity h is known as the *convection heat
transfer coefficient*, which depends on the thermal properties of the fluid
(thermal conductivity, specific heat, density) as well as its flow rate and
viscosity. The units of h are $W\,m^{-2}\,°C^{-1}$ or $Btu/h\,ft^2\,°F$. Typical values for
the convection heat transfer coefficient are given in Table 6.1.

Table 6.1. Typical Values of Convection Heat Transfer Coefficient
(for Forced Convection)

Coolant	$h\ (W\,m^{-2}\,°C^{-1})$	$h\ (Btu/h\,ft^2\,°F)$
Gases	25–500	5–100
Water	10,000–40,000	2000–8000
Sodium	25,000–250,000	5000–50,000

The rate of conduction of heat out of a surface A of the fuel element is

$$q = hA(T_c - T_f) \qquad (6.41)$$

or

$$q = \frac{T_c - T_f}{1/hA} \qquad (6.42)$$

By analogy with the terms in the denominator of equations (6.25) and (6.38), the term $1/hA$ represents the thermal resistance for convective heat transfer from the cladding of the fuel to the bulk coolant. Equation (6.25), for example, may then be extended to cover the whole heat transfer from the plate-type fuel element in terms of the difference in temperature between the centers of the fuel and the bulk coolant, yielding for the rate of heat flow out of each face the expression

$$q = \frac{T_m - T_f}{(t/2k_f A) + (c/k_c A) + (1/hA)} \qquad (6.43)$$

or

$$q = \frac{T_m - T_f}{R_{mf}} \qquad (6.44)$$

where R_{mf} is the total thermal resistance from the center of the fuel to the bulk coolant.

Equation (6.44) also applies to the cylindrical element, with the total thermal resistance given by (see equation 6.39)

$$R_{mf} = \frac{1}{4\pi k_f H} + \frac{\ln(1 + c/R)}{2\pi k_c H} + \frac{1}{2\pi h H (R + c)} \qquad (6.45)$$

where the surface area A_c of the cladding has been taken as $2\pi H(R + c)$.

6.4.2. Axial Variation of Fuel, Clad, and Coolant Temperatures

In deriving the equations for the heat flow out of fuel elements, we have so far assumed that the rate of heat generation per unit volume is constant in the axial direction. In practice, of course, this is not the case; for the unreflected reactor with uniform coolant density the flux, and hence the power production rate, will have a cosine dependence in the vertical direction. We will now derive expressions for the variation of cladding, fuel, and coolant temperatures in the vertical direction for the case where the volumetric heat production rate varies along the length of the fuel element.

We consider a core of height H, made up of vertical rod-type elements. The coolant is assumed to flow upwards through the core, parallel to the fuel elements. Each fuel element is considered as having a certain fraction of the total coolant flow associated with it. The core cross section may be divided as shown in Fig. 6.8 in order to define the flow area for each element. The mass flow rate per element is given the symbol $\dot{m}$.

For simplicity we consider the fuel rod which lies along the axis of the core. If the extrapolation distance is small enough to be ignored in comparison with H, the flux variation along the length of the element is of cosine form and, if the fuel distribution is uniform, the variation of heat generation rate per unit volume is then

$$q_z''' = q_0''' \cos \frac{\pi z}{H} \tag{6.46}$$

where, as before, $z = 0$ at the central plane of the reactor and q_0''' is the maximum heat production rate per unit volume of fuel in the reactor. For a fuel rod at a radial distance r from the central axis, the expression (6.46) has to be multiplied by $J_0(2.405r/R)$ to give the appropriate heat generation rate per unit volume.

Consider a small section of the element, of length dz, at distance z above the central plane. The heat generation rate in this segment is $q_z''' A_s \, dz$, where A_s is the cross-sectional area of the fuel. In equilibrium, the heat generation rate will be equal to the rate at which heat is taken up by the coolant, which is given by $\dot{m} c_p \, dT_f$, where c_p is the specific heat at constant pressure for the fluid and dT_f is the rise in bulk coolant temperature over the length dz (it is

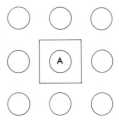

Fig. 6.8. Definition of equivalent cooling area for fuel element. The square area encloses the flow associated with the element A.

assumed that no boiling takes place anywhere in the channel). We therefore have

$$q_z''' A_s \, dz = \dot{m} c_p \, dT_f \tag{6.47}$$

Integrating equation (6.47) from the bottom of the core to the height z, and substituting from (6.46) gives

$$q_0''' A_s \int_{-H/2}^{z} \cos \frac{\pi z}{H} \, dz = \dot{m} c_p \int_{T_{fi}}^{T_{fz}} dT_f$$

where T_{fi} and T_{fz} are the bulk coolant temperatures at core inlet and height z, respectively.

Performing the integration, we obtain

$$T_{fz} = T_{fi} + \frac{q_0''' A_s H}{\pi \dot{m} c_p} \left(1 + \sin \frac{\pi z}{H} \right) \tag{6.48}$$

In particular, the temperature at core outlet $(z = H/2)$ is

$$T_{f0} = T_{fi} + \frac{2 q_0''' A_s H}{\pi \dot{m} c_p} \tag{6.49}$$

In order to derive the variation of cladding temperature with height, we note that the rate of heat generation in the section dz of the fuel is equal to the rate of heat transfer to the coolant. Using equation (6.40), we have

$$q_z''' A_s \, dz = q_z'' C \, dz = h C (T_{cz} - T_{fz}) \, dz \tag{6.50}$$

where h is the convection heat transfer coefficient and C is the circumference of the fuel element.

Hence,

$$T_{cz} = T_{fz} + \frac{q_z''' A_s}{hC} \tag{6.51}$$

and, substituting for T_{fz} from equation (6.48), we have

$$T_{cz} = T_{fi} + q_0''' A_s \left[\frac{1}{hC} \cos \frac{\pi z}{H} + \frac{H}{\pi \dot{m} c_p} \left(1 + \sin \frac{\pi z}{H} \right) \right] \tag{6.52}$$

In order to derive an expression for the axial variation of the central fuel temperature, T_{mz}, we start by equating the rate of heat production in the section dz of the fuel to the rate of transport of heat to the coolant. This is given, by a variation of equation (6.44), as

$$q_z''' A_s \, dz = \frac{T_{mz} - T_{fz}}{R_{mfz}} \tag{6.53}$$

where R_{mfz} is the total thermal resistance associated with the section dz of the fuel. It is obtained by substituting dz for H in the expression for the total thermal resistance of the whole element, R_{mf}, as given by equation (6.45), i.e., $R_{mfz} = (H/dz)R_{mf}$.

Multiplying equation (6.53) by H/dz yields

$$q_z''' A_s H = \frac{T_{mz} - T_{fz}}{R_{mf}}$$

Hence,

$$T_{mz} - T_{fz} = A_s H R_{mf} q_0''' \cos \frac{\pi z}{H} \tag{6.54}$$

which leads to

$$T_{mz} = T_{fi} + \frac{q_0''' A_s H}{\pi \dot{m} c_p} \left(1 + \sin \frac{\pi z}{H} \right) + q_0''' A_s H R_{mf} \cos \frac{\pi z}{H} \tag{6.55}$$

The variation of the bulk coolant, cladding, and maximum fuel temperatures, given by equations (6.49), (6.52), and (6.55), is illustrated in Fig. 6.9. The fuel and cladding temperatures reach their peak values some distance above the plane of maximum heat generation rate because of the progressive heating of the coolant as it passes up through the core.

Example 6.2. A heavy-water-moderated reactor has fuel in the form of natural uranium metal rods of diameter 1 in., with cladding of 0.04-in. aluminum. The heat generation rate per unit volume of a fuel element at the center plane of the reactor is 1.85×10^7 Btu/h ft³. The maximum fuel temperature is 690°F and the heat transfer coefficient to the coolant is 5500 Btu/h ft² °F. If the thermal conductivities of uranium and aluminum are 18.5 Btu/h ft °F and 131 Btu/h ft °F, respectively, calculate the bulk coolant temperature at the center plane.

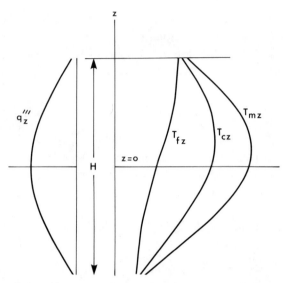

Fig. 6.9. Axial variation of heat generation rate per unit volume (q_z'''), bulk coolant temperature (T_{fz}), cladding temperature (T_{cz}), and center-line fuel temperature (T_{mz}).

From equation (6.54), with $z = 0$, we have

$$T_m - T_f = A_s HR_{mf} q_0'''$$

where A_s, the cross-sectional area of the fuel element, is equal to 5.455×10^{-3} ft^2. The product HR_{mf} is given by equation (6.45) as

$$HR_{mf} = \frac{1}{4\pi k_f} + \frac{\ln(1 + c/R)}{2\pi k_c} + \frac{1}{2\pi h(R + c)}$$

Using the values given for the thermal conductivities k_f and k_c, and the heat transfer coefficient, h, we find

$$HR_{mf} = 5.04 \times 10^{-3} \text{ h ft}^2 \text{ °F/Btu}$$

Hence from equation (6.55)

$$\begin{aligned}
T_f &= T_m - A_s HR_{mf} q_0''' \\
&= 690 - (5.455 \times 10^{-3} \times 5.04 \times 10^{-3} \times 1.85 \times 10^7) \\
&= 690 - 509 = 181°\text{F}
\end{aligned}$$

□

6.5. Heat Transfer by Boiling Coolant

The considerations in the previous section apply only when the liquid coolant is not permitted to boil, as, for example, by the pressurization to above the critical pressure which is characteristic of the pressurized water reactor. Allowing the coolant to boil has the advantage of relaxing the requirements on circuit pressure and of increasing the heat removal capacity of the coolant, since heat may now be absorbed in causing a change of phase as well as in raising the temperature of the liquid. The reactor steam cycle is greatly simplified by dispensing with the need for ancillary steam generators. On the other hand, the existence of varying voidage within the core may give rise to problems of reactor stability. Care must also be taken that the more complicated processes of heat transfer with a boiling coolant do not lead to a mismatch of heat flux and cooling capacity which could result in overheating and consequent damage to the fuel elements.

The prediction of the conditions under which such a mismatch might occur is a difficult task when the system under consideration is a bundle of fuel rods past which a boiling coolant is being pumped, since in this case the analysis is complicated by factors such as the nonuniform axial heat flux and the mixing of the coolant flow in the subchannels between the rods. We shall start by considering a simpler system, where water is being pumped through a vertical tube which is uniformly heated so that the heat flux (q'') is constant along the tube. The behavior of the coolant is illustrated in Fig. 6.10. It is assumed that the water entering at the base of the tube is subcooled, i.e., that its temperature is below the saturation temperature corresponding to the pressure of the coolant. Over the lower part of the tube, the heat taken up by the coolant is therefore used simply to increase its temperature. The region AB, over which no boiling takes place, is called the *single-phase region*.

At the point B, although the bulk temperature of the water is still lower than the saturation temperature, bubbles of vapor begin to form at *nucleation centers* (imperfections on the tube surface). Since the bulk water is below saturation temperature, the bubbles escaping into the main coolant flow recondense. The process occurring in the region BC is called *subcooled nucleate boiling (local boiling)*. At the point C, the bulk liquid reaches saturation temperature, and the bubbles leaving the wall no longer recondense. The type of flow here is referred to as *bubbly flow* and the boiling regime is called *saturated nucleate boiling (bulk boiling)*. Towards the upper part of the region CD, coalescence of the bubbles leads to a condition known as *slug flow* where the coolant is predominantly in the form of large bubbles of vapor.

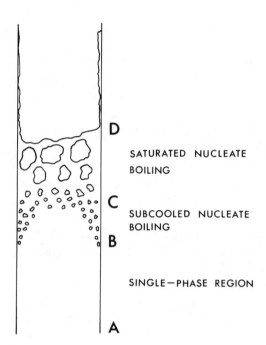

D

SATURATED NUCLEATE
BOILING

C

SUBCOOLED NUCLEATE
BOILING

B

SINGLE—PHASE REGION

A

Fig. 6.10. Flow regimes for coolant pumped through a heated tube.

At the point *D*, the coolant enters a regime known as *annular flow*, where a thin film of water flows along the wall of the tube while the center of the channel is filled with high-velocity vapor carrying small droplets of water with it.

There are a number of terms, such as *critical heat flux, burnout* and *departure from nucleate boiling*, which are used to describe the breakdown of the heat transfer process at sufficiently high heat fluxes. To illustrate these, consider what happens as the temperature of the tube wall is steadily raised from an initial low value. Figure 6.11 shows a graph of the logarithm of the heat flux as a function of the temperature of the wall of the tube. At low heat fluxes, the heat is transferred entirely in the non-boiling regime and goes to increase the sensible heat of the liquid (region *a*). As the temperature of the wall is raised, the heat flux increases to the point where nucleate boiling begins. Thereafter, the rate of heat transfer increases more rapidly with wall temperature, partly on accouunt of the heat taken up in the liquid–vapor transition and partly because the agitation produced by bubble formation leads to more effective coolant mixing (region *b*).

With further increase of heat flux a situation is reached where the bubble density becomes so great that the bubbles coalesce and blanket a portion of the heating surface with vapor (*film boiling*). This condition is known as the *departure from nucleate boiling* (DNB) or *boiling crisis*. Since the vapor is a poor heat transfer medium, the heat flux decreases as the wall temperature increases (region *c*). The heat flux reaches a minimum value when the whole of the wall surface is covered with a film of vapor. Thereafter, as the wall temperature increases further, radiation becomes an effective means of heat transfer, and the heat flux consequently begins to rise (region *d*).

In practice, the curve shown in Fig. 6.11 can be obtained only if the wall temperature can be controlled by some means, so that it is the independent variable. When the heat transfer process being considered is that from a reactor fuel bundle to the coolant, the parameter which is varied is the heat production rate in the fuel pins. In this case, once departure from nucleate boiling is reached, any further increase in heat production rate drives the fuel pin into region *d* (from *P* to *Q*), thus leading to a rapid increase in the pin temperature and possible damage to the cladding and fuel of the pin. This phenomenon is known as *burnout*, and if the temperature rise along the path PQ is sufficient to produce a rupture of the pin, the heat flux corresponding to the point P is called the *burnout heat flux*. If burnout does not occur over the path *PQ*, a further increase in heat flux takes the pin into the region *e*,

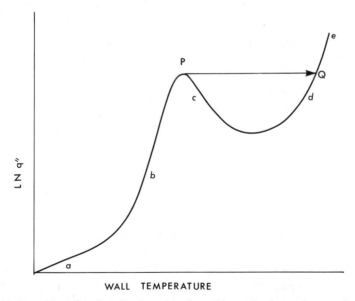

WALL TEMPERATURE

Fig. 6.11. Logarithm of the heat flux from the tube wall into a flowing coolant as a function of the wall temperature.

where the wall temperature continues to rise with increasing heat flux until the burnout heat flux is reached and burnout occurs.

In a power reactor, it is important that the conditions in the coolant channels are such as to provide an adequate margin against burnout occurring in any of the fuel pins. The prediction of the onset of burnout is complicated by the number of parameters which may be involved and the difficulty of evolving a physical theory for such a complex process. In the absence of a basic theory, the approach that has been used is to derive empirical correlations between the burnout heat flux and the more important parameters which control its onset. These correlations are based on a very large number of experimental determinations of the burnout flux for different types of geometry (pin, annulus, and pin bundle, for example) and for a variety of coolants. Apart from water, the main coolants for which experimental work has been done are the Freons, liquids of low specific heats which have the advantage that the departure from nucleate boiling occurs at wall temperatures which are not high enough to cause physical burnout.

For the case of liquid flowing through a round tube with a uniformly heated wall, for example, it is found that the burnout heat flux, ϕ_{BO}, can be represented by an expression of the form

$$\phi_{BO} = f(P, D, L, G, \Delta h) \tag{6.56}$$

where P is the pressure, D the internal diameter of the tube, L the heated length, G the mass velocity of the coolant (mass per unit time per unit area), and Δh is the enthalpy of subcooling at the channel inlet.

The *enthalpy of subcooling* is the amount by which the *specific enthalpy* of the coolant entering the channel is below the specific enthalpy of the saturated liquid at the system pressure. The specific enthalpy (enthalpy per unit mass) is defined by the relation

$$h = U + PV \tag{6.57}$$

where U is the internal energy per unit mass of the coolant, and P and V are its pressure and specific volume (volume per unit mass).

The only one of the five parameters listed above for which the burnout flux shows an approximately linear dependence is the inlet subcooling. A typical set of experimental data illustrating the linear relation between burnout flux and inlet subcooling for water in a uniformly heated tube is shown in Fig. 6.12. A commonly used correlation formula for burnout flux is

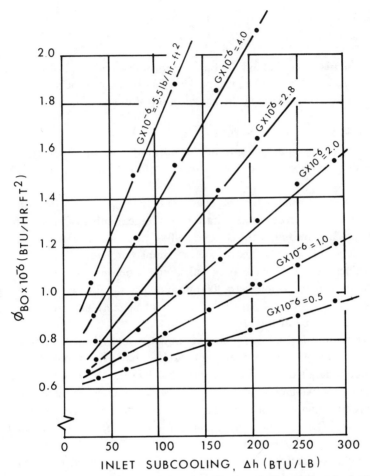

Fig. 6.12. Variation of burnout flux with inlet subcooling for a uniformly heated tube [from R. V. Macbeth, *Adv. Chem. Eng.* **7**, 207 (1968)]. $P = 1000$ psi, $D = 1.475$ in., $L = 77$ in.

that due to Barnett, which may be written as

$$\phi_{BO} \times 10^{-6} = \frac{A' + 0.25D(G \times 10^{-6})\,\Delta h}{C' + L} \qquad (6.58)$$

where the parameters A' and C' are functions of G and D of the form

$$A' = Y_0 D^{Y_1}(G \times 10^{-6})^{Y_2}$$
$$\times\, [1 + Y_3 D + Y_4(G \times 10^{-6}) + Y_5 D(G \times 10^{-6})] \qquad (6.59)$$

$$C' = Y_6 D^{Y_7} (G \times 10^{-6})^{Y_8}$$
$$\times [1 + Y_9 D + Y_{10}(G \times 10^{-6}) + Y_{11}D(G \times 10^{-6})] \quad (6.60)$$

The reasoning behind the burnout formula (6.56) is known as the *local-conditions hypothesis*. The burnout heat flux is found to be a function of the values of four variables at the actual point of burnout, i.e.,

$$\phi_{BO} = f(P, D, G, X) \qquad (6.61)$$

where X is the steam quality at the burnout position. This expression is equivalent to (6.56), since X is a function of the heated length and the inlet subcooling.

Numerical values for the coefficients $Y_0, Y_1, \ldots, Y_{11}$ are quoted by Macbeth, based on computer analysis of a large number of measurements of burnout flux in uniformly heated tubes with water coolant. Since the dependence of burnout flux on the remaining parameter P is very complex, the values of the Y_i have been calculated independently for each of the pressures, ranging from 15 to 2000 pounds per square inch absolute (psia), for which experimental data exist. Tables of the coefficient are given in the references quoted at the end of this chapter.

In dealing with the much more complex case of a bundle of fuel pins in a coolant channel, there are two approaches that may be taken. One is to attempt a detailed computer analysis by dividing the cross-sectional area available for coolant flow into a number of interacting subchannels and solving the equations of mass, heat, and momentum conservation for each subchannel. The program should allow for mixing of flow between subchannels based on the pressure difference developed across the subchannel boundaries as the flow moves up the core. From a knowledge of the resulting flow rate and enthalpy for each channel as a function of vertical position, a comparison may be made with the burnout correlation for some simpler geometry to predict the burnout heat flux for that channel. A number of subchannel models are under development and good progress is being made in using these to predict burnout in complex geometries.

A simpler approach is the whole channel model, which involves a correlation of the burnout power of the whole cluster with that of a system of simpler geometry, such as an annulus consisting of a heated rod (of diameter D_I) placed concentrically in an unheated shroud tube (of diameter D_O). Barnett has shown that for the latter case, the burnout flux is given, for pressures around 1000 pounds per square inch (psi), by the expression

$$\phi_{BO} \times 10^{-6} = \frac{A' + B' \Delta h}{C' + L} \qquad (6.62)$$

where

$$A' = Y_0 D_{HE}^{Y_1}(G \times 10^{-6})^{Y_2}\{1 + Y_3 \exp[Y_4 D_{HY}(G \times 10^{-6})]\}$$
$$B' = Y_5 D_{HE}^{Y_6}(G \times 10^{-6})^{Y_7}$$
$$C' = Y_8 D_{HY}^{Y_9}(G \times 10^{-6})^{Y_{10}}$$

In the expressions above, D_{HE} and D_{HY} are, respectively, the heated and hydraulic equivalent diameters, defined by the relations

$$D_{HE} = \frac{4 \times \text{flow area}}{\text{heated perimeter}} = \frac{D_0^2 - D_I^2}{D_I} \qquad (6.63)$$

$$D_{HY} = \frac{4 \times \text{flow area}}{\text{wetted perimeter}} = D_0 - D_I \qquad (6.64)$$

The values for the coefficients $Y_0, Y_2, \ldots, Y_{10}$ are again obtained by fitting to experimental data for burnout in annuli at a pressure of 1000 psia.

For the rod cluster, Barnett used the same correlation, but converted the cluster to an "equivalent annulus" by using empirical equivalences for the dimensions of the annulus. The diameter of the inner rod of the annulus was taken to be the same as that of the fuel pins, while the equivalent diameter for the shroud tube was given by the expression

$$D_0 = [D_R(D_R + D_H^*)]^{1/2} \qquad (6.65)$$

Here,

$$D_H^* = \frac{4 \times \text{flow area}}{\pi D_R S^*} \qquad (6.66)$$

where S^*, which takes account of the variation of heat flux between the rods in the bundle, is given by

$$S^* = \sum_{\text{rods}} \frac{\phi_r}{\phi_m} \qquad (6.67)$$

where ϕ_r/ϕ_m is the ratio of the heat flux on the rth pin to that on the pin of highest rating.

The effectiveness of the annulus correlation in predicting burnout in fuel bundles is illustrated for a typical bundle in Fig. 6.13. It will be seen that the burnout flux can be predicted within an accuracy of 10% for all three mass flow rates. This degree of accuracy is typical of most of the fuel bundle cases investigated.

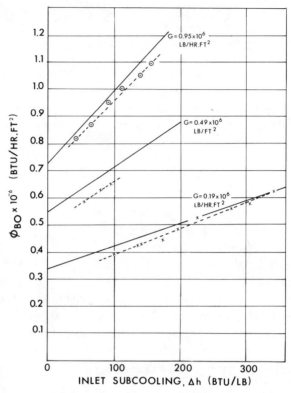

Fig. 6.13. Comparison of burnout data for a 19-rod fuel bundle with Barnett's annulus correlation (from P. G. Barnett, AEEW-R463). The experimental points are compared with the continuous line based on the annulus correlation.

All the correlations mentioned above apply to situations which are simpler than those found in an operating reactor. All of them, for example, assume that the heat flux is constant along the length of the tube or pin, whereas in a reactor fuel element the heat flux will vary in a cosine or skewed cosine fashion. One consequence is that, whereas for uniform heating the burnout will always occur at the exit end of the coolant channel, in the cosine case the burnout may occur at some other position. It has been found that, for heat flux profiles which are symmetrical and show only a moderate deviation from a uniform distribution (maximum to average heat fluxes of 1.4 or less), the total power required to produce burnout is the same for the nonuniform as for the uniform case (assuming that all other conditions, such as geometry and inlet subcooling, are the same). For more extreme variations, however, some form of local-conditions correlation is necessary.

6.6. Nuclear Power Reactor Cooling Systems

6.6.1. Choice of Coolant

The principal coolants which have been used for nuclear reactors are (1) liquids (e.g., water, heavy water, organic fluid), (2) gases (e.g., carbon dioxide, helium), and (3) liquid metals (e.g., sodium).

Water, owing to its ready availability and good heat transfer characteristics, is the most commonly used coolant. Along with ordinary water, we may consider heavy water, which has similar thermodynamic properties. Both of these have the advantage of being excellent moderators but, as mentioned earlier, the relatively high neutron absorption of light water necessitates the use of enriched fuel. Heavy water, on the other hand, has a low enough cross section to permit the use of natural uranium fuel; this advantage has to be balanced against its high cost of production.

Water-cooled reactors may be divided into two types, the boiling water reactor (BWR) and the pressurized water reactor (PWR), depending upon whether the design does, or does not, permit the water to boil on its passage through the core. At typical coolant operating temperatures (of around 300°C), the PWR circuit has to be maintained at a high pressure (around 2250 psi) in order to prevent boiling at this temperature. The BWR, on the other hand, may be operated at about half this pressure. The circuit pressure of the heavy water reactor (HWR) is similar to that of the PWR, since boiling of the coolant is avoided in this case also.

The use of certain organic coolants, such as terphenyls, has been studied in some detail. These have the advantage of permitting operation at lower pressures, on account of their lower vapor pressure as compared with water. They have the additional advantage of being less corrosive than water at reactor temperatures. Their main disadvantage is a tendency to decompose under irradiation and high temperature.

For gas-cooled reactors, the coolant which has been most widely used is carbon dioxide, which is readily available and has a low neutron absorption cross section. The poor heat transfer capability of a gas as compared to a liquid coolant requires the gas-cooled reactor to be operated at a high pressure (about 600 psi) in order to provide the required heat removal capacity.

The third type of coolant, the liquid metal, has found its main use in the fast breeder reactor, where liquid sodium is the commonly accepted coolant, since water or organic liquids are ruled out because of their high moderating power. Sodium has the advantage of low moderation and neutron absorption, coupled with the excellent heat transfer capability which is needed for a reactor with the very high power density of the fast breeder. The high boiling point also eliminates the need for operation at high pressure. On the other hand, the high radioactivity induced by neutron irradiation, and the high chemical activity of sodium require elaborate design arrangements to avoid the possibility of explosive interactions between radioactive sodium and the water in the steam-raising plant.

6.6.2. Water-Cooled Reactors

We shall first consider the pressurized water reactor, since the in-core conditions are simpler, although the overall coolant circuit is more complicated owing to the need for external steam generators to provide steam for the turbine. The cooling circuit for a PWR is shown diagrammatically in Fig. 6.14. The *primary coolant* is pumped through the reactor core, taking up heat generated by fission in the fuel elements, and then passes through a *heat exchanger* where the heat from the coolant is transferred to a *working fluid* (water) which is converted to steam which is taken off to drive the turbine. After passing through the turbine, the steam is converted back to water in the condenser and pumped back to the heat exchanger. The condenser is normally cooled by a supply of cold water from a source such as a river or lake.

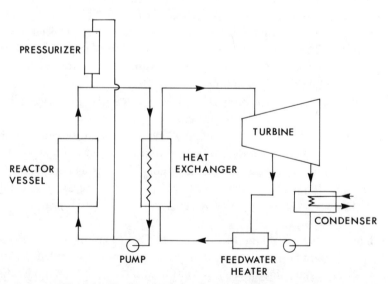

Fig. 6.14. Cooling circuit for a pressurized water reactor.

The thermal efficiency (η) of a nuclear power plant is defined by the ratio

$$\eta = \frac{\text{rate of electrical energy output}}{\text{rate of energy generation in core}} \tag{6.68}$$

The unit of energy in both cases is the megawatt. The abbreviations MWe and MWt are used where it is necessary to distinguish between electrical and thermal power output.

It should be noted that not all of the energy released in the fission process is given up to the coolant passing through the core. Of the average energy release per U^{235} fission of 205 MeV (see Section 2.3), the 12 MeV associated with the neutrinos is simply lost from the system. This loss is partially compensated for by the energy deposited by γ rays arising from stray neutrons undergoing (n, γ) reactions in the reactor components; this process contributes an extra 7 MeV of energy. Only about one third of the total energy arising from γ rays emitted from the fuel is absorbed in the core, the remainder being taken up in the coolant, reflector, shield, and pressure vessel. As a result, about 184 MeV (90%) of the recoverable energy per fission is deposited in the core, while another 16 MeV or so is deposited outside the core.

For the simplest type of coolant cycle, the reversible Carnot cycle, the thermodynamic efficiency is defined by the relation

$$\eta = 1 - \frac{T_R}{T_A} \tag{6.69}$$

where T_A is the temperature at which heat is added to the working fluid and T_R is the temperature at which it is rejected by the fluid. The cycle shown in Fig. 6.14 is more complicated than the simple Carnot cycle, but the efficiency in this case also increases as the ratio T_R/T_A decreases.

Since there is little flexibility in the choice of T_R, which is set by the characteristics of the external water supply for the condenser, the achievement of a high thermal efficiency makes it necessary to operate with steam at the highest practicable temperature. This in turn means that the primary coolant should also be at a high temperature. In order to prevent boiling at this temperature, the primary circuit must be run at a pressure well above the saturation pressure of water at the temperature of the coolant. The most basic limitation on the primary coolant temperature is set by the condition that the fuel pins must remain at a temperature below that at which damage might occur to the cladding or the fuel itself. In practice, in a water-cooled reactor, this limitation restricts the maximum coolant outlet temperature to a value of slightly over 300°C. The corresponding primary circuit pressure is around 2250 psi (157 kg cm^{-2}).

The pressure in a PWR is maintained at the desirable level by a surge chamber known as a *pressurizer* (Fig..6.15). This is most commonly a tank partly filled with water, the volume above the water being occupied by steam. The water can be heated by electrical heaters and the steam can be condensed by a spray of cooler water located at the top of the tank.If a positive pressure surge occurs in the reactor cooling circuit, water from one of the cold legs of the system is sprayed into the pressurizer, causing some of the vapor to condense and so limiting the pressure rise. Similarly, in the event of a negative pressure surge, the heaters are switched on to increase the pressure. The actions of the spray and the heaters are aided by the automatic ballasting action of the pressurizer, since a positive pressure surge, for example, will result in an expansion of the cooler primary coolant into the pressurizer, result in a partial vapor condensation which will also limit the pressure rise.

A typical heat exchanger, or *steam generator*, for a PWR is illustrated in Fig. 6.16. The generator is divided into a lower and an upper section, known as the evaporator and the steam drum. The former consists of a U-tube heat exchanger and the latter houses moisture-separating equipment. Each of the U tubes is welded to a tubesheet; the space under the tubesheet is divided in two by a vertical divider plate which separates the inlet from the outlet flow. The high-temperature primary coolant enters by the primary inlet, flows through the U tubes, and leaves by the primary outlet.

The feedwater for the steam circuit enters the generator through a nozzle located on the upper shell and is distributed by a feedwater ring into the downcomer annulus between the tube wrapper and the steam generator shell. The feedwater mixes with the recirculation flow and flows down to enter the tube bundle just above the tubesheet. The steam generated as the secondary water flows up through the tube bundles passes to a set of centrifugal moisture separators, located in the upper shell. It finally passes through steam dryers which increase the steam quality to at least 99.75% (0.25% moisture content).

The heat balance equations for the PWR may be written in terms of the *specific enthalpy*, h, the enthalpy per unit mass of the water flowing through the core (in Btu lb^{-1}). The specific enthalpy is a thermodynamic function equal to the sum of the internal energy per unit mass and the product pV, where V is the volume per unit mass. When heat is added to a substance at constant pressure, the heat absorbed appears as an increase in the enthalpy.

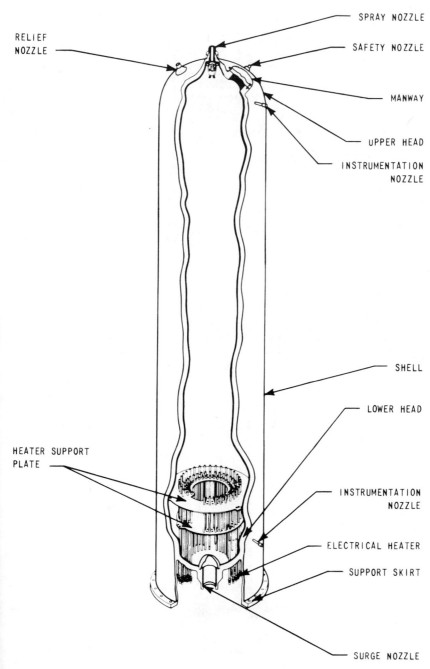

Fig. 6.15. Pressurizer for a pressurized water reactor (courtesy Westinghouse Electric Corporation).

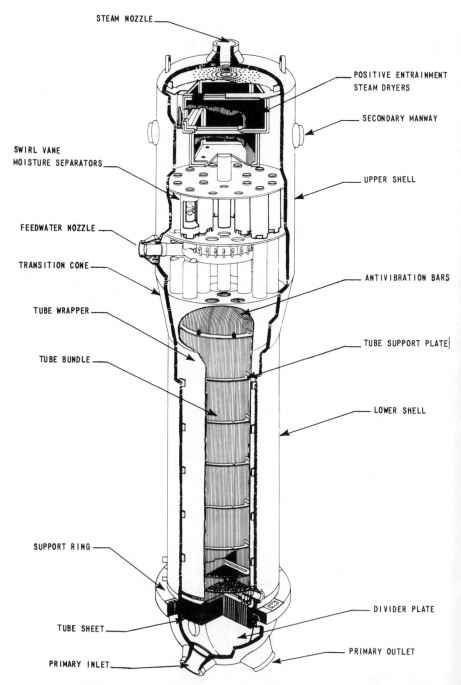

Fig. 6.16. Typical PWR steam generator (courtesy of Westinghouse Electric Corporation).

Values of the specific enthalpy of water as a function of temperature and pressure are given in Tables A.7–A.8.

Assuming no heat losses from the system, the heat generation rate in the core, Q, is simply equal to the rate of flow of coolant multiplied by the increase in the specific enthalpy of the coolant between entry and exit. Hence,

$$Q = \dot{m}_i(h_0 - h_i) \tag{6.70}$$

where Q is the heat generation rate in core (Btu h^{-1}), $\dot{m}_i$ the coolant flow rate (lb h^{-1}), h_0 the specific enthalpy of coolant at outlet (Btu lb^{-1}), and h_i is the specific enthalpy of coolant at inlet (Btu lb^{-1}).

Example 6.3. The flow rate through the core of a PWR operating at a pressure of 2250 psi is 120×10^6 lb h^{-1}. The inlet and outlet temperatures are 450°F and 510°F, respectively. Calculate the thermal power of the reactor (in megawatts).

According to Table A.7, the specific enthalpies at inlet and outlet are 430.2 and 499.6 Btu lb^{-1}, respectively. From equation (6.70), the thermal power is

$$Q = \dot{m}_i(h_0 - h_i) = 120 \times 10^6 \times (499.6 - 430.2)$$
$$= 8.328 \times 10^9 \text{ Btu } h^{-1}$$

Using the conversion factor given in Table A.2, we have

$$Q = 8.328 \times 10^9 \times 0.2931 \times 10^{-6} \text{ MW}$$
$$= 2440 \text{ MW} \qquad \square$$

The boiling water reactor is shown diagrammatically in Fig. 6.17. In this case, subcooled water enters at the bottom of the core and flows past the fuel elements. Some of the liquid is converted to steam as it goes through the core, and the water–steam mixture leaving at the top passes to steam separators in which vanes cause a centrifugal motion which separates the steam from the water. The steam passes through steam dryers to the turbine and the separated water flows downward in the space between the pressure vessel wall and the *core shroud* which surrounds the core. This water is driven by a number of *jet pumps* located in the annular space outside the core shroud. The jet pumps in turn are driven by a flow of water provided by external *recirculation pumps*. The design of the jet pump is such that the external water flow creates a suction which entrains the reactor water which is driven down

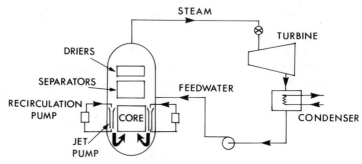

Fig. 6.17. Cooling circuit for a boiling water reactor.

to the plenum underneath the core. The introduction of the internal jet pumps has the advantage of reducing the required external flow, which is less than one third of the total recirculation flow through the reactor.

The coolant flow pattern for the BWR is shown in simple outline in Fig. 6.18. The total water mass flow rate is $\dot{m}_i$ lb h^{-1} flowing through the core, where some of the water is converted to steam, and the steam flows off to the turbine at a mass flow rate of $\dot{m}_g$ lb h^{-1}. The feedwater returns from the condenser at a mass flow rate of $\dot{m}_w$ lb h^{-1}, and mixes with the saturated recirculation flow of $\dot{m}_f$ lb h^{-1}. The mass balance equations for the core flow are therefore

$$\dot{m}_w = \dot{m}_g \qquad (6.71)$$

$$\dot{m}_i = \dot{m}_f + \dot{m}_w = \dot{m}_f + \dot{m}_g \qquad (6.72)$$

Assuming no heat loss from the system, the heat generated within the core essentially goes into converting feedwater into saturated steam at a rate of $\dot{m}_g$ per hour. The heat balance equation for the core may then be written as

$$Q = \dot{m}_g(h_g - h_w) \qquad (6.73)$$

where h_g is the specific enthalpy of the saturated steam (at reactor pressure), h_w is the specific enthalpy of the feedwater, and Q is the total heat generation rate in the core (Btu h^{-1}).

The total heat generation rate may also be written in the form

$$Q = \dot{m}_i(h_f - h_i) + \dot{m}_g h_{fg} \qquad (6.74)$$

where h_f is the specific enthalpy of the saturated recirculation liquid and h_i is the specific enthalpy at core inlet. The quantity h_{fg} is the specific enthalpy of evaporation. The equation simply states that the core heat production rate is being used to raise $\dot{m}_i$ lb of water per hour from the inlet enthalpy to the saturation enthalpy and also to convert $\dot{m}_g$ lb/h of water to steam at saturation.

Equation (6.72) may be written in a different form by introducing the *steam exit quality* x, which is defined by

$$x = \frac{\dot{m}_g}{\dot{m}_g + \dot{m}_f} = \frac{\dot{m}_g}{\dot{m}_i} \tag{6.75}$$

Substituting in equation (6.74) gives

$$Q = \dot{m}_i[(h_f + xh_{fg}) - h_i] \tag{6.76}$$

The specific enthalpy at inlet, h_i, may be expressed in terms of the exit quality by starting with the heat balance equation for the mixture of recirculation and feedwater flows

$$\dot{m}_i h_i = \dot{m}_f h_f + \dot{m}_w h_w \tag{6.77}$$

Using equations (6.71), (6.72), and (6.75), we find that

$$h_i = (1 - x)h_f + xh_w \tag{6.78}$$

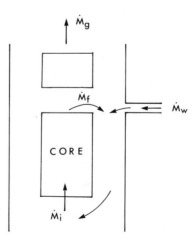

Fig. 6.18. Mass flow rates for boiling water reactor.

The inlet enthalpy is related to the temperature of the inlet water, T_i. The *inlet subcooling*, ΔT_i, is the difference between the inlet temperature and the saturation temperature, T_f, i.e.,

$$\Delta T_i = T_f - T_i \tag{6.79}$$

Example 6.4. Calculate the inlet subcooling and the inlet and steam mass flows for a BWR with the following operating parameters: total heat generation rate, 3300 MW; steam exit quality, 0.070; pressure, 1100 psi, and feedwater temperature, 350°F. Using standard steam tables (see Tables A.7–A.8), we find that

$$h_g = 1188.3 \text{ Btu lb}^{-1}$$
$$h_f = 557.4 \text{ Btu lb}^{-1}$$
$$h_w = 321.8 \text{ Btu lb}^{-1}$$
$$h_{fg} = 631.0 \text{ Btu lb}^{-1}$$
$$T_f = 556.45°F$$

From equation (6.78), the specific enthalpy at inlet temperature is

$$h_i = (0.93 \times 557.4) + (0.07 \times 321.8) = 540.9 \text{ Btu lb}^{-1}$$

From Table A.7, the corresponding inlet temperature is

$$T_i = 543.55°F$$

The inlet subcooling is therefore

$$\Delta T_i = T_f - T_i = 556.45 - 543.55 = 12.9°F$$

The steam mass flow rate is derived from equation (6.73), i.e.,

$$Q = \dot{m}_g(h_g - h_w)$$

Now,

$$Q = 3300 \text{ MW} = 3300 \times 3.412 \times 10^6 = 1.126 \times 10^{10} \text{ Btu h}^{-1}$$

Hence,

$$1.126 \times 10^{10} = \dot{m}_g(1183.3 - 321.8)$$

giving

$$\dot{m}_g = 1.299 \times 10^7 \text{ lb h}^{-1}$$

The inlet mass flow rate is given by equation (6.76), i.e.,

$$Q = \dot{m}_i[(h_f + xh_{fg}) - h_i]$$

or,

$$1.126 \times 10^{10} = \dot{m}_i\{[557.4 + (0.07 \times 631.0)] - 540.9\}$$

Hence,

$$\dot{m}_i = 1.856 \times 10^8 \text{ lb h}^{-1} \qquad \square$$

6.6.3. Gas-Cooled Reactors

Gas-cooled reactors have been extensively developed in the United Kingdom and have also been used in several other countries. Current designs of gas-cooled reactor use an indirect closed cycle, as illustrated in Fig. 6.19. Carbon dioxide at a pressure of around 600 psi (42 kg cm^{-2}) is pumped through the core of the reactor and then to a heat exchanger (boiler) with water as the secondary coolant. The steam produced in the heat exchanger is used to drive the turbine.

Carbon dioxide has the advantages of low cost and low neutron absorption cross section. The use of gas as coolant permits much higher core

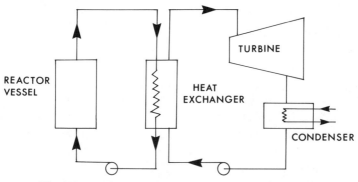

Fig. 6.19. Cooling circuit for a gas-cooled reactor (indirect cycle).

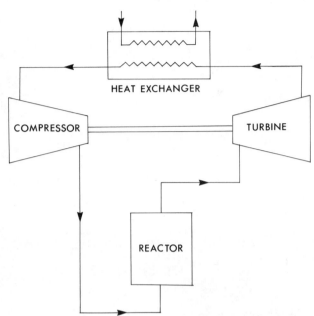

Fig. 6.20. Cooling circuit for a gas-cooled reactor (direct cycle).

outlet temperature [650°C for the new designs, compare with around 320°C for a light water reactor (LWR)], thereby increasing the thermal efficiency of the plant to a value in excess of 40% (compared with around 33% for the LWR). Disadvantages include the poorer heat transfer capability of the gaseous coolant and the relatively higher power demand for pumping a gas as compared to a liquid.

The coolant outlet temperatures which can be achieved with gas cooling are high enough to raise the possibility of eliminating the secondary circuit and passing the high-temperature coolant directly to a gas turbine. The layout of a direct cycle of this type is illustrated in Fig. 6.20. In this case, the coolant gas expands through the turbine and then rejects heat in the heat exchanger before being compressed and pumped back to the reactor.

The high gas temperature required for a gas turbine results in carbon dioxide being unsuitable as a coolant, owing to its radiation-enhanced reaction with graphite at these temperatures. The favored coolant for high-temperature gas-cooled reactors is helium, which is chemically inert and has an essentially zero neutron absorption cross section. The requirement for high coolant temperature results in the fuel having to withstand temperatures well in excess of those in the earlier designs of gas-cooled reactor. This has led to the development of coated-particle dispersed fuels, as discussed in Chapter 5.

6.6.4. Liquid Metal Cooling

Cooling by a liquid metal, such as sodium, has been universally adopted for fast breeder reactors. The main reason is that so many potential coolants are unsuitable because of the need to minimize moderation of the fast neutron spectrum in the core. Water (light or heavy) and organic coolants are therefore eliminated. Consideration has been given to the use of steam or helium as fast reactor coolants, but the heat transfer characteristics of a liquid metal are so much more favorable, particularly for a reactor of high power density, that neither steam nor gas cooling has been adopted for current fast breeder designs.

Of the liquid metals which might possibly be used, sodium turns out to be the most suitable. It has a relatively low thermal neutron capture cross section (0.53 b), and remains liquid over a relatively wide range of temperature (98–883°C). The fact that it solidifies at temperatures below 98°C means that auxiliary heating must be available to keep it liquid when the reactor is out of operation. On the other hand, the high boiling point has the advantage of permitting high coolant outlet temperatures to be attained without requiring high circuit pressure. The high thermal conductivity of sodium has the effect of reducing the temperature gradient in the coolant channels as compared with those of a water-cooled reactor, where heat transfer is largely by convection rather than conduction.

Two disadvantages of sodium are the fairly high radioactivity induced by neutron capture in Na^{23} and its violent reaction on coming in contact with water. The process of heat transfer from the sodium primary coolant to the water in the steam generation loop must therefore be carefully designed to reduce the probability of sodium–water interaction and to ensure that, even if this were to occur, there would be no large-scale release of radioactive sodium to the environment. In practice, this implies the use of an intermediate sodium circuit between the sodium core coolant and the water. The heat removed from the core is transferred in a sodium-to-sodium heat exchanger to the intermediate circuit, and the sodium in this circuit then transfers heat to water through a sodium–water heat exchanger (see Fig. 7.2). The prevention of sodium–water interactions in the latter has proved to be one of the main problems in a commercial-scale fast breeder reactor; in most other respects sodium has turned out to be a very tractable coolant.

6.6.5. Superheat and Reheat

The condition of the working fluid in the heat exchanger of the reactor circuit shown in Fig. 6.14 can be illustrated by the temperature–entropy

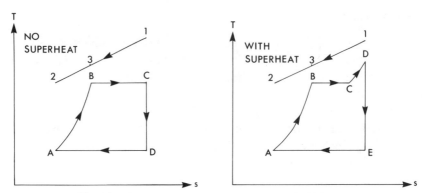

Fig. 6.21. Temperature–entropy diagram for a Rankine cycle.

diagram for a Rankine cycle, given in Fig. 6.21. The secondary water entering the heat exchanger takes up heat from the primary coolant in the form of sensible heat along the path AB and then as heat of evaporation along BC. The saturated vapor then expands through the turbine (CD) and is condensed and returned to the heat exchanger (DA). Assuming that the heat exchanger is of the counterflow type, where the primary and secondary coolants flow in opposite directions, the temperature variation of the primary coolant is given by the line $1 \rightarrow 2$.

For high thermal efficiency, it is desirable that the cycle should approximate to a reversible one. The cycle can be reversible only if the temperature difference between the primary and secondary coolants is everywhere equal to zero, a condition impossible to attain in practice, since there has to be a finite difference between the two fluids in order to promote heat transfer between them. In the diagram, the smallest temperature difference between the primary and secondary fluids occurs between the points B and 3; this gap is called the *pinch point*.

In order to approach a reversible cycle, the temperature difference (ΔT) at the pinch point should be as small as possible. Since the heat transfer rate in the exchanger is also proportional to the surface area of contact between the two fluids, it is possible to compensate for a smll ΔT by increasing the size of the heat exchanger. There are, of course, obvious limitations to the practical size of the unit, so that with this type of cycle there will always be an appreciable difference between the average temperatures of the two coolants.

One way of reducing the average temperature difference is by *superheating* of the saturated steam, as shown by the line CD in the right portion of Fig. 6.21. This not only increases the thermal efficiency of the cycle, but has the added advantage of reducing the water content of the steam in the turbine and thus minimizing the problem of turbine blade erosion. Superheating is normally done in a special section of the steam generator, but some work has also been done on using the reactor itself as a source of superheat, by passing the saturated steam through special coolant channels in the core. The latter effort has been largely abandoned because of the increased design complexity which it entails.

Another technique used to improve efficiency and reduce water content in the turbine is *reheating*. The turbine is divided into high-pressure and low-pressure sections; the wet steam leaving the high-pressure section is passed through a reheater unit in the heat exchanger, where its temperature is increased and the water content reduced before it goes on to the low-pressure section of the turbine. This is illustrated in Fig. 6.22, which also shows a typical subdivision of the heat exchanger into three sections, known as the *economizer, evaporator,* and *superheater*; in the first of these, the working fluid is taken up to evaporation temperature. Also shown in the figure is the *feedwater heater*; in this, steam bled from the turbine is used to heat the feedwater returning to the heat exchanger. This process, known as *regeneration*, has the effect of increasing the efficiency of the working fluid cycle.

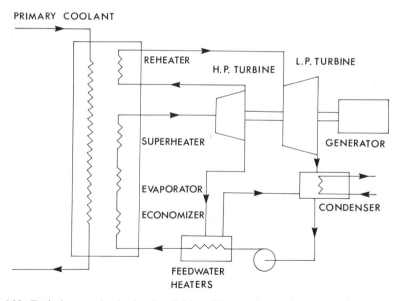

Fig. 6.22. Typical steam circuit, showing division of heat exchanger into economizer, evaporator, superheater, and reheater.

Problems

6.1. A PWR fuel element is 12 ft long and has a fuel diameter of 0.366 in. Find the total heat generation rate for the element if the heat generation rate per unit volume at the central plane is 6.52×10^7 Btu/h ft^3.

6.2. A PWR has a core in the form of a cylinder of diameter 11 ft. It contains 193 fuel assemblies, each with 204 pins. The fueled length is 12 ft and the fuel diameter 0.366 in. The maximum heat generation rate per unit volume at the core central plane is 9.0×10^7 Btu/h ft^3. Calculate the heat generation rate per unit cross-sectional area of the core at a distance of 3 ft from the core axis.

6.3. Calculate the total power (in megawatts) which is being generated in the reactor described in the previous problem.

6.4. A plate-type fuel element consists of a uniform fuel-bearing region of thickness 0.1 in. If the heat generation rate per unit volume in the element (assumed uniform throughout) is 7.80×10^7 Btu/h ft^3, and the center temperature is 650°F, calculate the temperature of the outer surface of the plate. Take the thermal conductivity of the element as 18.5 Btu/h ft °F.

6.5. A plate-type element consists of a 100×5-cm sandwich of fuel of thickness 5 mm, with a cladding of thickness 1.2 mm. The thermal conductivities of the fuel and cladding are 32 and 19 W/m °C, respectively. If the maximum fuel temperature is 370°C and the surface temperature is 315°C, calculate the rate of heat conduction out of each surface of the plate.

6.6. A cylindrical fuel rod consists of a fuel region of diameter 0.4 in., with a zircaloy cladding of thickness 0.025 in. If the temperature at the center line of the fuel is 3400°F and the heat generation rate is 4.25×10^7 Btu/h ft^3, calculate the temperature at the surface of the fuel. Take the thermal conductivity of the (oxide) fuel to be 1.1 Btu/h ft °F.

6.7. For the fuel rod above, calculate the temperature at the outer surface of the cladding, if the thermal conductivity of the cladding is 10.2 Btu/h ft °F.

6.8. A plate-type fuel element consists of a 0.2-in.-thick fuel region with 0.025-in. cladding. The thermal conductivities of the fuel and cladding are 18.5 and 10.5 Btu/h ft °F, respectively. The convection heat transfer coefficient to the coolant is 6500 Btu/h ft^2 °F. If the maximum fuel temperature is 800°F and the bulk coolant temperature is 650°F, calculate the rate of heat flow per unit surface area of the element.

6.9. The fuel elements of a PWR have a length of 3.66 m and a fuel region diameter of 10.2 mm. The coolant inlet temperature is 260°C and the flow associated with each element is 1.23×10^3 kg h^{-1}. The maximum heat generation rate in the fuel is 4.40×10^8 W m^{-3}. Taking the specific heat of the coolant as 1.35 J/g °C, calculate the bulk coolant temperature at the center plane of the reactor.

6.10. The fuel elements in the previous problem have cladding of thickness 0.5 mm. If the convection heat transfer coefficient to the coolant is 40,000 W/m^2 °C calculate the surface temperature of the cladding at the reactor centre plane.

6.11. The maximum heat generation rate at the center of a PWR fuel element is 4.9×10^7 Btu/h ft^3. The length of the element is 12 ft, the fuel diameter 0.32 in., and the cladding thickness 0.025 in. The mass flow rate per element is 2.80×10^3 lb h^{-1} and the coolant inlet temperature is 500°F. The reactor pressure is 2200 psi. The convection heat transfer coefficient is 7000 Btu/h ft^2 °F and the thermal conductivities of the fuel and cladding are 1.1 Btu/h ft °F and 10.5 Btu/h ft °F, respectively. The specific heat of the coolant at the operating pressure is equal to 1.15 Btu/lb °F. Calculate the maximum fuel temperature in the element at a height of 3 ft above the core center plane.

6.12. In the previous problem, find the axial location and the magnitude of the maximum cladding temperature.

6.13. Water at a pressure of 1000 psi flows through a uniformly heated tube of the same diameter and heated length as the tube in Fig. 6.12. If the flow rate is 2.8×10^6 lb h^{-1}, and the inlet temperature is 415°F, estimate the burnout flux.

6.14. The total coolant flow rate for a PWR operating at a power of 3150 MW is 1.47×10^8 lb h^{-1}. If the coolant inlet temperature is 540°F, calculate the outlet temperature.

6.15. A BWR operates at a power of 3000 MW and a pressure of 1100 psi. The total coolant flow rate through the core is 1.70×10^8 lb h^{-1} and the inlet temperature is 540°F. Calculate the mass of steam generated per hour.

6.16. A BWR operating at a pressure of 1100 psi generates steam at a rate of 9.35×10^6 lb h^{-1}. If the feedwater temperature is 410°F, at what power (in megawatts) is the reactor operating?

6.17. A PWR of cylindrical core shape, operating at a power of 3400 MW, contains 193 fuel bundles, each with 264 fuel pins. The diameter of the fuel region is 8.1 mm and the fuel height is 3.66 m. The maximum heat generation rate in the core is 5.74×10^8 W m^{-3} and the total coolant flow rate is 6.44×10^7 kg h^{-1}. If the coolant inlet temperature is 290°C, calculate (a) the average outlet temperature and (b) the coolant outlet temperature in the hottest channel (assume that the coolant flow is equally distributed in the channels). Take $c_p = 1.4$ J/g °C.

6.18. For the previous problem, calculate the maximum fuel temperature at the center plane of the core if the cladding thickness is 0.635 mm. The thermal conductivities of the fuel and cladding are 1.9 and 18 W/m °C, respectively, and the convection heat transfer coefficient 37,000 W/m^2 °C.

7

General Survey of Reactor Types

7.1. Introduction

Over the more than 40 years since the first nuclear fission reactor was constructed numerous designs of reactor have been evolved by variation of the basic parameters such as fuel type, moderator, and coolant. One possible classification is by intended use, e.g., research, plutonium production, electricity generation, or propulsion units for submarines or surface ships. In this chapter we will concentrate on power reactors, both on account of their practical importance and because of the complexity in engineering design introduced by the need to convert the energy released by nuclear fission into a mechanical or electrical output. Many of the characteristics of the various reactor types have been touched on in earlier chapters, but the objective in the present chapter is to provide a systematic summary of the main classifications of reactor prior to the more detailed descriptions to be given in the following chapters.

All power-producing reactors require a fissile material, which may be U^{233}, U^{235}, or Pu^{239}, and a coolant to remove the heat from the fuel. In most cases the heat transported by the coolant is used to produce steam to drive a turbine. In this respect the nuclear power plant is similar to a conventional fossil-fueled station, but with the reactor acting as the heat source (see, e.g., Fig. 6.17).

From the nuclear viewpoint, the most general subdivision of reactor types is on the basis of whether or not a moderator is deliberately introduced in order to slow the neutrons down to thermal energies. A basic distinction may therefore be made between *thermal reactors,* where a moderator is introduced, and *fast breeder reactors,* where the only moderation taking place is the relatively minor effect produced by neutron collisions with the coolant, fuel, and structural components of the reactor.

7.2. Classification of Thermal Reactors

These may most conveniently be classified on the basis of type of moderator. The practical moderators, which satisfy the requirement of good slowing-down power combined with acceptably low neutron absorption, are hydrogen (as ordinary, or light water), deuterium (as heavy water), beryllium, and carbon (as graphite). All but beryllium have been used on a commercial scale; beryllium, while offering advantages such as compactness and low neutron absorption, has proved too expensive for use in power reactor systems. The main thermal reactor types, based on type of moderator, are listed in Table 7.1. The table is restricted to solid-fueled reactors; an additional type is the molten-salt breeder reactor (MSBR), where the fuel is continuously circulated in the form of a molten salt through a fixed graphite moderator.

The two *light water reactors* (LWRs) are the *pressurized water reactor* (PWR) and the *boiling water reactor* (BWR). The systems are similar in that both employ light water as both moderator and coolant, which necessitates the use of fuel enriched to about 2.0%–2.5% due to the appreciable neutron absorption of the water in the core. They are also alike in that in both cases the core is enclosed within a single large steel *pressure vessel* (see Fig. 7.1). The distinction between the two designs is also illustrated in the figure. In the

Table 7.1. Main Types of Thermal Reactor for Power Production

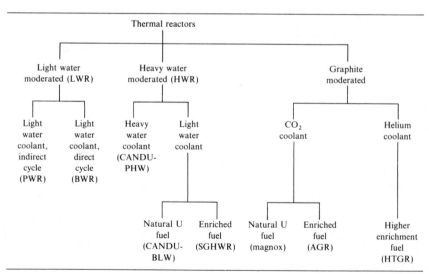

PWR, the coolant circuit is maintained at a high enough pressure (approximately 158 kg/cm^2) that no significant boiling can take place at the coolant operating temperature of around 320°C. The coolant leaving the core is circulated through the tubes of *heat exchangers* where it transfers heat to produce steam in a secondary water circuit maintained at a lower pressure (about 55 kg/cm^2). The steam from the steam generators passes through the turbine and is then changed back to water in the condenser, which is cooled by water from an external source such as a lake or river.

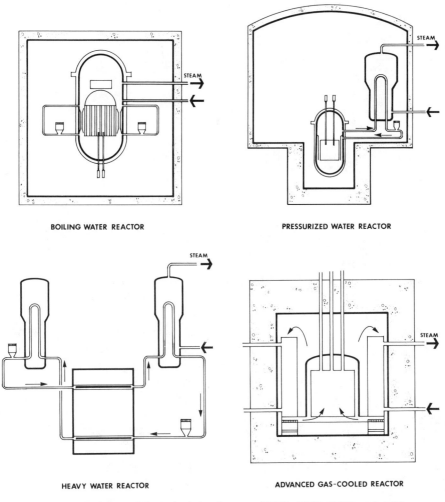

BOILING WATER REACTOR   PRESSURIZED WATER REACTOR

HEAVY WATER REACTOR   ADVANCED GAS-COOLED REACTOR

Fig. 7.1. Comparison of the main design features of BWR, PWR, HWR, and AGR.

REACTOR HEAT EXCHANGER TURBINE GENERATOR

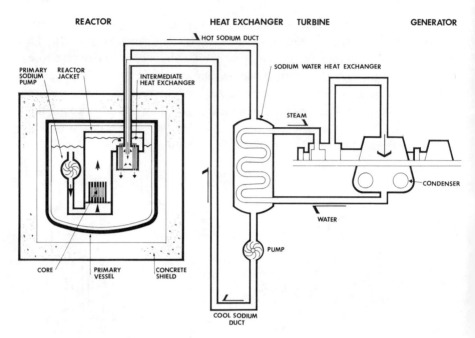

Fig. 7.2. Design features of a liquid-metal-cooled fast breeder reactor (courtesy United Kingdom Atomic Energy Authority).

When a heat exchanger is used to transfer heat from the primary coolant to a secondary water circuit, the system is said to operate in an *indirect cycle*. In the BWR, on the other hand, the primary circuit pressure is lower (about 73 kg/cm²) than in the PWR, and large-scale boiling takes place in the core, the steam produced being fed directly to the turbine. This *direct cycle* mode of operation has the advantage of simplifying the heat exchange system, as shown in Fig. 7.1. The development and comparative merits of the two designs of BWR will be dealt with in Chapter 9.

The main commercial development of the *heavy water moderated* designs (HWRs) has been based on the *pressure tube* concept, where the coolant is circulated past the fuel in pressure tubes which keep it separate from the heavy water moderator, which can therefore be contained in a low pressure tank. The only large-scale power program based on the HWR is the Canadian CANDU-PHW (*pressurized heavy water*) program. The reactor operates in an indirect cycle, as shown in Fig. 7.1. Variants employing light water as coolant are the CANDU-BLW (*boiling light water*) and the United Kingdom SGHWR (*steam-generating heavy water reactor*).

The range of *gas-cooled graphite-moderated reactors* represents a progression from the early natural uranium-fueled reactors of the British *magnox* type (so-called from the alloy used for the fuel cladding), through the enriched uranium *advanced gas-cooled reactor* (AGR) to the *high-temperature gas-cooled reactor* (HTGR) which uses the coated-particle fuel discussed in Chapter 5. Figure 7.1 shows the AGR design where the heat exchangers are enclosed in a prestressed concrete pressure vessel. The series has been characterized by a steady development towards higher gas outlet temperature and therefore higher thermodynamic efficiency. The higher temperature of the HTGR necessitated a move from carbon dioxide to helium as the coolant gas.

In the *fast breeder reactor* (FBR) the moderator is eliminated in order to keep the neutron energies high enough to achieve a favorable value of η and thus permit the possibility of breeding. The currently favored coolant for the fast breeder reactor is a liquid metal (sodium), although alternatives such as steam or helium have been explored in the past, and there is still some interest in the potentialities of a *gas-cooled fast breeder reactor* (GCFBR). The design of a typical *liquid metal fast breeder reactor* (LMFBR) is shown diagrammatically in Fig. 7.2. The relatively small core is surrounded by a *blanket* containing natural or depleted uranium fuel elements. In order to prevent the possibility of widespread contamination resulting from the radioactive sodium of the primary coolant circuit coming into contact with the water of the steam-generating system, the two circuits are isolated from one another by an intermediate heat exchanger where primary coolant gives up heat to a secondary sodium coolant. The model shown is of the "pool type," where the core, together with the primary pumps and the intermediate heat exchangers, is suspended in a large vessel containing the liquid sodium.

8

The Gas-Cooled Graphite-Moderated Reactor

8.1. General Characteristics of the Gas-Cooled Graphite-Moderated Reactor

The graphite-moderated reactor has a longer history than any other type, since the first critical assembly constructed under the direction of Enrico Fermi at Chicago in December 1942 was a natural uranium, graphite-moderated reactor. The principal use of graphite-moderated reactors for power generation, however, has been in the United Kingdom and France. In both cases, the original motivation was the advantage of constructing a system combining natural uranium fuel with a readily available and relatively inexpensive moderator.

Gas-cooled graphite-moderated reactors can be divided into three main types, representing progressive stages of development of the system. These are (i) natural uranium metal, CO_2 cooled (e.g., the U.K. magnox and the French reactors of the G2 and EDF1 type), (ii) enriched uranium oxide, CO_2 cooled (the U.K. advanced gas-cooled reactor), and (iii) higher-enrichment ceramic fuel, helium-cooled, high-temperature (e.g., the OECD Dragon high-temperature reactor, the West German thorium high-temperature reactor, and the U.S. Fort St. Vrain HTGR).

All these reactors share the common feature of a dual coolant cycle, although the high-temperature reactor has the potential for operating on a direct cycle with a helium gas turbine.

The development of the magnox reactor line in the United Kingdom arose out of an original requirement for a plutonium producer based on natural uranium fueling, which led to the construction of the Windscale reactors for that specific purpose. The use of natural fuel restricted the possible moderators to beryllium, heavy water, or graphite. The last of these was selected on grounds of availability. Cooling by light water, as in the U.S. Hanford plutonium production reactors, was rejected both on the grounds of complexity and the adverse effects on the neutron economy of the reactor. The original plutonium-producing reactors at Windscale were cooled by forced circulation of air at atmospheric pressure, but the much greater heat removal rates required for economic operation of a reactor designed for power production necessitates the use of high-pressure coolant gas. Carbon dioxide was selected as coolant on the basis of its relative cheapness, low capture cross section, and negligible interaction with the graphite at the temperature characteristic of a reactor with uranium metal fuel. The low absorption property is important, not only to the neutron economy, but also to prevent the possibility of a major reactivity excursion in the event of sudden depressurization of the primary circuit. The compatibility of coolant and moderator allows the whole core to be enclosed in a single pressure vessel.

As a moderator, graphite is characterized by a reasonably good slowing-down power ($\xi\Sigma_s \sim 0.06$ cm^{-1}) and a low absorption cross section (3.4 mb), giving a moderating ratio $\xi\Sigma_s/\Sigma_a \sim 220$. The diffusion length for pure graphite is approximately 54 cm, with the consequence that the typical graphite-moderated reactor is considerably less compact than one moderated by light water. The large size is particularly a feature of the natural uranium reactor, where the leakage has to be made very small if an adequate reactivity margin is to be achieved. A magnox reactor is also considerably larger than a D_2O-moderated system, such as CANDU, of the same output.

The combination of natural uranium fuel and graphite moderator requires a marked heterogeneity of the fuel elements in order to minimize the effective resonance integral. The typical element is a metal rod of 28-mm diameter and about 1 m in length. The rods are fitted with finned cans made from the magnesium alloy (magnox), from which the reactor type derives its name. The low absorption cross section of magnesium ($\sigma_a = 63$ mb) permits the can to be fitted with the substantial finning which is necessary to achieve adequate heat transfer with a gaseous coolant.

The first magnox station to be commissioned was the four-reactor Calder Hall station, which came on line in 1956. In the course of the next 11 years the magnox program was expanded to a total capacity of some 5000 MWe generated by 24 reactors in ten power stations in the United Kingdom. All but two of the stations employed large, spherical, steel pressure vessels for containment of the reactor core, but the last units of the series, at Oldbury and Wylfa, have prestressed concrete pressure vessels of the type pioneered by the French gas-cooled reactors at Marcoule. With this design, the reactor and its heat exchangers form an integral unit within the pressure vessel.

The thermal efficiency of the natural uranium metal reactor is restricted to a value of around 30 % by the limitations on coolant temperature imposed by the properties of the fuel and canning. The problems of heat removal from the large-diameter metal fuel rods, and the limited ability of the uranium to accommodate damage due to irradiation, restrict the fuel rating and attainable burn-up to figures of the order of 5 MW/tonne and 3600 MW d/tonne, respectively. Greater thermal efficiency, coupled with higher ratings and fuel burn-up, can be achieved by changing to uranium dioxide fuel and a stronger canning material, such as stainless steel. The improved performance is obtained at the expense of having to enrich the fuel. This development of the graphite-moderated concept is represented by the advanced gas-cooled reactor (AGR), which forms the second generation of the U.K. nuclear power system. Like the magnox reactor, the AGR has a fixed graphite moderator and is cooled by CO_2 gas.

The extension of the graphite reactor to still higher power densities requires a radical redesign, to a form known as the high-temperature gas-cooled reactor (HTGR). The AGR is limited in its potential thermal efficiency by the following factors:

i. The heterogeneous nature of the fuel assembly restricts the power density, since all the heat is produced in an isolated and relatively small fraction of the core volume. The low thermal conductivity of UO_2 leads to a large temperature differential between the center of the fuel pin and the coolant. The limitation on power density is made more stringent by the need to protect the fuel from excessive central temperatures in the event of a possible power excursion, since the low thermal capacity of the fuel assemblies makes them respond fairly rapidly to changes in power.

ii. The radiation-enhanced reaction of CO_2 with graphite becomes increasingly severe as the coolant temperature increases, and effectively limits its use to temperatures below 600°C.

iii. Dimensional changes in the permanent graphite core structure, as a result of fast neutron irradiation, impose a limit on the total irradiation dose which it can sustain. This in turn implies a restriction on the power density in the core.

The HTGR concept has its origin in an attempt to avoid these limitations by the use of dispersed fuel and a helium coolant. The fuel and fertile material are in the form of very small oxide particles, individually coated with impervious material and dispersed in a graphite matrix. The consequent increase in heat transfer area reduces the temperature difference between fuel and coolant, while the elimination of metallic cladding removes a further limitation on fuel rating. The whole of the core graphite is usually incorporated in the fuel elements and is removed at fuel discharge, thus removing the restriction which would be imposed by fast neutron irradiation of the graphite. The adoption of helium coolant eliminates the problem of chemical interaction.

The success of the HTGR is crucially dependent on the provision of impervious fuel coatings capable of withstanding high-temperature conditions throughout the irradiation life of the fuel elements. The early development work on the HTGR has involved extensive fuel testing in helium-cooled high-temperature reactor experiments. The reactors concerned were the Gulf General Atomic 40-MWe plant at Peach Bottom, Pennsylvania, the OECD 20-MWt Dragon reactor at Winfrith Heath, Dorset, England, and the AVR pebble-bed reactor at Jülich, West Germany. The design of coated-particle fuels, and the encouraging results of fuel testing in these reactors, have been discussed in Chapter 5.

As an illustration of the range of graphite-moderated reactor designs, the following sections describe the magnox reactor at Wylfa Head, North Wales, the advanced gas-cooled reactor station at Hartlepool and the Fort St. Vrain design of high-temperature reactor. As a general indication of the advances in design in the gas-cooled reactor field, Table 8.1 lists some of the main parameters of the three types for comparison purposes.

8.2. The Magnox Natural Uranium Graphite-Moderated Reactor

The general design features of the magnox line of reactors will be illustrated by describing the last of the series to be constructed in the United Kingdom, the twin-reactor 1180-MWe station at Wylfa Head, North Wales. The Wylfa station commenced generation in January 1971.

The early magnox reactors were enclosed by large spherical steel pressure vessels, connected by ducts to external boilers. The construction of a steel vessel of the size needed to enclose the large core of a magnox reactor presents a considerable engineering problem and the scale of the later stations was such as to approach the practical limit in steel pressure vessel construction. A major advance was achieved with the introduction of the concrete pressure vessel, where the whole pressure circuit, including core, circulators, and boilers, was entirely contained within a prestressed concrete vessel. The high integrity offered by the massive concrete vessel, and in particular its resistance to any mode of catastrophic failure, have eased siting requirements to the extent that the later AGR stations, which also are equipped with prestressed concrete pressure vessels, are being located close to

Table 8.1. Comparison of Operating Parameters for Typical Magnox, AGR, and HTGR Designs

Parameter	Magnox (Wylfa)	AGR (Hartlepool)	HTGR (Fort St. Vrain)
Output (MW)	590	625	330
Core diameter (m)	17.4	9.3	5.95
Core height (m)	9.2	8.2	4.75
Fuel loading	595 tonne natural U	120 tonne 2.3% U-235	0.87 tonne U-235, 19.5 tonne Th (initial loading)
Core outlet temperature (°C)	414	648	785
Average power density (MW m^{-3})	0.86	3.4	6.3

areas of high population density. While the concept offers considerable advantages from the viewpoint of compactness, economy, and safety, much effort has been required to overcome the problem of protecting the concrete from the possibility of exposure to the hot coolant gas. This has involved the development of water-cooled steel liners on the inside of the pressure vessel and in the penetrations for the circulators and boilers.

The layout of the reactor circuit for Wylfa is shown in Fig. 8.1. The spherical internal surface of the concrete pressure vessel in 29.3 m in diameter and the minimum concrete thickness is 3.3 m. The external profile is in the form of a succession of stepped cylinders. The prestressing is applied by three sets of tendons. One consists of external hoop tendons, anchored to 16 vertical ribs on the external walls of the vessel. A second set runs in vertical planes through the side walls, and the third set runs across the upper and lower caps of the vessel. The four single-stage axial flow gas circulators are mounted in penetrations in the walls of the pressure vessel, discharging into a common plenum chamber and thence through an inlet annulus to the underside of the core. After passing through the core the CO_2 gas, at 414°C, flows via the outlet annulus to the boilers, which are situated just outside the side shield of the reactor. The circuit pressure is 400 psi (28 kg cm^{-2}).

The need to keep the pressure vessel as small as possible puts a premium on compactness of design for the steam-generating plant. This leads to the adoption of a single-pressure steam cycle with a compact once-through boiler, in which the economizer, boiler, and superheater are arranged in one continuous length, without gaps. An advantage of the design is the avoidance of the additional penetrations of the vessel that would be required for the

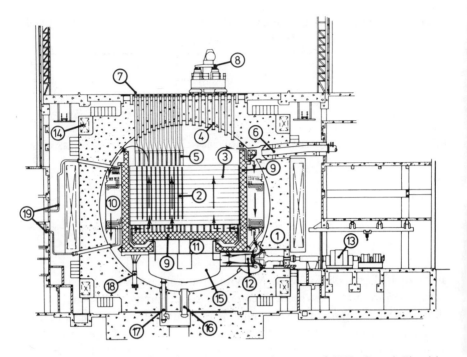

Fig. 8.1. Sectional view of Wylfa magnox reactor (courtesy of U.K. Central Electricity Generating Board). 1, Reactor pressure vessel; 2, fuel elements; 3, graphite moderator; 4, charge standpipes; 5, guide tube assemblies; 6, safety relief valve penetration; 7, pile cap; 8, charge machine on transporter; 9, neutron shield; 10, boiler; 11, radial grid; 12, gas circulator; 13, gas circulation moter drives; 14, pressure vessel prestressing cables; 15, core gas inlet plenum; 16, vessel man access; 17, CO_2 penetration; 18, structural support columns; 19, boiler steam and feed pipework.

connections in a double-pressure system. A steam-to-steam reheater is provided to achieve suitable steam dryness at the turbine exhaust.

The internal surface of the pressure vessel is protected by a steel liner of 19-mm thickness, which also serves as a gas-tight containment for the reactor. Numerous cooling water pipes are welded to the external surface of the liner for heat removal purposes. Neutron and γ shielding for the boilers is supplied by a pair of concentric steel tanks, with the interspace between them filled with steel plates and carbon blocks.

The core and its associated shielding rest on a radial grid which is supported on rollers on the pressure vessel. The natural uranium fuel elements are located in vertical channels in a structure of graphite blocks 17.4 m in diameter by 9.2 m high. The channels are arranged on a square pitch of 19.7 cm, the total number of channels being 6150. The fuel is in the form of uranium metal rods of 28-mm diameter, eight rods of just over 1 m in length being stacked vertically in each channel. The rods are canned in Al-80 magnox, the surface being fitted with so-called "herringbone" finning (see Fig. 8.2). The element surface is divided into quadrants by a series of lugs which serve to centralize the element in its channel and restrain its tendency to bow. The fins on the element surface are aligned to produce gas flow patterns in adjacent quadrants which are mirror images rather than identical to each other, thereby improving the stability of the element in the gas stream.

Wylfa, like all the magnox series, is refueled on load. The fueling machine operates on the top face of the reactor, access to the core being provided by standpipes passing through the top cap. Each of the standpipes

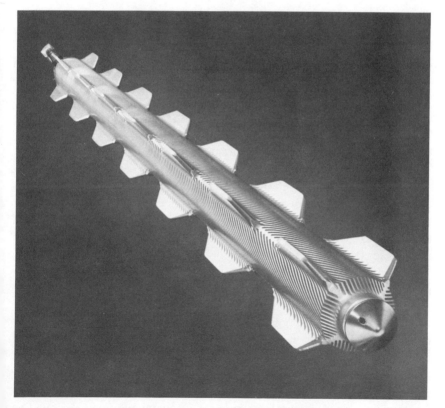

Fig. 8.2. Fuel element for magnox reactor, showing herringbone finning (courtesy of United Kingdom Atomic Energy Authority).

incorporates a rotating charge chute which enables a number of fuel channels to be serviced from that standpipe. Unlike the earlier magnox stations, Wylfa does not employ a water pond for storage of irradiated fuel elements. These are stored in vertical sealed tubes in an atmosphere of CO_2, the tubes being cooled by natural convection of air.

The core reactivity variations due to burn-up are compensated for by flattening absorbers which are loaded by the refueling machine. Sector control is provided by 32 mild steel automatic rods controlling the outlet gas temperature in each of 16 zones. In addition boron steel rods with a total reactivity worth of around 6.5% are used to compensate for the gross reactivity changes from the shutdown to the full-power condition.

Problems encountered in the operation of the magnox stations included (a) early difficulties with the complex refuelling equipment, (b) vibration of components due to gas glow, and (c) corrosion of gas circuit steel components in the carbon dioxide atmosphere. The last of these has proved the most serious, since it has led to the imposition of a reduction in power output in order to maintain the gas outlet temperature at a value where the corrosion rate is low enough to assure an economic life for the reactor. The problem is confined to small components, such as nuts and bolts made from low-silicon mild steel, and there is no danger to the more massive components of the pressure circuit.

8.3. The Advanced Gas-Cooled Reactor

As mentioned earlier, the advanced gas-cooled reactor (AGR) represents the second generation of reactors in the United Kingdom nuclear power program, which is planned to provide a total generating capacity of 8600 MWe in seven twin-reactor stations. Typical features of the planned reactors of the AGR type include the following:

i. Enriched uranium oxide fuel in long 36-pin clusters, with stainless steel cladding, with an average rating of around 12.5 MWt per tonne of uranium.

ii. Carbon dioxide coolant at about 600 psi (42 kg cm^{-2}) pressure, with maximum core outlet temperature of around 650°C, giving a thermal efficiency of over 40%.

iii. Steam at 2300 psi at 540°C, compatible with modern steam-raising plant.

iv. Prestressed concrete pressure vessel containing core, surrounding shields, boilers, and gas circulators, with internal steel dome to provide a

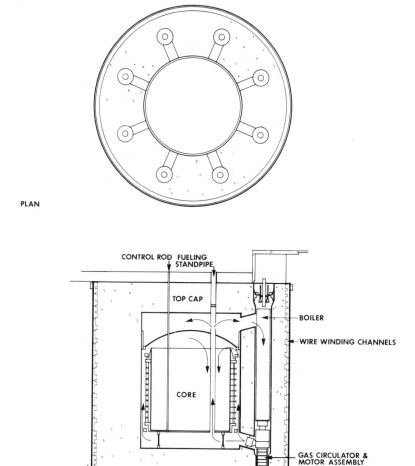

Fig. 8.3. Coolant circulation system for Hartlepool advanced gas-cooled reactor (courtesy of Nuclear Engineering International).

reentrant flow of coolant gas for maintaining the core graphite at a temperature where stored energy and dimensional changes due to irradiation are kept to a minimum.

   v. On-load refueling with a machine of simpler design than that used in the magnox stations; maximum fuel burn-up of 18,000 MW d/tonne.

   To illustrate the AGR design, the 625-MWe unit of the twin-reactor station at Hartlepool, in the northeast of England, has been chosen. One of the most interesting features of the design adopted by the contractors, the U.K. National Nuclear Corporation, is the pod-boiler concept, where the boilers are located in cylindrical cavities within the walls of the concrete pressure vessel, in contrast to the more conventional layout where both the reactor core and the boilers are grouped together within the central void.

   The arrangement of the reactor core and boilers is shown in Fig. 8.3. The pressure vessel is in the form of a cylinder, of external dimensions 96 ft (29.3 m) in height by 85 ft (25.9 m) in diameter. The eight boilers are placed in circular cavities of 9-ft (2.75-m) diameter within the 21-ft-thick walls of the pressure vessel. The cavities run the full height of the vessel and are joined by gas ducts to the main core void. The gas circulators are mounted below the boilers. The advantages of the pod-boiler design include the following:

   i. There is an increased accessibility for maintenance or possible removal of boilers compared with the case where they are enclosed in the same space as the reactor core.

   ii. A special radiation shield between core and boilers can be eliminated since this role is served by the intervening concrete of the pressure vessel wall.

   iii. Since the steam and water feed lines are taken to each boiler through the closure head at its top, no internal welds or connections have to be made inside the vessel itself, and the need for horizontal penetrations is eliminated.

   iv. The absence of ducts on the cylindrical outer faces of the vessel permits the adoption of a relatively simple form of prestressing in the form of circumferentially wound external cables.

   These advantages are offset to some extent by the increased thickness needed for the pressure vessel wall and an increase in the total area of the insulating liner which has to be fitted to protect the inner surface of the pressure vessel from the hot circulating gas.

   The core void, of diameter 43 ft (13.1 m) is surmounted by the concrete top cap, of thickness 18 ft (5.5 m). Since the refueling system is one which permits individual access to every channel, the top cap is penetrated by a large number of standpipes, making it impracticable to pass prestressing tendons through the section above the core. Prestressing is therefore accomplished by circumferential tendons located in the outer region of the cap. Axial prestressing of the main pressure vessel is by means of vertical tendons running through the walls, while circumferential prestressing is applied by bands of wire wound under tension into preformed channels in the vessel walls.

The gas circulation system is also illustrated in Fig. 8.3. The basic features of the design are set by the requirement for maintaining the core graphite at a temperature which minimizes dimensional changes and stored energy; this is achieved by allowing gas at inlet temperature to flow downwards through the core in the annular space between the graphite sleeves surrounding the fuel elements (see Fig. 8.4) and the moderator bricks which comprise the main core structure. The main circulators discharge gas at the inlet temperature of 290°C into a plenum space below the core. The gas flow then divides, about 60% flowing into the fuel channels and the remainder up the space between the outer edge of the core and the inside of the pressure vessel. In order to separate the hot gas leaving the fuel channels from the cooler gas flowing up at the periphery, a steel dome spans the space above the core, providing complete separation of the two gas streams. The peripheral flow is discharged into the space below the dome, and then

Fig. 8.4. Fuel element for advanced gas-cooled reactor. (Copyright United Kingdom Atomic Energy Authority/Courtesy British Nuclear Fuels Limited.)

provides the reentrant flow down through the core graphite. At the bottom of the core it mixes with the direct flow from the circulators and is circulated through the fuel channels. The cooling effect of the reentrant flow allows the core graphite to be maintained within the temperature range of approximately 325°–500°C, in which irradiation-induced effects are relatively small.

The flow of hot gas from each fuel channel passes into the space above the dome through a guide tube and standpipe assembly which forms an extension of the channel. The standpipes pass through gas-tight seals in the hot dome and the gas at outlet temperature is discharged into the hot-box above the dome through outlet ports in the standpipe wall. The outlet gas, at a temperature of 650°C, then flows through the top ducts into the boiler inlets. The upper surface of the dome is insulated to protect the steelwork from the gas at outlet temperature, and this, coupled with the cooling effect of the cooler gas below the dome, maintains the steel temperature at below about 370°C. Careful design of the dome, and of the circumferential skirt which connects it to the vessel liner, are essential, since a major failure of the dome would lead to loss of cooling to the core.

A major problem with concrete pressure vessels is the protection of the concrete from the hot coolant gas. Since the maximum temperature to which the concrete may be exposed is of the order of 60°C, all internal surfaces of the vessel must be lined with heat-resistant material. The main void is lined with steel, generally of thickness 3/4 in. (1.9 cm), except in the region of the penetrations in the top cap, where the thickness is increased to 1 in. (2.54 cm). The liner is tied to the concrete by closely pitched studs. Water is circulated through square-section pipes welded to the outer face of the liner in order to provide adequate cooling, and the inner face is covered with stainless steel foil or ceramic fiber insulation. Similar liners are provided for the boiler pods and their interconnecting ducts. Cooling is arranged for the fuel channel standpipes by having these of triple concentric construction, the space between the inner and intermediate tubes being filled with insulation and that between the intermediate and outer forming a cooling jacket through which water is circulated.

The boilers are of the once-through type, with a steam reheater located above the superheater of the high-pressure boiler section. The tubular heating surface is arranged in the form of finned tube multistart helical coils. The coolant gas, entering through the duct from the hot dome at 570 psi pressure and 650°C temperature is drawn through the boiler by the centrifugal-type gas circulator mounted vertically below the boiler. The pod-boiler design allows for easy installation of factory-built boiler units, the only on-site work required being the fitting of the top closure and the feed and steam connections. The steam conditions obtainable from the AGR permit the use of standard turbines, as used in modern fossil stations, thus reducing capital costs.

The core structure, comprising the fuel region, graphite reflectors and shields, is supported on pillars resting on the bottom cap of the pressure vessel. The core contains a total of 324 fuel channels on a square pitch of 457 mm. The core graphite is in the form of a column of annular graphite bricks into which the fuel elements are inserted. The lowest layer of the stack comprises the bottom neutron shield which attenuates the neutron flux at the circulator outlet ducts to a level such that activation of the circulator and boiler components is at an acceptably low level. The shield is made up of graphite blocks with steel spiral scatter plugs in the gas entry holes to the fuel channels. Above this is the core, of height 8.2 m and diameter 9.3 m, surrounded by the radial and top and bottom graphite reflectors. Finally, there is a top neutron shield, again composed of graphite, which limits activation of components above the core to a level permitting access during a shutdown.

The graphite comprising the core itself is a doubly impregnated isotropic material, specially developed to minimize dimensional changes arising from neutron irradiation. As an additional safeguard against this effect, the support structure for the core is designed in such a way that any bulk movement of the assembly is determined by the steel supports rather than the graphite itself. The graphite stack is made up of layers of bricks, each column being keyed radially to its neighbor, and the stack being linked at each layer to a peripheral support structure which is itself keyed to the vessel wall at the top and bottom of the stack. The moderator consists of columns of circular graphite bricks, each with a concentric channel to accommodate a fuel element; these circular bricks alternate with smaller bricks used to accommodate the control rods which are distributed between the fuel channels in a one-in-four array. The control rods themselves are made up of four sections, joined end-to-end with flexible joints, giving insertion characteristics superior to those of a single rod.

The standard AGR fuel cluster is a 36-pin bundle, containing hollow pellets of slightly enriched UO_2, clad in stainless steel with low-profile radial finning. The cluster is surrounded by a graphite sleeve (Fig. 8.4) to form the complete fuel element. At one stage, beryllium was considered as a possible canning material, in the interests of improved neutron economy, but the difficulty of working the material led to the choice of stainless steel as the cladding. The basic fuel element is 1041 mm in length, and eight elements are stacked end-to-end on a central tie-bar to form the complete fuel assembly which is loaded and unloaded as a unit. While on-load refueling is not as important to economic operation as it is for the natural uranium magnox system, it has been adopted for the AGRs to maintain high availability. The refueling rate is low (about three channels per week) and the fuel management scheme adopted for Hartlepool involves no requirement for routine axial or radial shuffling. Coupled with the facility of individual channel access, the result is that it is possible to operate with a relatively simple design of refueling machine. The approach to equilibrium operation will follow the reduced initial loading principle; the initial core will contain less than the full complement of elements, and the vacant positions will be filled as the core reactivity falls with burn-up.

Despite the successful operation and extensive development program which had been carried out on the prototype AGR at Windscale, which has been in operation since 1962, unexpected problems have led to appreciable delays in the implementation of the AGR installation program, particularly on the first of the scheduled stations, Dungeness B. Most of the problems encountered arose from the scaling up from the 33-MWe prototype to the full-scale power plants of 625-MWe capacity. Among the problems which arose were the following:

   i. Noise and vibration effects were found to be much more severe than in the prototype, largely owing to the increase of pressure from 20 to 43 kg cm^{-2}. The main areas affected were the circulators, the adjustable gags in the fuel element stringers, and the cover plates for the thermal insulation of the vessel walls. The latter was particularly serious, since breakaway of the metal, followed by erosion of the insulation, would have exposed the concrete to the high-temperature coolant, with a temperature far in excess of the maximum which it is capable of withstanding. All these vibration problems required extensive modifications to the components involved.

   ii. Potential corrosion problems in the boilers, both on the steam and carbon dioxide sides, led to changes in the construction materials, and also to a need for derating the units to reduce operating temperatures. The complex mixture of materials in the once-through boilers makes it difficult to optimize the temperature of operation in this region.

   iii. The correct composition of the coolant has also given concern, since the methane added to minimize graphite corrosion leads to carbon deposition on the fins of the fuel cladding, with potential corrosion. An acceptable composition for the gas seems to have been achieved, but complete confidence in the avoidance of corrosion will come only after extended operation of the new stations. A further complication has been the need to drill vertical channels in the graphite blocks in order to ensure adequate diffusion of the methane inhibitor into the bulk graphite.

The first two of the AGR stations, Hinkley Point B and Hunterston B, came on line in 1976. The success of the program will depend on the avoidance of further corrosion or vibration damage, particularly in the extremely complicated region above the reactor core, where repair work would be exceptionally difficult.

8.4. The High-Temperature Gas-Cooled Reactor

The Fort St. Vrain generating station, constructed by Gulf General Atomic at a site near Denver, Colorado, is the first full-scale commercial reactor of the high-temperature type to come into operation. Based on the experience gained with the earlier prototype 40-MWe reactor at Peach Bottom, Pennsylvania, the Fort St. Vrain station consists of a single reactor of 842-MWt capacity, with a net output of 330 MWe. The reactor attained criticality in February 1974, and commenced regular power operation in 1979. The main features of the system include a uranium–thorium fuel cycle, based on coated-particle fuel, use of graphite for the fuel cladding and for the moderator in the core and reflector, helium coolant with an outlet temperature of 770°C, once-through modular steam generators with integral superheaters, and prestressed concrete pressure vessel.

The core, steam generators, and helium circulators are enclosed within a reactor vessel of concrete reinforced by bonded reinforcement steel and prestressed by steel tendons. The top head houses a number of penetration channels which are used for refueling and as housings for the control rod drives. The walls of the vessel have an internal water-cooled liner of carbon steel. The liner forms a gas-tight seal and acts as the primary containment for the reactor, while the secondary containment is provided by the concrete vessel itself. All penetrations of the walls have two independent closures to maintain the principle of the double containment.

The arrangement of the primary circuit of the reactor is illustrated in Fig. 8.5. The cooling system is divided into two loops, each of which has a six-module steam generator and two steam-driven helium circulators. The gas from all four circulators discharges into a plenum underneath the core support floor. The full flow then passes up the outside of the core to the core inlet plenum above the core. It then flows downward through the core, where it is heated to a temperature of 780°C, to the steam generators, to produce superheated and reheated steam. The steam turbines of the circulators are

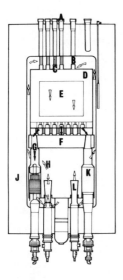

Fig. 8.5. Coolant circulation system for Fort St. Vrain high-temperature gas-cooled reactor (courtesy of General Atomic Company and Nuclear Engineering International). A, Fueling penetrations; B, control rods; C, orifice valves; D, reflector; E, reactor core; F, core floor; G, hot helium; H, cold helium; J, pressurized concrete reactor vessel; K, steam generator; L, helium circulator.

driven by cold reheat steam from the main turbine. Each of the steam generator modules incorporates reheater, superheater, and evaporator-economizer sections.

The fuel region of the core is roughly a cylinder of height 4.75 m and diameter 6 m (see Fig. 8.6). This is surrounded by graphite reflectors of thickness 1 m at the core top and 1.2 m at bottom and sides. The core contains a total of 247 vertical columns of fuel, each column consisting of six individual elements stacked on top of each other. The elements are hexagonal in section, with a width across flats of 0.36 m and a length of 0.79 m. For refueling purposes, the core is divided into regions, which, apart from a few at the core edge, contain seven columns of elements. The seven columns in a refueling region rest on a single hexagonal block, as shown in Fig. 8.7. The blocks are supported on the steel-lined and water-cooled concrete core support floor. The whole graphite structure is surrounded by a steel core barrel which provides lateral support for the fuel and reflector columns.

The high fuel rating of the HTGR design makes it impractical to have a permanent structure of graphite in the core and inner reflector regions. The part of the radial reflector immediately adjacent to the core consists of removable blocks which are replaced when the associated fuel region is refueled; the axial reflector is similarly replaced along with the fuel column. Since all the moderator in the active core region forms part of the fuel elements themselves, the core moderator is automatically replaced after the same exposure as the fuel.

The fuel is disposed in the form of rods in 210 vertical channels running through each hexagonal graphite block. The element also incorporates 108 vertical channels for the coolant to pass through, the coolant channels in each element being aligned with those in the elements above and below. The fuel and fertile material is in the form of coated particles distributed in a coke filler. The structure of the coated particle has been illustrated in Fig. 5.4. The kernel of the fuel particle contains a mixture of uranium and thorium dicarbides, while the fertile kernel contains only thorium dicarbide. Each type has a four-layer coating. The innermost layer is a porous pyrolytic carbon which acts to absorb fission recoils and provides space for the buildup of gaseous fission products. The next is of high-density pyrolytic carbon, followed by a layer of silicon carbide, which is highly impervious to volatile solid fission products, such as strontium and cesium. A final layer of isotropic

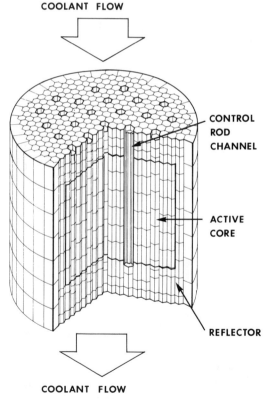

Fig. 8.6. Core layout for Fort St. Vrain high-temperature gas-cooled reactor (courtesy of General Atomic Company).

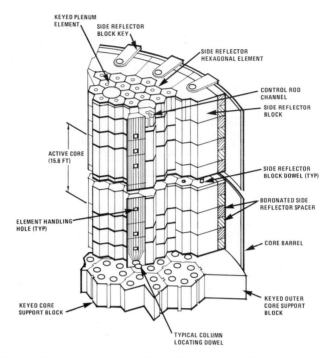

Fig. 8.7. Structural arrangement for graphite in Fort St. Vrain high-temperature gas-cooled reactor (courtesy of General Atomic Company and Nuclear Engineering International).

pyrolytic carbon is added to increase the strength of the composite coating, and to protect the silicon carbide from external chemical attack. The fertile particle is typically about twice the diameter of the fissile. Burnable poison in the form of boron carbide may be included either within the fuel holes or in separate channels within the fuel element, in order to compensate for reactivity loss due to fuel depletion or fission product buildup.

The central column of each of the seven-element fuel regions has three cylindrical holes running through the top reflector and core region. Two of these channels are occupied by control rods containing boron carbide, which are driven by motors located in the refueling penetrations in the top cap of the pressure vessel. The third channel is available as a receptacle for boron carbide spheres which can be released to fall into the core under gravity as an emergency shutdown system. The reactor has a total of 74 control rods, arranged to move in pairs.

Refueling is off-load, one sixth of the fuel in the core being replaced at a time. For St. Vrain will be operated initially without recycle of the U-233 bred in the fertile particles, which will be stored for future use when the necessary facilities become available. The average burn-up is planned to be 100,000 MW d/tonne (corresponding to a 20% burn-up of heavy metal atoms in the fuel particles). Tests carried out in the Peach Bottom and other reactors have given confidence in the stability of the coated-particle fuel to withstand burn-up of this order.

The core is divided into four radial zones, of different uranium loadings, in order to flatten the power distribution; similar zoning of the fertile material minimizes the variation of radial form factor with burn-up. The fuel loading is also divided into two axial zones, with the higher concentration in the upper part of the core in order to shift the peak rating towards the cooler inlet gas. Since, for the equilibrium cycle, the ages of the fuel elements in adjacent refueling regions are different, there are appreciable discontinuities in the power distribution at the region boundaries, increasing the overall radial peaking factor.

Important safety features of the Fort St. Vrain design include coated-particle fuel of proven integrity in normal operation and affording good resistance to temperature transients and to chemical attack in the event of accidental ingress of air or water vapor; double containment provided by the prestressed concrete pressure vessel and its steel liner; slow transient response due to large thermal capacity of core structure with distributed fuel; negative prompt fuel temperature coefficient due to Doppler broadening of the thorium resonances; and normal control rod scram action supplemented by reserve system of boron carbide spheres.

It should be noted that Fort St. Vrain represents a movement away from the original highly homogeneous concept for HTGR fuel, where the fuel rods contained the whole of the core graphite, to a more heterogeneous design, where the fuel is concentrated into a smaller portion of a graphite block which also incorporates the cooling channels. This change facilitates fuel loading by giving a mechanically stronger fuel element and has the additional advantage that the bulk of the graphite is in a lower temperature range where the effects of fast neutron irradiation are less serious. On the other hand, the reversion to a higher concentration of fissile material in the fueled region itself has the effect of increasing the heat flux and temperature at the fuel location. The economic advantage of a high loading of heavy metal atoms has led to the development of coated particles with a kernel diameter as large as 800 μm.

An interesting variant of the HTGR design is the pebble-bed concept, first demonstrated in the 15-MWe AVR reactor built by Brown Boveri/Krupp at Jülich, West Germany. The fuel is a mixture of U^{235} and thorium, incorporated as coated particles in fuel elements in the form of 6-cm-diameter graphite spheres, which are contained in a large reactor vessel, through which they are continuously circulated at a slow rate. The advantages of the pebble-bed design include a simple core structure, without the requirements of rigidity inherent in the conventional design, and unaffected by the usual problems of thermal expansion and irradiation-induced dimensional changes. The typical pebble-bed fuel element results in the fuel being intimately mixed with the majority of moderator in the core, giving moderate temperature gradients and low thermal stresses in the fuel element. The continuous fuel circulation guarantees uniform burn-up of fuel and makes it possible to run with a relatively low excess reactivity. The prismatic type used in Fort St. Vrain, however, has the advantage of permitting the adoption of a multizone enrichment scheme. One disadvantage of the pebble-bed reactor is the requirement that the control rods be driven directly into the pebble bed, thereby involving complicated and expensive drive mechanisms.

Following the successful operation of the AVR, which went to full power in 1968 and has been operating with a low fission product concentration in the coolant gas at an outlet temperature of 850°C for some years, work is proceeding on a 300-MWe prototype plant known as the THTR (thorium high-temperature reactor). The fuel region of this reactor will contain a total of 675,000 spherical elements of diameter 6 cm. Within a 0.5-cm-thick outer skin, each element contains about 33,000 particles, of kernel diameter 400 μm, in a graphite matrix. The spheres are contained in a core region of diameter 5.6 m and height 6 m, through which they gradually circulate under gravity, passing out through a withdrawal tube at the bottom of the core. They then run through a sorting system which removes any damaged elements and measures the sphere burn-up, returning those selected for recirculation to the top of the reactor vessel through pneumatic tubes. Normal control is by rods situated in the side graphite reflector, while shutdown is by rods driven into the core, backed up by injection of boron trifluoride. The power density is around 5 MW m^{-3} and the helium temperature at the steam generator inlet is 750°C. The helium is at a pressure of 40 kg cm^{-2} within a prestressed concrete pressure vessel.

The high operating temperature of the HTGR makes it particularly suitable as a source of process heat for a variety of industrial applications. Among the possible uses are the production of liquid and gaseous hydrocarbon fuels from solid fossil fuels, such as coal and lignite. On a longer time scale, there is interest in the production of hydrogen from water, where the HTGR offers better conversion efficiency than can be obtained with current electrolytic techniques, with the advantage that the production method is independent of the cost of hydrocarbon sources. Interest is also being shown in the use of HTGRs in conjunction with steel production. In this case the reactor would serve the dual purpose of providing the power required for an electric arc furnace and also the hydrogen needed for reduction of the ore, using a steam–methane reforming method.

While present designs of HTGR already offer good thermal efficiency, with steam conditions comparable with those of modern fossil-fueled stations, the likelihood of achieving much higher helium outlet temperatures, in excess of 1000°C, raises the prospect of direct cycle operation with a gas turbine. This mode of operation not only reduces capital cost because of system simplification, but also offers the prospect of a considerable increase in power conversion efficiency. The fact that waste heat is rejected at a relatively high temperature makes dry cooling economically feasible, with a resulting flexibility in plant location. The corrosion problems that arise when the gas turbine is used in conventional power systems are avoided in this case since inert helium is the working fluid.

The advantages of the HTGR may be summarized as follows: high thermal efficiency, leading to reduced heat rejection to the environment; high temperature, making possible the use of dry cooling towers and enabling the reactor to be employed for process heat applications; low activity release rate in normal operation and large safety factor associated with the prestressed concrete pressure vessel, making the reactor suitable for urban siting. The safety of the system is based largely on inherent features such as the slow response to loss of coolant, due to the large heat capacity of the graphite core. The use of a single-phase coolant avoids the problems associated with change of phase under accident conditions in a liquid-cooled reactor such as the PWR. Until large-scale experience of the operation of commercial HTGRs has been accumulated, however, it is difficult to forecast to what extent these advantages will compensate for unknowns such as the costs associated with the fuel cycle.

9

The Light-Water-Moderated Reactor

9.1. General Characteristics of the Light-Water-Moderated Reactor

Much of the early development work on power reactors concentrated, of necessity, on the use of natural uranium. This demanded the selection of high-grade moderator, of low-neutron cross section, either graphite, as in the United Kingdom magnox system, or heavy water, as in the Canadian CANDU reactor. The United States, on the other hand, had available from an early date adequate supplies of enriched uranium from the diffusion plants constructed for military purposes. The high specific fission cross section of the enriched fuel enabled an adequate thermal utilization to be achieved even when a relatively low-grade moderator of high absorption cross section, such as ordinary water, was used. Having chosen water as moderator on account of its excellent moderation properties and ready availability, it was natural to employ it also as the coolant, because of its good heat transfer properties. Reactor development in the United States over the past two decades has been largely devoted to the water-moderated, water-cooled type of reactor, and the last few years have seen an increasing installation of this type in other countries. As of 1980, the 142 light water reactors in operation were producing a total annual output of 109 GWe, out of a total world nuclear generating capacity of about 130 GWe.

Owing to the ample U.S. reserves of alternative fuels, the rate of installation of commercial nuclear plants in that country was lower initially than in the United Kingdom. The first type of light water reactor to be utilized for power production was the pressurized water reactor (PWR), which had its origin in the nuclear submarine propulsion program. The first commercial plant of this design was the Shippingport reactor, which commenced full-power operation in December 1957 at its full output of 141-MWe net power. The basic design feature of the PWR is that the reactor pressure is maintained at a value high enough to prevent boiling of the coolant taking place at the temperature of operation. The production of steam takes place in a secondary circuit, necessitating the provision of heat

exchangers and associated ancillary equipment. The reason for the pro-
hibition of in-core boiling of the coolant was the desire to avoid un-
predictable changes in heat transfer efficiency and sudden reactivity variations due to changes in the voidage fraction in the core. At a later date,
however, experiments carried out in the BORAX-I test reactor in Idaho
showed that it was in fact possible to operate perfectly safely with boiling
taking place in the core. This led to the development of a second line of light
water reactor, the *boiling water reactor* (BWR), where the pressure is lower
than in the PWR and saturated steam is produced in the reactor core. This
simplification makes possible the use of a direct coolant cycle, the steam
generated passing directly to the turbine. The first commercial plant of the
BWR type was the 210-MWe Dresden-I, which commenced operation in
August 1960. The distinction between the two types of light water reactor has
been illustrated in Fig. 7.1.

Certain features are common to both types of light water reactor. The
principal common features include the following:

i. The core, which contains enriched fuel, is immersed in water, which
acts as moderator and is circulated past the fuel elements for heat removal
purposes. For all except a few relatively low-power reactors, forced
circulation is necessary to give an adequate rate of heat removal.

ii. The high effectiveness of water as a moderator, together with its good
specific heat transfer characteristics as compared with a gas coolant, leads to
a relatively compact core. The typical power density for a modern BWR, for
example, is of the order of 55 kW l^{-1} and for a PWR is around 100 kW l^{-1}
compared to about 2.7 kW l^{-1} for an AGR. The fraction of the station
output which has to be fed back to pump the coolant round the primary
circuit is also considerably less for the two water-moderated reactors.

iii. The high pressure needed to produce steam at a temperature great
enough to give a reasonable thermal efficiency (for the BWR) or to suppress
core boiling (for the PWR) means that the core has to be enclosed in a strong
pressure vessel. The PWR typically operates at a pressure of about 2250 psi
(157 kg cm^{-2}) with a coolant temperature of 320°C, while the corresponding pressure and temperature for the BWR are 1040 psi (73 kg cm^{-2}) and
285°C, respectively.

iv. Water at the temperature used is highly corrosive, and although
there is ample experience of the corrosion problem from conventional
stations operating at similar temperatures, little of this is relevant to the very
limited range of materials which have sufficiently low neutron absorption
cross sections to be acceptable for use in the core region of a reactor. In
addition to the adverse effects of corrosion on the reactor materials

themselves, it is important to minimize the carry-over of activated corrosion products to areas of the primary circuit such as the heat exchangers or even, in the case of the BWR, to the turbine itself.

v. Compared to a fossil-fueled boiler, the steam produced is at relatively low pressure and temperature and, unless superheating is employed, the thermal efficiency is limited to a value of around 32%.

vi. A major breach of the primary circuit would lead to a very rapid loss of coolant. To prevent damage to the core, a highly reliable emergency cooling system must be provided to remove the residual heat.

One of the main distinctions between the two reactor types is the simplification of the coolant circuit made possible in the BWR by the use of a direct cycle. The lower pressure is also an advantage in terms of the pressure vessel and containment requirements. The existence of in-core boiling improves the natural convection, so that the pumping requirements are much less for the BWR than for the PWR. On the other hand the use of a boiling heat transfer regime requires the maintenance of a power operating margin sufficient to eliminate the possibility of fuel element damage due to dryout at the fuel surface. Another disadvantage of the BWR is the higher radiation level at the turbine and other external plant due to carry-over of activity from the core. The control system of the BWR has to be designed to deal with the reactivity variations associated with changes in the core voidage, but this effect has the advantage that it can be utilized to provide the plant with automatic load-following characteristics. It should be noted that the distinction between the BWR and the PWR is not today as clear as it was in the original designs, since some degree of boiling is now permitted in PWRs.

The BWR can be operated on either a direct or a dual coolant cycle. In the first, all the steam generation takes place within the reactor vessel, the steam being fed to the turbine and the condensate returned to the reactor. In its simple form the direct-cycle BWR is not inherently load following, since an increase in steam demand lowers the reactor pressure, with a consequent increase in core voidage, leading to a fall in reactivity, rather than the desired rise. The dual-cycle variant is one method of providing automatic load following. In this design the steam supply to the turbine is only derived in part from the direct boiling in the core. In addition, hot water from the steam drum is subcooled, before being returned to the core, by passing through a heat exchanger where supplementary steam is generated, this steam being fed to a low-pressure stage of the turbine (Fig. 9.1). The turbine governor is used to regulate the flow of this secondary steam into the turbine. In this case, increase of turbine steam demand results in a greater subcooling of the inlet water to the core and hence to a reduction in voidage and a consequent

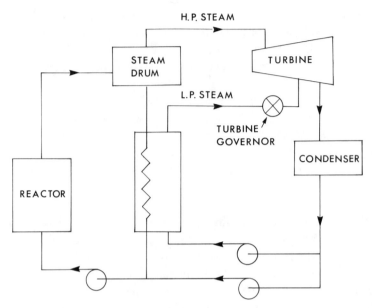

Fig. 9.1. Cooling circuit for dual-cycle boiling water reactor.

increase of reactivity. The automatic load following is obtained at the expense of an increased circuit complexity, and in some cases it may be preferable instead to modify the direct-cycle BWR for load following by arranging that an increased demand from the turbine leads to an increase in the recirculation rate through the core, with consequent reduction of voidage.

The low thermal efficiency associated with the modest temperature of the saturated steam produced in both BWRs and PWRs, coupled with the enhanced turbine corrosion rate due to the high moisture content, has led to attempts to superheat the reactor steam. The superheating may be produced either externally, by an auxiliary source which may or may not be nuclear, or in special superheating channels within the same reactor (integral superheat). A typical example of the latter technique was the Puerto Rican Boiling Nuclear Superheat Reactor (BONUS), in which the core was divided into two regions, the saturated steam being produced in the central zone and then passed through an outer zone for superheat purposes. The use of integral superheat presents a number of problems. These include the following:

i. The correct ratio of power production between the boiling and superheat channels must be maintained. Owing to the lower heat transfer capacity of the steam in the superheat region as compared with the water in the other channels, the power density there must be considerably lower than in the boiling region.

ii. The steam used for heat transfer from the superheater elements must, of course, be physically isolated from the water coolant in the rest of the core. Careful design is required to minimize heat losses from the superheat channels to the water and also to ensure that no sudden increase of reactivity can occur due to a sudden entry of water into these channels. Typically, the achievement of these objectives involves the use of a double-walled steam tube with an insulating interspace. Designs involving annular fuel element geometry have also been suggested.

iii. The more rigorous corrosion conditions imposed by the high temperature of the steam in the superheat channels require the use of more resistant materials, such as stainless steel, for the coolant tubes and fuel cladding in these channels, with a consequent reactivity penalty.

In general, there has been less emphasis in recent years on the use of superheat, because of the problems outlined above. It is possible that, with increasing incentive for higher thermal efficiency for environmental reasons, there may be a renewal of interest in its introduction.

9.2. Development of the Light Water Reactor

In comparison with the first light water plants, Shippingport (141 MWe) and Dresden-I (210 MWe), current unit sizes have advanced to around 1200 MWe. Even with the greater compactness achieved by such advances as higher power density, the pressure vessel for a 1200-MWe BWR is very large; typically the diameter is more than 6 m and the height of the order of 22 m, with a vessel thickness of 15–18 cm. The possibility of using prestressed concrete pressure vessels, of the type currently in use for gas-cooled reactors, has been considered, but the possible advantages of the scaling-up of power output have to be balanced against the potential difficulties of a new technology and the advantages of standardizing on the current 1200 MW design.

The standard fuel for both BWRs and PWRs is now UO_2 pellets clad in zircaloy and arranged in a square array. Stainless steel cladding, which had been adopted in the earlier LWRs, was abandoned on the basis of its undesirably high neutron absorption and because of an unacceptable stress corrosion rate. A recent trend has been towards a greater number of fuel pins in the element, in order to reduce the linear rating for a given power output from the cluster. The larger PWR sizes now have as the standard fuel pin arrangement a 17×17 array, while the BWR uses an 8×8 pattern. A modern PWR has a typical power density of $100 \, \text{kW} \, \text{l}^{-1}$, with a maximum linear rating of $44 \, \text{kW} \, \text{m}^{-1}$, while the BWR has a similar linear rating with a power density of around $56 \, \text{kW} \, \text{l}^{-1}$. The potential benefit of greater subdivision of the fuel may be realized either in higher per-channel ratings or in increased margins of safety, particularly in the event of a loss of coolant accident; these advantages have, of course, to be balanced against the increased fabrication costs of the more complicated fuel assemblies.

The fuel pins themselves are now full length rather than segmented, prepressurized with helium to minimize the compressive stresses in the cladding and reduce creep induced by the coolant pressure. Allowance for the buildup of gaseous fission products takes the form of an end plenum. As a result of the improvements in fuel element design, the standard burn-up is 33,000 MW d/tonne for the PWR and 27,000 MW d/tonne for the BWR. The limit to the desirable burn-up level is now set by economic rather than material limitation considerations.

A similar development has taken place in the structural design of the fuel cluster. The original PWR element, for example, incorporated small stainless steel ferrules, brazed together for rigidity, to support and space the fuel pins. This method had the disadvantage that the brazing process annealed the cladding to the extent that a relatively thick cladding was necessary to ensure adequate mechanical strength. This drawback was overcome in later designs by the replacement of the brazed spacers by spring grid assemblies, in which the fuel pins were free to expand axially through grid cells consisting of cantilever spring assemblies, thus allowing for differential thermal expansion. The grid restraining force exerted by the grids on the fuel rods is adjusted so as to minimize fretting, but at the same time to avoid overstressing the cladding at the points of contact.

The early PWRs made use of cruciform control rods of the same basic design as is still employed in the BWR. A significant improvement came with the introduction of rod cluster control (RCC), where the control poison is distributed in the form of small rods which are inserted in thimbles distributed throughout the fuel assemblies (Fig. 9.4). Apart from achieving a greater unit effectiveness of the control poison, the RCC concept eliminates the problem of flux peaking which occurs when the cruciform type of control rod is withdrawn, leaving a water gap behind it. The sizeable negative temperature coefficient of both the PWR and BWR, arising from the reduction of moderator density with temperature, means that a considerable control capacity must be built in to deal with the power coefficient in addition to the compensation required for burn-up and fission product poisoning. The control rod requirement has been eased by the increasing use of burnable poison and chemical shim, the latter using boron dissolved in the coolant.

While the burn-up experience with LWR fuel has on the whole been encouraging, problems have arisen in some PWR fuel owing to the phenomenon of densification of the fuel pellets (see Chapter 5). This effect can lead to a partial collapse of the fuel cladding, which was first observed in the Ginna PWR in 1972. Radiation-enhanced sintering leads to fuel pellet densification and shrinkage, which in turn, when coupled with cladding creep under coolant pressure, can lead to pellet hang-up, leaving interpellet gaps in the fuel column. Gradual cladding collapse can then follow in the region of the gaps. The gradual nature of the collapse, however, makes it unlikely that actual failure of the cladding will result. The problem, which has been confined to unpressurized fuel pins in PWRs, which are particularly susceptible owing to the lower fuel density and higher fast fluxes as compared with BWRs or the SGHWR, can be dealt with by such expedients as increasing the initial fuel density and achieving a more uniform distribution of porosity and a more complete sintering during the production of the fuel.

One of the areas where the most marked development has taken place is in the design of the containment and emergency cooling systems for the light water reactor. The presence of a water coolant under high pressure leads to the requirement for a containment structure capable of withstanding the pressure surge that would occur in the event of a major breach of the primary circuit, which would cause immediate vaporization of the coolant. Early designs for light water reactors, such as Dresden-I, incorporated a concrete biological shield around the reactor vessel and a large steel containment sphere enclosing the vessel, the external steam drum, steam generators, and primary coolant pumps. The outer containment sphere, which has a diameter of some 57 m and a thickness of 3–3.5 cm, will withstand the pressure transient caused by a complete breach of the primary circuit and prevent any serious subsequent release of radioactivity to the atmosphere. The problems involved in the fabrication of large containment vessels of this type have led to a search for more compact containment methods involving means of pressure suppression by condensation of the escaping steam by water or ice. Some idea of the developments in BWR design which have led to a considerable gain in compactness and containment efficiency may be gained by comparing the Dresden-I layout with a relatively recent plant such as Grand Gulf (see Fig. 9.9).

It should also be noted that important advances have been made in simplifying the primary circuit of the reactor itself. The experience with

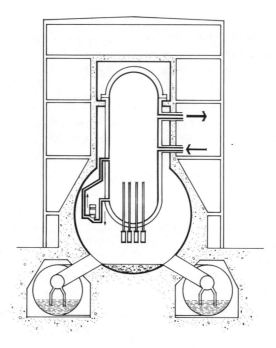

Fig. 9.2. Bulb-and-torus design of BWR containment.

Dresden-I showed that it was possible to operate with a steam quality high enough that external risers and an external steam drum were no longer necessary. Modern BWRs incorporate steam separators and driers mounted above the core inside the pressure vessel itself. A further advance is the adoption of internal jet pumps, mounted on the inside of the pressure vessel, for circulation of the coolant through the core. These are driven by external loop pumps, but the fraction of the core recirculation flow which is required to drive the jet pumps is only about one third of the whole, leading to a reduction in the number and size of the external loops, with a consequent reduction in the severity of the potential effects of breaches of the primary circuit.

The shielding and containment for a modern BWR typically comprises a steel or concrete inner containment (the drywell) connected by underwater ducts to a suppression pool full of water, the purpose of which is to condense the escaping steam in the event of a circuit breach. Two layouts which have been adopted are the bulb-and-torus design (Fig. 9.2) and the more recent weir wall construction described in detail later for the Grand Gulf BWR. The standard PWR containment consists of a concrete structure with either a steel liner or a separate inner steel shell, with pressure suppression being provided by either cold water sprays or an ice condenser.

An aspect of water reactor technology which has received considerable attention in recent years is the emergency core cooling system (ECCS). Although the control circuits are arranged to shut the reactor down immediately following a loss-of-coolant accident, the residual heating of the fuel rods due to the energy release from their radioactive content requires that they be supplied with a reliable source of cooling for extended periods after the accident if excessive fuel temperatures, possibly leading to melting of the core, are to be avoided. The emphasis on preventing major damage of this kind has led to the installation of multiple spray cooling and emergency circulation systems. The discussion of the effectiveness of these in preventing the possibility of major releases of activity is deferred to Chapter 12.

9.3. The Pressurized Water Reactor

As an illustration of the design aspects of a modern pressurized water reactor, we may consider the two-unit Sequoyah plant of total net capacity 2280 MWe, which is being built by Westinghouse at a site near Chattanooga, Tennessee. Full power operation of the first of the two reactors commenced in 1981.

The core is contained within a cylindrical pressure vessel with a hemispherical bottom head, the total height of the vessel being 12.6 m and its diameter 4.4 m (Fig. 9.3). The cylindrical removable upper head of the vessel contains the control rod drive mechanisms. The core consists of 193 square-section fuel elements which are supported between upper and lower grid plates and surrounded by a stainless steel shroud. The water entering the pressure vessel through the four inlet nozzles flows downward in the annulus between the core shroud and the vessel wall. It then flows upward through the core and out through one of the four outlet nozzles to the steam generators. The external coolant circuit consists of four loops, each with its steam generator and circulator, the latter being of the vertical, single-stage, shaft seal design (see Fig. 12.3). A pressurizer is connected to one of the loops to maintain the circuit pressure at 2250 psi (157 kg cm^{-2}).

The steam generators are divided into two sections, the evaporator below and the steam drum above. The evaporator consists of a U-tube heat exchanger. Water of the primary coolant, at a temperature of 321°C, flows through the tubes, which are made from Inconel. The flow in the secondary circuit is upward through the tube bundle, the steam–water mixture passing

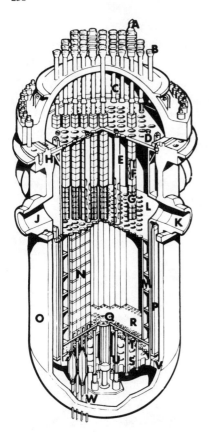

Fig. 9.3. Sectional view of the Sequoyah pressurized water reactor (courtesy of Nuclear Engineering International). A, Control rod drive head adaptors; B, instrumentation ports; C, thermal sleeves; D, upper support plate; E, support column; F, control rod drive shaft; G, control rod guide tube; H, internals support ledge; J, inlet nozzle; K, outlet nozzle; L, upper core plate; M, baffle and former; N, fuel assemblies; O, reactor vessel; P, thermal shield; Q, access port; R, lower core plate; S, core support; T, diffuser plate; U, lower support column; V, radial supports; W, instrumentation thimble guides.

to centrifugal moisture separators and steam driers mounted in the top section of the steam generator shell before leaving for the turbine. The flow in the secondary side of the steam generator is by thermal convection. The steam leaving the generator is at a temperature of 291°C, at a pressure of 60 kg cm^{-2}, and has a quality of at least 99.75%.

The single pressurizer serves all four coolant loops. It serves to hold the primary circuit pressure constant and to allow for expansion and contraction of the coolant volume caused by changes of load. An electric immersion heater of 1800-kW capacity is used to raise the pressure to, and maintain it at, the required value during negative pressure surges caused by an increase in load demand. In the event of a positive surge, caused by a reduction in load demand, a spray system, fed from the cold leg of one of the reactor coolant loops, condenses steam in the pressurizer to avoid tripping the pressure relief valves with which it is fitted.

The fuel assembly for the Sequoyah reactor is shown in Fig. 9.4. The fuel rods are arranged in a 17×17 array to form a square-section element with a side dimension of 21.4 cm. The overall height of the assembly is 407 cm. Of the 289 possible rod locations, 264 are occupied by fuel pins, 24 by the rod cluster control guide sheaths and the remaining position is for in-core instrumentation. The fuel assembly is bounded at top and bottom by square nozzles which control the inlet and outlet coolant flow. The stainless steel nozzles are welded to the zircaloy guide sheaths to form a basic structural skeleton for the element, which is strengthened by a set of seven grid assemblies which are attached to the sheaths at intervals along their length, giving lateral support to the fuel pins. The grids consist of an assembly of spring clips interlocked to form an egg-crate arrangement, providing a rigid support and spacing of the fuel rods. The magnitude of the grid restraining force is high enough to minimize fretting, but at the same time low enough to allow axial expansion of the fuel assembly without imposing a restraint force sufficient to cause buckling or distortion of the pins. The grids are fabricated from Inconel 718 because of its high corrosion resistance.

The fuel is in the form of pellets of slightly enriched UO_2 contained in tubes of cold-worked zircaloy-4. The pellets, of diameter 8.2 mm, are dished at the ends to allow for the differential thermal expansion due to the temperature profile across the pellet. Following the technique pioneered by Westinghouse, the rods are pre-pressurized with helium to minimize the compressive clad stresses and creep induced by the coolant pressure. Adequate void volume is provided to take up the differential expansion between fuel and cladding and to allow for accumulation of fission products. There is no differential enrichment of fuel within an individual assembly, but different enrichments are used for the three radial regions into which the core is divided to improve the radial form factor.

The control system for the reactor incorporates the rod cluster control (RCC) concept, which has replaced the cruciform control elements used on the earlier PWRs. The control rod poison is distributed uniformly in the form of small-diameter rods which are inserted into the sheaths located within the fuel clusters. The 24-rod assembly is coupled to a drive shaft which is actuated by a drive mechanism mounted on the reactor vessel head. Compared with the large cruciform assemblies, the RCC system gives an increased reactivity worth per unit weight of absorber, coupled with a more uniform power distribution and a greatly reduced flux perturbation effect due to the water gaps which are created by movement of the control rods out of the core. The absorber material in the control rods is silver–indium–cadmium in the form of extruded rods which are sealed within stainless steel tubes.

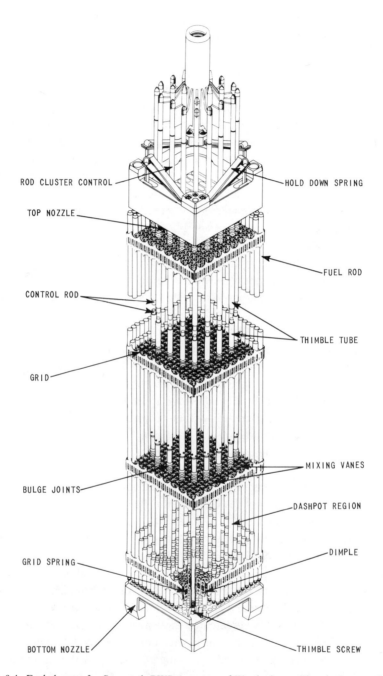

ROD CLUSTER CONTROL

HOLD DOWN SPRING

TOP NOZZLE

FUEL ROD

CONTROL ROD

THIMBLE TUBE

GRID

MIXING VANES

BULGE JOINTS

DASHPOT REGION

DIMPLE

GRID SPRING

BOTTOM NOZZLE

THIMBLE SCREW

Fig. 9.4. Fuel element for Sequoyah PWR (courtesy of Westinghouse Electric Corporation).

Chemical shim control, using boron dissolved in the coolant, is employed as a supplementary means of reactivity control, the concentration being adjusted so that normal operation is with the control rods almost completely withdrawn from the core. In addition to chemical shim, fixed burnable poison absorbers are loaded for the first cycle in order to guarantee that the moderator temperature coefficient will remain negative throughout the whole cycle. This poison is in the form of borated glass rods distributed throughout the core in vacant control rod guide thimbles.

The Sequoyah plant utilizes the ice condenser system of reactor containment. In this system, the potential energy release following on a major breach of the primary circuit is dissipated in melting a large quantity of ice stored in a number of modules surrounding the reactor. The effectiveness of this arrangement in limiting the pressure rise in the primary containment considerably eases the demands on the structural strength of the latter. The layout of the containment is shown in Fig. 9.5. The inner containment consists of a free-standing steel shell, with its base embedded in a steel-lined concrete base slab. Outside the shell is a separate reinforced concrete shield building. The ice condenser is made up of 24 modules which are arranged around the reactor, each module effectively comprising a large tube and shell heat exchanger. The ice bed is bounded on the outside by the steel containment shell and on the inside by a concrete wall. The ice is contained within wire mesh baskets of 30-cm diameter, each module of the condenser consisting of a 9×9 array of basket columns. The total height of the ice column is some 15 m. The condenser is insulated on all sides and is cooled by a continuous flow of refrigerated air.

In the event of a loss of coolant accident, the pressure rise in the lower compartment of the reactor containment opens the inlet doors in the bottom section of the condenser and all the steam–air mixture flows through the ice columns into the upper compartment. The rapid condensation of the steam in the ice condenser limits the maximum pressure in the containment to a value comfortably below the 11 pounds per square inch gauge (psig) (0.77 kg cm^{-2}) design pressure. The enhancement of the steam flow rate due to the rapid condensation process helps substantially to reduce the duration of the initial pressure transient. Following this transient, longer-term heat removal is provided by a double-containment spray system, which sprays water into the upper compartment of the reactor containment. The water from the melting ice, together with the spray water, collects in the sump of the reactor containment and is recirculated through heat exchangers. The combined effect of the ice and spray cooling enables a large long-term energy release to be accepted without overstressing the containment. The system is

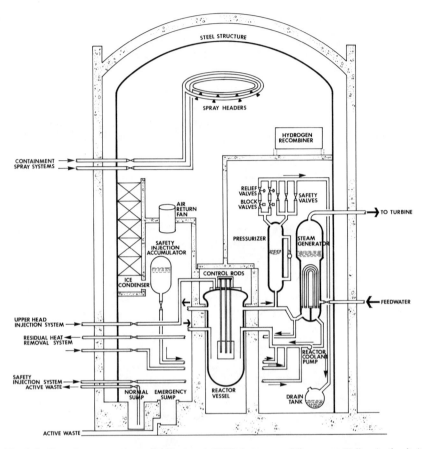

Fig. 9.5. Containment structure for Sequoyah PWR (courtesy of Tennessee Valley Authority).

backed up by an emergency safety injection system which supplies borated water to the reactor core, for prevention of excessive temperatures which would lead to clad melting and damage to the core structure.

As is typical for both PWR and BWR designs, the reactor is refueled off-load, one third of the fuel elements being replaced at a time. Refueling time is estimated at 30 days. For refueling, the upper head of the reactor vessel is removed and the refueling cavity above the vessel is flooded with water, which contains boric acid to ensure that the core remains subcritical at all times. The refueling cavity is connected by means of a fuel transfer tube to the spent fuel storage pit, which is permanently flooded. The fuel elements are removed from the core by a crane and placed on an underwater conveyer car which transfers them through the transfer tube to the storage pit.

9.4. The Boiling Water Reactor

An example of a modern boiling water reactor is the General Electric reactor of the BWR/6 type which is scheduled for construction at Grand Gulf, about 25 miles south of Vicksburg, Mississippi. The first reactor of the two-unit station is expected to achieve commercial operation in 1982 and the second in 1986. Each reactor will have a net output of 1250 MWe.

The layout of the core and pressure vessel is shown in Fig. 9.6. The core, steam separators, and driers are enclosed within a low alloy steel pressure vessel of diameter approximately 6.4 m and height 22 m, with a removable head. The wall thickness is nominally 6 in. (152 mm). The core is made up of

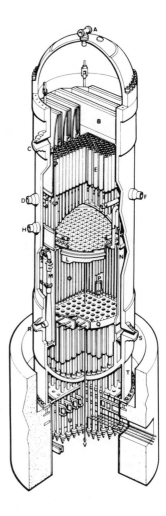

Fig. 9.6. Sectional view of the Grand Gulf boiling water reactor (courtesy of General Electric Company and Nuclear Engineering International). A, Vent and head spray; B, steam dryer; C, steam outlet; D, core spray outlet; E, steam separators; F, feedwater inlet; G, feedwater sparger; H, L.P. coolant injection inlet; J, core spray pipe; K, core spray sparger; L, top guide; M, jet pump; N, core shroud; O, fuel assemblies; P, control blade; Q, core plate; R, jet pump water inlet; S, recirculation water outlet; T, vessel support skirt; U, control rod drives; V, in-core flux monitor.

individual assemblies which rest on an orificed core support plate. A top guide plate above the core provides lateral support for the tops of the fuel elements. The core is surrounded by a stainless steel shroud, which, together with the wall of the pressure vessel itself, forms an annulus down which the recirculating coolant flows before entering the inlet plenum beneath the core support plate, whence it passes up through the core. The flow is driven by 24 jet pumps located in the annulus between the shroud and vessel wall. The driving flow for the jet pumps is provided by two external centrifugal pump loops, the flow through the loops being approximately one third of the total core flow. This system gives coolant recirculation with a minimum of external loops and avoids the use of moving parts for the components of the system within the vessel.

After passing through the core, the coolant steam–water mixture enters the bank of centrifugal steam separators mounted above the core, where the water is separated out by vortex action and flows down to join the recirculation flow through the annulus. The steam passes upwards into the steam driers in which the moisture is further reduced, and thence to the turbine. The steam leaving the core is at a temperature of 286°C, at a pressure of 1040 psi (73 kg cm^{-2}). The total thermal output from the core is 3833 MWt. Recirculation flow control is used to provide automatic load following; power changes of up to 25% of full power can be accommodated in this way.

The control rods are of the cruciform type and are driven hydraulically into the core from underneath. They occupy alternate spaces between fuel elements (Fig. 9.7). The presence of partially inserted rods helps to offset the higher power density in the lower part of the core due to the axial variation of coolant density in the BWR. The location of the rod drive mechanisms below the core has the additional advantage that the control system remains operational even when the vessel head is removed for refueling purposes. The hydraulic mechanism allows for very rapid insertion of the rods for emergency shutdown, while velocity limiters are fitted to minimize the rate of reactivity addition in the unlikely event of a rod dropping under gravity. The absorbing element in the rods is in the form of boron carbide powder in stainless steel tubes, each arm of the cruciform containing 18 of these tubes.

The core is made up of a total of 800 fuel assemblies, arranged to form an array of roughly circular cross section, as shown in Fig. 9.8. Each of the assemblies consists of an 8 × 8 square array of fuel pins, with zircaloy-2 cladding, surrounded by a square-shaped channel of zircaloy-4. The assembly has tie plates at both ends, the lower of which has a nosepiece which fits into the fuel support and distributes the coolant flow to the rods. The use

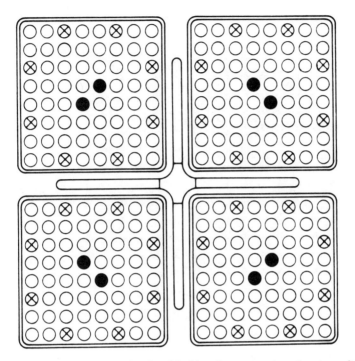

Fig. 9.7. Grand Gulf BWR: control rod and fuel bundle cross sections (courtesy of General Electric Company and Nuclear Engineering International). ○, Fuel rod; ●, water rod; ⊗, tie rod.

of separate fuel channels allows each bundle to be individually orificed, thereby permitting the flow rate in the channel to be adjusted to match the fuel rating in the bundle. The fuel is in the form of pellets of diameter 10.6 mm. Each assembly contains rods of several different enrichments in order to minimize the local variation throughout the cluster. Two of the rods in the center region of the cluster are "water rods," not containing fuel, which provide extra moderation and thereby reduce the flux depression in the middle of the assembly. Reactivity compensation for fuel burn-up is provided by burnable poison (gadolinia) added to the fuel pellets. The fuel assemblies have a common average enrichment, which for the equilibrium cycle will be in the range 2.4–2.8%.

Refueling involves the transfer of the spent fuel from a water-filled reactor well above the core to a storage pool in the adjoining fuel building. The reactor containment building is linked to the fuel building through a 0.91-m-diameter horizontal tube of length 9 m, through which the spent fuel

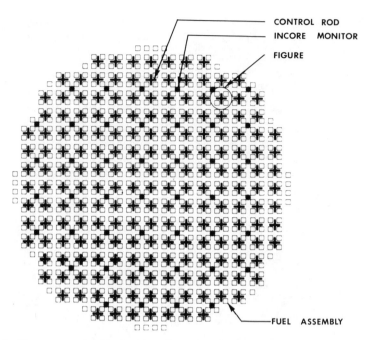

Fig. 9.8. Core cross section for Grand Gulf BWR (courtesy of General Electric Company and Nuclear Engineering International).

is transported in a special carrier. The fuel storage and transfer areas are separated from the reactor well by a water-tight gate so that the well may be drained to allow the removal of the reactor vessel head. After this has been done, the gate is opened to connect up the reactor well and the fuel transfer area, so that the whole fuel transfer operation may be carried out underwater. Time between reloadings can be varied between one and two years. For a one-year reload cycle, about one fourth of the core is discharged, the required outage time being estimated at around 30 days. If desired, the cycle time can be stretched out by a gradual reduction of the feedwater temperature at the nominal end of the core life, the resultant reduction in core voidage increasing the reactivity and allowing an extension of operation at the expense of some loss in power output.

The containment is of the pressure suppression type, in which a large volume of water is used to cause a rapid reduction of pressure by condensation of steam in the event of a loss of coolant accident. The protection against release of radioactivity into the atmosphere comprises three separate barriers: a concrete drywell around the reactor vessel, a water suppression pool, and a steel-lined prestressed concrete containment building. The drywell is connected by means of horizontal vent passages to the annular suppression pool which runs between the drywell and the steel containment vessel (see Fig. 9.9). A weir wall on the inside of the vent area of the drywell maintains a water seal for the pool. A major breach of the coolant circuit would lead to a rapid increase of pressure in the drywell which would depress the water level in the drywell–weir wall annulus and drive the steam–water mixture through the vents into the suppression pool. Complete condensation of the steam would take place, the displaced air from the drywell being accommodated within the steel-lined containment building, which forms the basic barrier against release of fission products to the atmosphere.

The reactor is equipped with a number of emergency core cooling systems. The core isolation cooling system is activated by low water level in the pressure vessel, and injects water into the vessel from the condensate storage tank, the suppression pool, or the steam condensed in the residual heat removal heat exchangers. In addition, a high-pressure core spray, again fed from the condensate storage tank or suppression pool, sprays water over

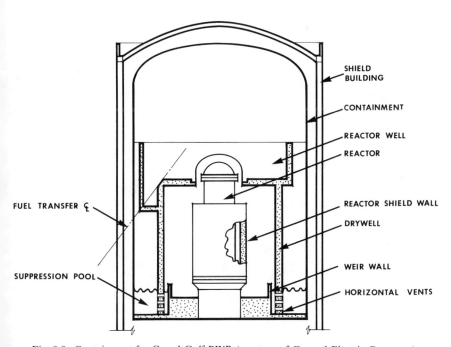

Fig. 9.9. Containment for Grand Gulf BWR (courtesy of General Electric Company).

the core. This system will operate over the full range of reactor pressures, and is supplemented by a low-pressure core spray system at low pressures. Owing to the strong natural circulation capacity of the BWR, adequate core cooling will be maintained so long as the core is completely covered with water. The large suppression pool has an energy absorption capacity of 4–5 h of decay heat, if necessary, and is backed up by a variety of long-term heat removal systems.

IO

The Heavy-Water-Moderated Reactor

10.1. General Characteristics of the Heavy Water Reactor

The incentive to use heavy water as moderator arises from the excellent neutron economy made possible by the low neutron cross sections of both deuterium and oxygen (0.5 and 0.27 mb, respectively). In combination with its slowing-down power of 0.175 cm^{-1}, the low cross section of D_2O leads to a moderating ratio of 6000, making heavy water, by a wide margin, the most desirable moderator on purely theoretical grounds. The fast and slow leakages from a heavy water reactor tend to be roughly equal, in contrast to the situation for a light water system, where the fast leakage is much the greater.

The neutron lifetime is also greatly different from that of a light water reactor. It can be as large as 50 ms, in comparison with a typical light water value of about 0.1 ms. The prompt response to a change of reactivity is therefore much slower for the heavy water system. Another phenomenon which affects the kinetic response of a heavy water reactor is the production of photoneutrons by γ radiation from fission product decay. As mentioned in Section 1.8.1, γ rays of energies in excess of 2.23 MeV can interact with deuterium to produce neutrons through the (γ, n) reaction. Since the photoneutron periods, determined by the β-decay periods of their γ-emitting precursors, are generally much longer than the delayed neutron periods, the longer-term kinetic behavior of a heavy water reactor can be considerably more sluggish than that of other types.

The excellent neutron economy associated with the D_2O moderator leads to a highly efficient use of the U^{235} content of the fuel; in comparison with a light water reactor, twice as much energy is generated per unit of U^{235} consumed, with typically some 50% of this coming from Pu^{239} generated from U^{238}. The discharged fuel also has around twice the Pu^{239} content of that discharged from a light water system.

The economics of a D_2O reactor are basically a matter of balancing the

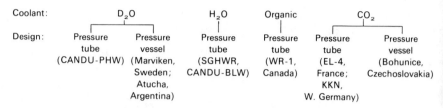

Fig. 10.1. Types of D_2O-Moderated Reactor.

better fuel utilization against the higher capital cost of the system, due in part to the relatively high cost of the moderator.

Because of its good neutron economy, the heavy water reactor is an obvious candidate for the role of thermal near-breeder or advanced converter, using the U^{233}–Th^{232} fuel cycle. The value of the cycle depends on the high thermal η value of U^{233} (approximately 2.29) and the greater absorption cross section of Th^{232} as compared with U^{238}. While the production of plutonium in a natural uranium D_2O reactor is about 2.7 g of Pu^{239} per kilogram of uranium, the equilibrium production of U^{233} is typically 16 g per kilogram of thorium.

The variety of heavy water reactors which exist may be classified on the basis of coolant (D_2O, H_2O, organic liquid, or CO_2 gas) and of basic design (pressure tube or pressure vessel). The subdivisions of reactor types constructed to date, with examples of each, are illustrated in Fig. 10.1.

The distinction between the pressure tube and pressure vessel design is shown schematically in Fig. 10.2. In the former, the fuel elements are located in individual tubes through which the high-pressure coolant is pumped; these tubes pass through a separate tank containing the D_2O moderator, which can thus be maintained at a relatively low pressure and temperature. In the pressure vessel design, the heavy water moderator is within the high-pressure containment (although it may be physically separated from the coolant by relatively thin tubes which do not have to withstand a pressure differential).

One of the main advantages of the pressure vessel design is the avoidance of the highly complicated pipework associated with the large number of individual cooling pipes of the pressure tube reactor. The main disadvantage is the requirement for a heavy pressure vessel to contain the pressurized moderator. This may well impose an economic limitation on the size of pressure vessel reactor that is practicable with natural uranium as fuel. Among the advantages of the pressure tube system are the following: flexibility in choice of a coolant other than D_2O; readier demonstration of safety in the event of pressure system rupture, arising from the subdivision of

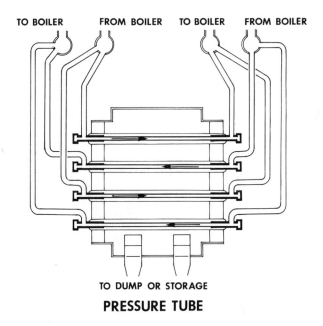

PRESSURE TUBE

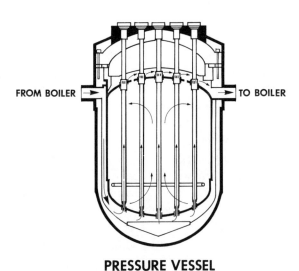

PRESSURE VESSEL

Fig. 10.2. Comparison of pressure tube and pressure vessel designs of heavy water reactor (courtesy of Atomic Energy of Canada Limited).

the pressure circuit; simpler scaling-up to larger reactor sizes on the basis of experience, since all that is basically involved is the addition of further fuel channels; simpler access for fuel changing, on account of the individual channel arrangement, facilitating on-load refueling, which is particularly advantageous for natural uranium reactors. On the other hand, the complexity of the individual channel layout involves a significantly higher capital cost for the pressure tube system, and one is also faced with the problem of providing pressure tubes which combine the property of low neutron absorption with the assurance of an adequately long life under irradiation in the reactor core. This problem has been approached by the development of various alloys of zirconium.

Similar advantages and disadvantages can be cited for the various *coolants* which may be used in a heavy water reactor. The use of D_2O itself for coolant has the advantage of retaining the optimum neutron economy in the system, at the expense of maximizing the total D_2O inventory, which is a significant factor (typically of the order of 15% for a CANDU-PHW) in the overall capital cost. The control of leakage, both of expensive D_2O and of radioactive tritium gas, demands high standards of engineering for all joints and seals in the high-pressure primary circuit. Employment of D_2O as coolant may be considered to imply an indirect cycle, with the associated capital cost of steam generators, since the attempt to operate on a direct cycle (on the Swedish Marviken reactor) has not been encouraging.

The use of H_2O as coolant in the pressure tube reactor has the advantage of direct cycle operation, thus eliminating the complication of heat exchangers and, in addition, further lowering capital cost by the reduction of D_2O inventory. (In the U.K. steam-generating heavy water reactor, for example, approximately 30% of the moderation takes place in the light water coolant.) Leakage from the cooling circuit becomes less of a problem, although there will be a higher activity level at the turbine, which is now coupled to the reactor directly, rather than through a heat exchanger. Among the disadvantages of operation with boiling coolant is the limitation in system power which is imposed by the requirement to maintain an adequate margin against dryout in the fuel channels. A further operational and safety consideration is the high positive void coefficient of reactivity which can arise with natural uranium fuel in an H_2O-cooled pressure tube reactor; this may be avoided by the use of enriched fuel and a relatively close fuel spacing to give a degree of undermoderation, as in SGHWR, where the void coefficient is close to zero. Another consideration with light water cooling, which has to be balanced against the capital cost savings due to the direct cycle, is the lower burn-up attainable owing to the poorer overall neutron economy.

Table 10.1. Characteristics of Heavy-Water Reactors

Pressurized Heavy Water Reactor (PHWR)

1. Low thermal efficiency
2. Capital cost about 10%–15% higher than light water reactor
3. Low fueling costs because of excellent neutron economy
4. Problem of preventing leakage from hot, pressurized D_2O coolant circuit
5. Tritiated atmosphere from hot D_2O leakage
6. Good plutonium producer

Boiling Light Water Reactor (BLWR)

1. Low thermal efficiency
2. Direct cycle reduces capital costs
3. Lower D_2O inventory than PHWR
4. Better steam conditions at turbine compared with PHWR
5. Higher activity level at turbine
6. With natural uranium fuel, positive void and power coefficients may complicate control
7. Appreciable power margin necessary to eliminate possibility of burn-out of fuel

Gas-Cooled Heavy Water Reactor (GCHWR)

1. High thermal efficiency
2. Lower power density than with liquid cooling
3. Higher fuel temperature demands use of more heat-resistant core materials, worsening neutron economy

Organic-Cooled Heavy Water Reactor (OCR)

1. High thermal efficiency
2. Requires high-density fuel
3. Coolant filtering required to prevent fuel tube fouling
4. Low D_2O inventory (about 1/5 that of PHWR)
5. Low activity can be achieved in primary circuit
6. Demanding conditions of operation of pressure tubes ($\sim 375°C$) and fuel cladding ($\sim 475°C$)

Owing to the restrictions on steam temperature and pressure imposed by the need to minimize the thickness of pressure tube material for reasons of nuclear absorption, the thermal efficiency is relatively low when either H_2O or D_2O is used as a coolant. A significant improvement in thermal efficiency may be obtained by the use of gas cooling with CO_2. The high fuel temperatures necessary to achieve this improvement impose more rigorous demands on the cladding and primary circuit materials, necessitating the use of elements which have higher absorption cross sections than the zirconium alloys acceptable for the cladding and pressure tubes of the water-cooled reactors. In addition, limitations on fuel and cladding temperatures make it difficult to obtain a power density comparable to that achieved with water cooling. The advantage of higher temperature operation to the overall system efficiency is partially offset by the higher auxiliary power demand for gas circulation.

High pressures are required with either light or heavy water cooling in order to attain temperatures high enough to give adequate thermal efficiency. An alternative method of improving efficiency is the use of a coolant of sufficiently low volatility to allow the reactor to operate at high temperature but moderate pressure. A suitable coolant of this type is an organic liquid such as a hydrogenated terphenyl oil, which allows the coolant outlet temperature to be raised to around 400°C, compared with the 300°C temperature characteristic of a CANDU-PHW, where the coolant circuit has to operate at a pressure of about 87 atm. High power density and a thermal efficiency in the range of 36–40% are obtainable with organic coolant. A further potential advantage is the low coolant activity, since neutron irradiation of the hydrocarbon coolant gives rise to a negligible amount of radioactivity. Another benefit is a great reduction in the D_2O inventory, and consequently in the size of the reactor, owing to the good moderation characteristics of the organic coolant; it is estimated that the total D_2O inventory per kWe for an advanced organic-cooled CANDU would be reduced by a factor of 5, relative to standard designs, as a result of changing from D_2O to organic coolant.

The use of organic coolant (terphenyl mixtures) in a D_2O-moderated system has been investigated in the Canadian WR-1 research reactor at Whiteshell, Manitoba. Experience has confirmed that zirconium-based alloys are suitable for extended operation in organic coolant provided there is careful control of the coolant chemistry to prevent excessive hydrogen uptake and fuel tube fouling.

For economic operation, the organic-cooled, D_2O-moderated reactor requires fuel of higher density than UO_2. Encouraging results have been obtained in WR-1 with uranium carbide (UC) and triuranium silicide (U_3Si). The consensus of opinion at the present time, however, appears to be that the cost advantage of organic cooling is not sufficient to justify diversion of the CANDU natural uranium reactor on to this type. The combination of organic cooling with the thorium fuel cycle, however, appears promising as a longer-term development of the CANDU system.

A short summary of the principal features of the four types of cooling discussed above is given in Table 10.1. This is followed by a detailed description of the CANDU-PHW reactor at Pickering Point in Ontario.

10.2. The CANDU Pressure Tube Heavy Water Reactor

The CANDU heavy-water-moderated and -cooled pressure tube reactor represents the main line of development of the Canadian nuclear program. Following experience with the Nuclear Power Demonstration Reactor (NPD), which began operation in 1962, and the prototype 200-MWe CANDU reactor at Douglas Point, commissioned in 1967, the major application of the concept has been in the four-unit Pickering A generating station in Ontario. This station, which is operated by the Ontario Hydro electrical utility, is one of the largest nuclear stations constructed to date, with a total gross capacity of 2160 MWe.

Apart from minor differences of detail, the four Pickering reactors are of identical design (see Table 10.2). All share the following features: natural uranium fuel, heavy water moderator, pressurized heavy water coolant, indirect steam cycle, and on-load bidirectional refueling for the horizontal pressure tubes. The heavy water moderator is contained in a horizontal cylindrical tank (*calandria*) of about 8-m diameter, constructed of austenitic stainless steel (Fig. 10.3). A total of 390 zircaloy-2 calandria tubes run through the calandria from one end face to the other. The zircaloy-2 or Zr–$2\frac{1}{2}\%$Nb pressure tubes, in which the fuel bundles are located, are separated from the calandria tubes by a sealed insulating annulus containing nitrogen. Each pressure tube is attached at the ends by rolled joints to stainless steel end fittings, which are supported in sliding bearings at the end faces of the calandria.

The heavy water coolant, which, as in a PWR, is pressurized to prevent

Table 10.2. Data for Pickering CANDU Reactor

Reactor type: CANDU-PHW	Length of fuel bundle: 49.5 cm
Moderator: Heavy water	Fuel pellet diameter: 14.33 mm
Coolant: Heavy water	Thickness of zircaloy sheath: 0.41 mm
Fuel: Natural uranium (UO_2)	Total mass of UO_2 in core: 105 tonnes
Diameter of calandria: 8.1 m	Average fuel irradiation: 8300 MW d/tonne
Overall length of calandria: 8.25 m	Average linear rating of fuel pin: 37.6 W cm^{-1}
Total D_2O content of moderator circuit: 284 tonnes	Total D_2O content of coolant circuit: 158 tonnes
Number of fuel channels: 390	Coolant temperature at channel inlet: 249°C
Lattice spacing: 28.6 cm	Coolant temperature at channel outlet: 293°C
Core radius: 318.5 cm	Mean pressure at outlet headers: 1280 psi
Core length: 595 cm	(90 kg cm^{-2})
Fuel bundles per channel: 12	Total heat output: 1744 MWt
Pins per fuel bundle: 28	Net electrical output: 508 MWe
	Overall station efficiency: 29.1%

boiling, is circulated through the pressure tubes and thence to the heat exchangers. The heat transport system is divided into two identical loops, each containing six tube-in-shell heat exchangers, where the heat is transferred to light water to produce steam. In addition to some engineering convenience, the division into two circuits has advantages from the safety viewpoint, as it effectively halves the voidage which would be produced by a gross failure of the primary containment system. The coolant circuit contains some 160 tonnes of heavy water, the inlet and outlet temperatures being about 250 and 293°C, respectively.

The 280 tonnes of heavy water moderator in the calandria is maintained nominally at atmospheric pressure and is continuously circulated through an external cooler to maintain its temperature at around 60°C. Inside the calandria there are a number of spray nozzles which provide a drenching spray to cool any parts not covered by the moderator.

The Pickering reactors are fueled with natural uranium fuel in the form of cold-pressed, sintered UO_2 pellets of diameter 14.3 mm, clad in collapsible zircaloy sheaths of 0.4-mm wall thickness. Each pellet is dished at one end to allow for axial expansion. The short fuel element cluster, of length 495 mm, is made up of 28 fuel pins (Fig. 10.4). Interpin and bundle-to-pressure tube spacing is provided by zircaloy spacers brazed to the sheaths. Each pressure tube contains 12 fuel bundles placed end-to-end. The fuel bundle is designed to give maximum neutron economy, and only the minimum of structural material is included. All related components, such as orifices, fuel handling hardware, and control and monitoring facilities, are part of the reactor capital equipment. As a result, the nonfuel component of the bundle accounts for only some 0.7% of its total neutron absorption.

The CANDU system, like the natural uranium magnox reactor, requires on-load refueling for optimum economy of operation. Remotely operated refueling machines are located at both ends of a given channel for insertion and removal of the fuel assemblies. A fuel shuffling scheme is used to equalize burn-up of the clusters, the two identical fueling machines being arranged to move fuel in opposite directions in adjacent channels, thus maintaining an axially symmetric neutron flux distribution. The burn-up attained is 8000 MW d/t or better, and the fuel residence time is only about 20%–30% of that which is typical for the fuel of a light-water-moderated reactor. The increased complexity associated with on-load refueling is partly compensated, therefore, by the less stringent requirements on the material of the fuel bundle, the sheathing for which is only about half the thickness necessary in a BWR or PWR. Despite the reduced thickness, failure rates due to corrosion or fretting are negligible. The thin sheath has the advantage of

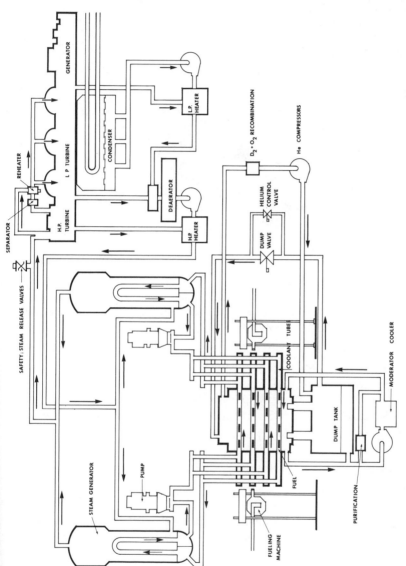

Fig. 10.3. Flow diagram for Pickering CANDU reactor (courtesy of Atomic Energy of Canada Limited).

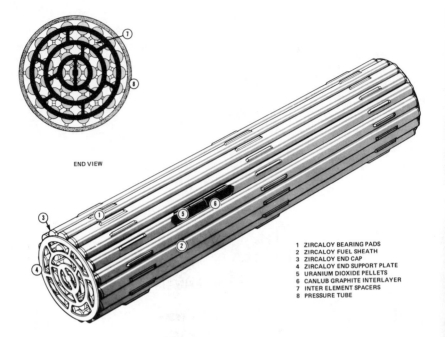

END VIEW

1 ZIRCALOY BEARING PADS
2 ZIRCALOY FUEL SHEATH
3 ZIRCALOY END CAP
4 ZIRCALOY END SUPPORT PLATE
5 URANIUM DIOXIDE PELLETS
6 CANLUB GRAPHITE INTERLAYER
7 INTER ELEMENT SPACERS
8 PRESSURE TUBE

Fig. 10.4. Fuel bundle for CANDU reactor (courtesy of Atomic Energy of Canada Limited). 1, Zircaloy bearing pads; 2, Zircaloy fuel sheath; 3, Zircaloy end cap; 4, Zircaloy end support plate; 5, uranium dioxide pellets; 6, canlub graphite interlayer; 7, interelement spacers; 8, pressure tube.

deforming under the coolant pressure to rest on the fuel pellets, thereby providing excellent heat transfer. Fuel densification effects are avoided, since the reduced need for voidage provision allows the use of high-density fuel in the first place.

The arrangements for reactivity control and shutdown of the reactor are as follows:

1. Once an equilibrium cycle has been achieved, the reactivity is maintained at a suitable value by on-load refueling. The average rate of refueling required is nine fuel bundles per day, to compensate for a reactivity decline due to burn-up of about 0.04 % per day.

2. Normal operating control of reactivity is provided by 14 zone control compartments located throughout the core. These are tubes which can be filled with ordinary water, thus reducing reactivity by increasing the neutron absorption. The requirement for stabilization of the core power level on a zone basis arises from the fact that the core radius is many times larger than

the diffusion length, and consequently the control system must be designed to prevent xenon-induced spatial instability.

3. An additional facility, used both for adjusting the flux shape for optimum peak-to-average ratio and to overcome the xenon transient following a trip from power, consists of 18 adjuster rods which may be inserted vertically between the calandria tubes. Cobalt is used as the absorber in the adjuster rods because of the commercial value of the radioisotope produced by irradiation. An alternative method of overcoming the xenon transient, which has been adopted for the subsequent Bruce generating station, is the use of booster rods containing enriched fuel.

4. Rapid shutdown is effected by dropping eleven cadmium-loaded shut-off rods operating through the top of the calandria tank, backed up if necessary by dumping of the moderator into a dump tank situated below the calandria. During normal operation, helium gas in the dump tank is maintained at a high pressure. Release of this pressure allows the moderator to fall under gravity into the dump tank. It is interesting to note that, in later CANDU designs, the use of moderator dump has been discontinued in favor of the rapid injection of poison (a salt of gadolinium) into the moderator.

5. For dealing with reactivity variations beyond the capacity of the other systems, for example, at the start of life, when the reactor is loaded with fresh fuel, boron can be added to the moderator in the form of a soluble oxide. The boron can be removed when necessary by ion exchange.

The void coefficient in CANDU is positive, yielding a reactivity change of about 0.75% for complete voiding of all channels in the equilibrium core. With this magnitude of coefficient, suitable subdivision of the primary circuit limits the rate and extent of the transient associated with any possible void increase to a level which can easily be covered by the shut-down rods.

The Pickering containment consists of a two-stage system. Each of the four reactors in the station is enclosed within its own concrete containment building, which is maintained slightly below atmospheric pressure. In the event of a sudden buildup of pressure, due to a major breach of the primary circuit, pressure relief valves would open and the steam and air mixture would be discharged to a "vacuum building" which serves all four reactors. This building, which has a volume of $8 \times 10^4\,m^3$, is constructed from reinforced concrete, and is normally maintained at an absolute pressure of 50 mm Hg. The vacuum building is capable of containing all the steam produced by the discharge of the entire primary coolant, but the process of containment is aided by a cooling spray which is automatically actuated at a preset pressure to cool the air and condense the steam. The water tank supplying the spray is located in the top of the vacuum building and has a capacity of two million gallons. The combination of reactor building and vacuum building is designed to contain all the energy release in the worst conceivable rupture of the primary circuit, the reactor building being exposed to a pressure transient of not more than 1.4 atm absolute, which is well within its capacity to withstand.

The choice of the CANDU type of reactor for the Canadian nuclear power program was initially motivated by the country's possession of an abundant supply of natural uranium and a large supply of hydroelectric power which could be used for the production of heavy water. Early problems with the prototype at Douglas Point included failure of coolant pumps, difficulties with the refueling machines, and an excessive rate of heavy water leakage, which was of concern not only on economic grounds but also on account of the associated tritium hazard. At one time the supply of heavy water presented a major problem. Prolonged delays in the commissioning of the Glace Bay production plant had created a critical shortage just at the point where maximum availability was needed for the startup of the Pickering reactors. This problem was solved in the short term by a complicated transfer of heavy water between the Pickering reactors and the other CANDU plants, coupled with the purchase of D_2O from foreign sources. The situation was eased by the commissioning of the Bruce heavy water plant in 1974, and by the early 1980s the number of production plants should be more than adequate to meet the anticipated CANDU installation rate.

The most significant problem that has arisen with the Pickering station has been the appearance of cracks in some of the zirconium–niobium pressure tubes in units 3 and 4. The initial cause of the cracking was found to be faulty rolling technique in the rolled joints connecting the pressure tubes to their end fittings. The cracks had been propagated by zirconium hydride deposition whenever the reactor was in the cold shutdown condition. Rectification of the problem necessitated the removal of the affected pressure tubes from the two Pickering units and also from two of the Bruce units under construction.

Apart from the pressure tube problem, now overcome, the performance of the Pickering reactors has been very encouraging, both as regards the commissioning schedule and the high availability since startup. This successful experience, combined with the Canadian emphasis on the efficient utilization of nuclear fuel resources, has strengthened the belief that CANDU is an appropriate choice for a major power program under Canadian conditions. Although the capital cost of the heavy water system is higher than that of a PWR or BWR, the uranium requirements are of the order of half those for a light water reactor of the same overall size. It is argued that the relatively low fuel costs of CANDU will be an increasingly important economic advantage in the 1990s, when there will be a high uranium demand, both for existing thermal reactors and for fueling the fast reactors which are expected to be coming into operation at that time.

The economics of the present CANDU system are based on the natural uranium, once-through, throwaway fuel cycle, that is, the fuel is passed through the reactor once before being discharged to permanent storage. No credit is currently allocated for the plutonium content of the irradiated fuel, although this may be worth recovering at a later date. The simple fuel cycle has the advantage of eliminating expensive enrichment costs and reprocessing costs, but in the longer term CANDU offers good development potential through the possibility of combining the use of thorium- and plutonium-based fuels with the high thermal efficiency of organic coolant. On a shorter time scale, improvements can be expected from the use of stronger zirconium alloys which are currently under development. These can be employed either to improve the neutron economy by reducing sheathing thickness or to increase the steam conversion efficiency. The latter objective will also be aided by the adoption of indirect cycle boiling, which will be standard in the later reactors of the CANDU type.

A development of the CANDU system which retains the advantages of the heavy water moderator while permitting the use of a direct coolant cycle is the CANDU-BLW (boiling light water) reactor, of which a prototype operated at Gentilly in Quebec from 1971 to 1980. One of the major concerns at Gentilly was the study of operation with positive void and power coefficients. Measurements during commissioning indicated a strongly positive power coefficient of the order of 10^{-4} per percent increase in power. In practice, no problem was encountered in controlling the reactor, since the long time constant of around 18 s associated with the large (19.5-mm-diameter) fuel pins in the 18-pin bundle led to an appreciable delay in the formation of voids following an increase in power. Close control of steam drum pressure was necessary, however, during transients, such as a turbine trip, to avoid overrapid variation of core voidage.

A feature of Gentilly is the use of enriched uranium booster rods to add reactivity during startup, in order to compensate for the negative reactivity effect of power increase. It was found that the insertion of the booster rods could result in a local reactivity imbalance which was amplified by the positive power coefficient. The result was the initiation of a side-to-side flux tilt, occurring in a time of the order of minutes, which, if uncorrected, could cause overrating of the fuel. To prevent this, the flux and power distributions in the core were continuously monitored and the control system was designed for automatic correction of any flux tilt which may occur.

Like CANDU-BLW, the United Kingdom SGHWR design has vertical fuel channels, but, unlike CANDU, uses fuel enriched to about 2.2%. A 100-MWe prototype of the SGHWR has been operating at Winfrith Heath, in Dorset, since 1968. The system had at one time been intended as the basis for the next phase of the United Kingdom nuclear power program, but, despite the excellent performance of the prototype, the SGHWR has been dropped from this role in favor of AGR or light water designs.

11

The Fast Reactor

11.1. Introduction

The formidable technical problems involved in the design of a full-scale commercial fast reactor have resulted in the introduction of this system lagging behind that of the thermal reactor by two to three decades. Despite these obstacles, there is a worldwide interest in the fast breeder because of its high potential utilization of uranium fuel, based on the high η value of Pu^{239} in the fast neutron spectrum, coupled with the bonus arising from the fast fission of U^{238} and Pu^{240} by the significant proportion of neutrons above the fission threshold energy (see Fig. 2.9). The production of more Pu^{239} (by conversion of U^{238}) than is consumed in the operation of the reactor leads to a dramatic reduction in the demands for fissile material, thus conserving uranium stocks and making the electricity generation costs almost insensitive to the price of uranium. The small size of the plant, due to the high power density of the fast reactor, is also seen as contributing to good economics by reduction of capital cost.

It is interesting to note that the first reactor to generate electricity was the U^{235}-fueled fast reactor EBR-I (Experimental Breeder Reactor I), which started operating in the United States in 1951. This reactor, cooled by molten sodium–potassium alloy, had an electrical output of 200 kW. Most of the effort devoted to fast reactors since that time has continued to be on the liquid-metal-cooled concept, the requirement for this highly efficient cooling method arising from the high rating characteristic of this reactor type. At the same time, however, some attention has also been given to the possibilities of gas cooling, using the technology developed for the high-temperature gas-cooled reactor.

Much of the earlier work on the fast reactor was carried out with pilot plants in the United States (EBR-I, Fermi, Sefor), the Soviet Union (BR-5 and BOR-60), the United Kingdom (DFR), and France (Rapsodie). Larger prototype plants are now operating, such as the Soviet BN-600, the French

Phenix, and the British Prototype Fast Reactor (PFR). Following successful experience with these prototypes, a number of countries plan to introduce full-scale commercial stations in the 1980s.

11.2. The Physics and Technology of Fast Reactors

The physics aspects of the fast reactor have been touched on at various points in the earlier chapters. On account of the considerable differences of emphasis which are necessitated by the differences between fast and thermal reactors, however, it is useful to summarize the physics of the former separately at this stage.

The principal distinction between the thermal and fast reactor is that whereas, in the former, moderating material is deliberately introduced in order to slow the neutrons down as rapidly as possible through the region of resonance capture, in the fast reactor the amount of moderating material is

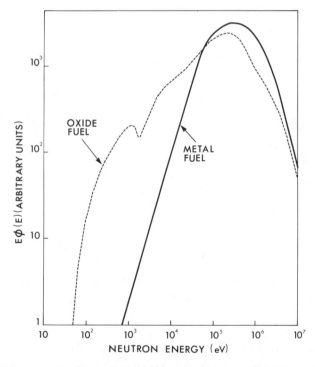

Fig. 11.1. Neutron spectra for small metal-fueled and large oxide-fueled fast breeder reactors [from T. D. Beynon, *Rep. Prog. Phys.* **37**, 951 (1974). Copyright Institute of Physics].

Table 11.1. Values of α, v, and η for U^{233}, U^{235}, Pu^{239}, and Pu^{241} in a Typical Fast Reactor Spectrum[a]

Parameter	U^{233}	U^{235}	Pu^{239}	Pu^{241}
α	0.108	0.244	0.231	0.114
v	2.57	2.50	2.96	3.04
η	2.32	2.01	2.40	2.73
$\eta - 1$	1.32	1.01	1.40	1.73

[a] From T. D. Beynon, *Rep. Prog. Phys.* **37**, 951 (1974). Copyright Institute of Physics.

reduced to the minimum. The neutron energy spectrum is consequently very different from that in a thermal reactor, the average energy of the neutrons causing fission being of the order of several hundred keV. The difference in spectrum may be clearly seen by comparing the typical fast reactor spectra shown in Fig. 11.1 with the thermal reactor spectrum of Fig. 3.2. In the absence of large amounts of moderator, the principal mechanism for the reduction of neutron kinetic energies below their emission energy is by inelastic collision with nuclei of the fuel and structural material. It is, in fact, the high inelastic scattering cross section of U^{238}, providing an efficient mechanism for reducing the neutron energies below the U^{238} fission threshold (1.4 MeV), that prevents a neutron chain reaction from being maintained in a mass of pure natural uranium.

Because of the effect of inelastic scattering in U^{238}, the spectrum softens as the proportion of fertile to fissile material is increased. Further softening is produced by the introduction of light elements, for example by the use of oxide or carbide fuel. As a result, the spectrum in a full-scale oxide-fueled fast reactor is considerably softer than that in the early compact cores of the EBR type. It may in fact be desirable deliberately to introduce a certain amount of additional moderation in order to improve the response of the reactor to power transients (see Section 11.3).

As was illustrated earlier (Fig. 4.1), the value of $\eta = v/(1 + \alpha)$ for both U^{235} and Pu^{239} is appreciably higher in a fast spectrum. The fission and capture cross sections themselves are of course much smaller at the higher energies (see, for example, Fig. 2.8 for U^{235}). Table 11.1 gives values of the important parameters of the fissile isotopes averaged over a typical fast reactor spectrum.

The superior performance of the plutonium isotopes, as evidenced by the value of $(\eta - 1)$, and in particular the advantage of the buildup of the higher isotope Pu^{241} as a result of neutron capture in Pu^{240}, are clearly seen. In

Table 11.2. Values of σ_f, σ_c, v, and Fission Threshold Energy for the Main Fertile Isotopes in a Fast Neutron Spectrum

	Th^{232}	U^{238}	Pu^{240}
σ_f (6)	0.010	0.048	0.408
σ_c(b)	0.240	0.202	0.354
v	2.54	2.70	3.17
E_f (MeV)	1.4	1.4	0.6

addition to its role in the production of Pu^{241}, the Pu^{240} also contributes appreciably to the neutron production by fast fission since, as shown in Table 11.2, it has both a higher v value and a lower threshold than, for example, the other fertile isotopes Th^{232} and U^{238}.

The factors which influence the breeding ratio of an FBR can be seen from the relation already derived in equation (4.7) for the conversion ratio in a thermal reactor. Owing to the unavoidable neutron losses from leakage and from capture in coolant, structural, and control materials, the breeding ratio for a typical sodium-cooled commercial FBR will be in the range 1.20–1.30.

The plutonium required for the initial fueling of a fast reactor power program would be available from the stocks built up in thermal reactors, such as CANDU and magnox, both of which are efficient plutonium producers. For a sustained and reasonably rapid growth in fast reactor generating capacity, however, the provision of adequate plutonium supplies will also demand a high breeding ratio in the fast reactors themselves. A measure of the efficiency of an FBR in increasing the inventory of fissile material is the *doubling time*, which is defined as the time taken to produce a quantity of surplus fuel sufficient to start up another reactor identical to the first.

In the simplest case, we consider a reactor where none of the fuel bred is reused in the course of the doubling time period. In this case, the doubling time is given by the relation

$$DT \text{ (yr)} = \frac{\text{initial loading of fissile fuel (kg)}}{\text{net rate of accumulation of fissile fuel (kg/yr)}} \quad (11.1)$$

Now the breeding ratio is

$$BR = \frac{\text{fuel produced per year}}{\text{fuel destroyed per year}} \quad (11.2)$$

Hence

$$\frac{\text{net fuel buildup per year}}{\text{fuel destroyed per year}} = \text{BR} - 1 \qquad (11.3)$$

The quantity BR $-$ 1 is known as the *breeding gain*. In the absence of any contribution to the fission rate from fast fission in the fertile isotopes, 1 MW d of power operation would be provided by the fission of almost exactly 10^{-3} kg of fuel (taking the recoverable energy per fission for Pu^{239} as 215 MeV). This corresponds to 10^{-3} $(1 + \alpha)$ kg of fuel destroyed, where α is the capture-to-fission ratio. Taking account of the contribution to power production by fast fission of the fertile isotopes, however, the rate of destruction of fuel will be reduced in the ratio

$$\frac{N_{\text{fi}}\sigma_{\text{fi}}^f}{N_{\text{fi}}\sigma_{\text{fi}}^f + N_{\text{fe}}\sigma_{\text{fe}}^f} = \frac{1}{1 + F}, \qquad \text{where } F = \frac{N_{\text{fe}}\sigma_{\text{fe}}^f}{N_{\text{fi}}\sigma_{\text{fi}}^f} \qquad (11.4)$$

Here N_{fi}, σ_{fi}^f are the number density and fission cross section of fuel and N_{fe}, σ_{fe}^f are the corresponding quantities for the fertile material. For simplicity, we consider a uniform reactor where the number densities are independent of position.

Hence 1 MW d at power results in the destruction of $10^{-3}(1 + \alpha)/(1 + F)$ kg of fuel, and the amount of fuel destroyed per year by operation at a power of P MW(thermal) at a load factor (fraction of time at full power) of L will be

$$\frac{365\,PL(1 + \alpha) \times 10^{-3}}{1 + F} \text{ kg}$$

Hence, from equation (11.3) the net rate of accumulation of fissile fuel is

$$\frac{0.365PL(1 + \alpha)(\text{BR} - 1)}{1 + F} \text{ kg/yr}$$

and the doubling time is then

$$\text{DT} = \frac{2.7\,M(1 + F)}{PL(1 + \alpha)(\text{BR} - 1)} \text{ yr} \qquad (11.5)$$

where M is the initial fuel inventory of the reactor (in kilograms).

It can be seen from equation (11.5) that the achievement of a short doubling time implies a high value for the power per unit mass of fissile material, P/M. This is equally desirable from the point of view of minimizing the inventory of the high-grade fuel required by the fast reactor; owing to the low fission cross section of fuel in a fast spectrum, the fuel has to be concentrated to achieve criticality, with a typical ratio of Pu^{239} to U^{238} of around 20%. The economic power density for an FBR is of the order of 500 kW/l, which is approximately ten times that of a typical thermal reactor. The resulting high specific heat removal requirement implies (a) a finely divided fuel geometry in order to maximize heat transfer area and (b) a coolant which must have excellent heat transfer characteristics, in addition to the basic requirement of low moderation. The favored coolants are liquid metals or possibly high-pressure helium.

Of the possible liquid metal coolants, lead has an inconveniently high melting point (327°C), while the boiling point of mercury (356°C at atmospheric pressure) is too low, as a high-pressure circuit would be required for its containment at the reactor operating temperature. Two other possible coolants are liquid sodium and lithium, both of which have high enough boiling points to permit operation at moderate pressure. Lithium, however, has a rather high capture cross section due to the Li^6 isotope (present as 7.5%), and it presents more corrosion problems at elevated temperature than does sodium. For these reasons, sodium is the preferred coolant for a full-scale fast reactor. Among its advantages are (i) high boiling point (881°C), allowing high operating temperatures and hence good thermal efficiency; (ii) low vapor pressure at working temperature, resulting in a low primary circuit pressure; and (iii) high thermal conductivity and good specific heat transfer rates without excessive pumping rate requirements.

On the other hand, liquid sodium presents a number of technical problems, and one of the main objectives of the fast reactor development program has been the proving of the practicability of using this coolant. The potential difficulties include the following:

i. The activation of sodium induced by neutron capture results in the formation of Na^{24}, a high-energy γ emitter with a 15-h half-life. The presence of Na^{24} gives rise to problems of shielding and access for maintenance.

ii. The possibility of an explosive reaction exists between sodium and water if these are allowed to come into contact, for example through a heat exchanger failure. This necessitates the provision of an intermediate heat exchanger, where the primary sodium gives up its heat to sodium in a secondary circuit, which in turn gives up heat to water in a sodium–water heat exchanger (see Fig. 7.2). The intermediate circuit acts as a buffer between the steam generator and the primary sodium, and prevents any possible explosive reaction in the former from propagating back into the reactor core itself.

iii. The reduction of corrosion demands cladding of the primary circuit and fuel in austenitic steels.

iv. Care is required to eliminate the possibility of gas entrainment in the sodium, leading to interruptions of heat transfer and to reactivity variations in the core. This is particularly the case with the pool design of coolant circuit (see Section 11.3), where baffles may have to be included to limit disturbance at the free coolant surface.

v. Since, unlike water or helium, sodium is not transparent, operations on fuel or in-vessel components have to be carried out "blind," increasing the complexity of fueling or maintenance.

The development of the prestressed concrete pressure vessel, together with the advances in high-temperature gas cooling associated with the helium-cooled HTGR system, has led to a revival of interest in the use of gas cooling for fast reactors. In addition to eliminating the complication of an intermediate heat exchange circuit, the use of helium avoids the spectrum softening effect associated with the sodium and thus leads to appreciably shorter doubling times. In order to minimize the spectrum effect with sodium cooling, it is important to keep the fraction of core volume occupied by the coolant as small as possible, resulting in a fuel element design with small interpin gaps, with a consequent concern about the effects of pin swelling or distortion on the coolant flow. The use of helium coolant eases design problems by permitting larger gaps between the fuel pins. Other advantages include its near-zero reactivity coefficient, chemical inertness, transparency, and freedom from neutron activation. On the other hand, the attainment of adequate heat transfer rates at the high ratings characteristic of the FBR requires a high gas pressure (up to 100 atm) and a high circulating power. The high coolant pressure necessitates the use of a vented fuel pin to avoid having a high differential pressure across the cladding. This, of course, has compensating features, since it eliminates clad stress due to fission gas buildup and reduces the activity held in the core by allowing the continuous extraction of fission products from the vented elements.

The safety aspects of the sodium- and gas-cooled FBRs are discussed in the following section. While, in the longer term, the gas-cooled breeder has features which make it worthy of further consideration, the main thrust of the fast reactor program in the immediate future will continue to be concentrated on the liquid metal system, partly because of the large investment which has already been made in this technology. Another long-term possibility is the use of N_2O_4 as coolant gas, the heat being taken up by dissociation of the

N_2O_4 to NO and O_2. The main interest in this cycle has been shown in the Soviet Union, and it is claimed to offer good heat transfer coupled with a rapid doubling time due to the hard neutron spectrum.

Owing to the high fabrication cost of the fuel elements, economic operation demands both a high cladding integrity and a considerably greater burn-up than is attainable in thermal reactors. The target burn-up for the FBR is around 100,000 MW d/tonne (corresponding to a burn-up of approximately 10 % of the fuel atoms), which is a factor of five greater than that of an AGR or three higher than a PWR. While the use of metal fuel would be preferable to achieve maximum density and minimize spectrum degradation, the high target burn-up makes it necessary to use ceramic fuel, in the form of mixed uranium–plutonium oxide or carbide. The greater thermal conductivity of the carbide compared with the oxide permits operation at a higher linear rating, but more irradiation experience is needed to provide assurance on factors such as compatibility between carbide fuel and cladding. The fuel pins are small, being only of the order of 5 mm in diameter. The fuel assembly is made up of a large number of pins, usually 300 or more, to facilitate reloading, which is a relatively frequent operation owing to the high fuel rating.

The leakage from a fast reactor is high because of the low absorption cross sections in the fast spectrum. It is therefore particularly advantageous to surround the core with a reflector. Low-mass materials are to be avoided because of the spectrum softening effect, but the relatively high scattering cross section of U^{238} makes it a very suitable choice as a reflector material. In addition to scattering neutrons back into the core, neutron capture in the U^{238} produces a large proportion of the Pu^{239} bred in the reactor. The uranium region surrounding the core is known as the *blanket*; it may consist of natural uranium but can also constitute a convenient way of utilizing the depleted uranium left over from an enrichment plant. Since an appreciable generation of power takes place in the blanket owing to the fast fission of the U^{238}, some of the sodium flow has to be diverted there for cooling purposes. In addition to the separate radial reflector, the fuel elements are desgined to have regions of natural uranium at either end, giving upper and lower blanket zones. The fuel pins in the radial blanket can be of larger diameter than those in the core owing to the lower power density.

A typical 1000 MWe FBR might have a core loading of some 4 tonnes of Pu^{239} and 16 tonnes of uranium, with about 25 tonnes of uranium in the blanket. One of the advantages of the fast spectrum is that plutonium containing an appreciable proportion of the higher isotopes Pu^{240} and Pu^{242} is perfectly acceptable as fuel. Indeed, the conversion of Pu^{240} to Pu^{241} is a

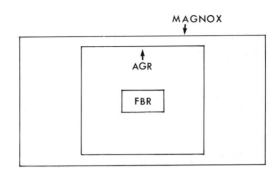

Fig. 11.2. Comparison of core sizes for magnox, AGR, and LMFBR designs.

useful means of minimizing the long-term change of the reactivity of the system. The FBR is ideal for utilizing the high burn-up plutonium produced, for example in a BWR, where the proportions of Pu^{240} and Pu^{242} in the discharged fuel are of the order of 24% and 5%, respectively. There is therefore no economic need to impose a restriction on the burn-up of fuel in existing thermal reactor stations which are producing the plutonium which will eventually be used for commissioning fast reactors.

The small size of the FBR core compared with that of relatively lowly rated thermal reactors with their large moderator volume may be seen from Fig. 11.2, where the dimensions of a 600 MWe FBR are contrasted with those of magnox and AGR systems of the same output. The overall dimensions of the fast reactor will be increased by 1–2 m when the axial and radial breeder zones are included.

Control of the early fast reactors tended to be by movement of fuel in or out of the core, but the mechanical complications of this method have led to its being superseded by ordinary control rods incorporating boron carbide or tantalum clad in stainless steel. The reduced effectiveness of the control material in the fast spectrum, due to the absorption cross section being so much lower than in a thermal reactor, is not of serious importance. The FBR does not need to incorporate a large built-in reactivity margin at start of life to allow for fuel depletion since the reactivity remains more constant owing to the compensating buildup of the bred fuel. In addition, fission product poisons are less important, accounting for only some 3%–4% reactivity penalty over the core life.

Although the calculation techniques discussed in Chapter 3 are applicable to fast reactors, the prediction of their characteristics is in general considerably more complex than for a thermal reactor, owing to the greater demands on the quality of the basic nuclear data. As discussed in Chapter 3, it is possible to divide the neutron spectrum in a thermal reactor into a relatively small number of groups, each group being dominated by a few important cross sections. It is possible to assume that the shape of the spectrum within a given group is invariant over a wide range of core compositions. In fast reactors, on the other hand, the competition between the various neutron reactions extends much more uniformly over the whole energy range, so that it is no longer possible to have a few cross sections dominating in each energy group. In addition, the spectrum shape within groups is now appreciably dependent on the core composition.

In principle, provided that accurate and detailed information on the relevant cross sections is available, it should be possible to predict the fast reactor parameters, such as critical loading or breeding gain, without recourse to adjustment of the data on the basis of comparison with experimental critical assemblies. While a great deal of effort, particularly in the United States, has been directed towards this goal, the time scale for the introduction of commercial fast reactors in other countries, such as the United Kingdom, has led to a need for adjustment of the basic cross section data by reference to measurements carried out on a range of test assemblies. The adjustment technique involves the measurement of quantities such as critical mass or relative reaction rates on an experimental assembly and the comparison of the measured values with predictions based on a calculation using about ten groups with averaged cross section data. The shape of the cross section variation within each group is based on differential cross section measurements, while the overall magnitude of each cross section is adjusted by comparison with the experimental assemblies.

In order to eliminate the complication of dealing with leakage, it is possible to construct a critical assembly where the central zone has been adjusted to have an infinite multiplication factor equal to unity. This zone is surrounded by a driver region containing highly enriched fuel. The condition of unit infinite multiplication factor (i.e., zero net neutron leakage) for the test zone is attained by uniform poisoning with boron until the situation is reached where the removal of a small section of the zone produces no effect on the neutron density of the reactor. This indicates that the composition of the test zone is equivalent to that of a critical homogeneous reactor of infinite size, since the replacement of any finite volume of such a reactor by a vacuum would leave the neutron density unchanged. Having achieved this condition, measurements are then made of the relative reaction rates of the various fissile and fertile isotopes, using activation foils of the appropriate composition. Particular attention has been given to the measurement of the capture-to-fission ratio of Pu^{239} and to the capture and fast fission cross sections of U^{238}.

The method may be illustrated by considering the measurement of α_9, the capture-to-fission ratio of Pu^{239}. Since we have $k = 1$ for the test region, the rate of neutron production within this region must be equal to the neutron absorption rate, i.e.,

$$v_9 F_9 + \sum_i v_i F_i = A_9 + \sum_i A_i \qquad (11.6)$$

where the quantities F and A refer to fission and absorption rates, respectively, and the summations are over all nuclei in the region, with the exception of Pu^{239}, which is considered separately.

Taking $A_i = F_i + C_i$, where C_i is the capture rate in the ith isotope, we have

$$F_9(v_9 - 1) + \sum_i F_i(v_i - 1) = C_9 + \sum_i C_i$$

Hence, we have

$$\alpha_9 = \frac{C_9}{F_9} = v_9 - 1 + \frac{1}{F_9}\sum_i F_i(v_i - 1) - \frac{1}{F_9}\sum_i C_i \qquad (11.7)$$

The ratios of the various capture and fission rates, relative to the Pu^{239} fission rate, are obtained by the standard methods, such as foil activation (see Section 1.8). Measurements of this kind are among those which have led to the abandonment of the concept of the steam-cooled fast reactor, which was at one time considered as a competitor for the liquid-metal- and gas-cooled systems. Measurements of α_9 showed that the degradation of the neutron spectrum had increased the capture-to-fission ratio to such an extent that the breeding ratio was no longer economic.

The current situation on fast reactor calculations is that, while a combination of basic differential cross section data, adjusted by correlation methods, with two-dimensional transport or three-dimensional diffusion codes, can give reasonable agreement for quantities such as critical mass or relative reaction rates in central regions of an FBR core, more precise data are still required for accurate prediction of differential effects such as the Doppler and sodium void coefficients. These two effects, together with the influence of the delayed neutron parameters for a fast reactor, are most conveniently dealt with in the following section, which considers the dynamic behavior of the system.

11.3. Dynamic Behavior and Safety Aspects of the Fast Reactor

Among the factors which affect the safety of the fast reactor are the following: (i) the high rating of the core; (ii) the high plutonium content of the core, which enhances the seriousness of a potential rupture of the primary circuit; (iii) the low value of the delayed neutron fraction associated with plutonium fueling, which reduces the reactivity margin from normal operation to prompt critical, and the short neutron lifetime, which would result in a very rapid rate of power rise if this margin were ever exceeded; (iv) the reactivity effects of changes of the sodium coolant volume in the core (sodium void coefficient); (v) the effect of core expansion and the Doppler coefficient in limiting the extent of possible transients; and (vi) the possibility of rapid propagation of fuel element damage due to a local blockage of the coolant flow.

Owing to the absence of moderator, the neutron lifetime l lies within the range 10^{-5}–10^{-7} s for the fast reactor, compared to a typical thermal reactor value of 10^{-3}–10^{-4} s. As discussed in Chapter 3, this fact in itself does not significantly affect the speed of response of the system to a positive change of reactivity, provided that its magnitude is less than the effective delayed neutron fraction, β. For positive step changes of reactivity, ρ, which are considerably in excess of β, however, the response of the fast reactor will be much more rapid. For large reactivity additions, equation (3.70) can be shown to yield a reactor period of approximately

$$T = \frac{l}{k_{\text{eff}}(\rho - \beta)} \tag{11.8}$$

where k_{eff} is the effective multiplication factor. For a fast reactor with a typical neutron lifetime of 10^{-6} s, a step reactivity addition of 1.2% to a critical reactor with an average delayed neutron fraction of $\beta = 0.0040$ would give a period of 1.25×10^{-4} s.

The effective delayed neutron fraction, β, is not only smaller for a plutonium-fueled fast reactor than for a uranium-fueled thermal system ($\beta = 0.0021$ for Pu^{239}, $\beta = 0.0065$ for U^{235}), but is also more difficult to predict accurately because of the relatively large number of isotopes (including fertile species such as U^{238} and Pu^{240}) which may be making a significant contribution to the delayed neutron population. Taking into account the effect of the fast fission in U^{238}, which has a delayed neutron fraction of around 0.0148, the average value of β for a large plutonium-fueled fast reactor is in the region of 0.0040.

In assessing the likely course of a power transient in a fast reactor, caused by the injection of a large positive reactivity by some appropriate mechanism, two effects are of crucial importance. These are described numerically by the sodium void and Doppler coefficients. The net effect of the removal of sodium from the core is given by the balance of three competing factors. In the first place, hardening of the neutron spectrum will give a positive reactivity effect, both because of the shift to a region of lower capture-to-fission ratio in the fuel isotopes and an increase in the fission rate due to the threshold effect in the fertile isotopes. An additional, but less important, reactivity addition arises from the reduction in neutron capture in the sodium itself. The third effect produced is negative and arises from the increase in neutron leakage due to the loss of the neutrons which would have been reflected back to the core by the sodium. For small cores, where the neutron leakage is high, the negative contribution of the last of these effects overcomes the other two, but for the large core of a commercial FBR the overall sodium coefficient will be positive.

It was at one time thought likely that, in order to ensure an adequate feedback mechanism for automatic limitation of power transients in fast reactors, it would be necessary to employ a core shape deliberately designed to produce a high leakage, such as a cylinder with a relatively low height-to-diameter ratio. The adoption of a high-leakage geometry brings with it the disadvantage of a reduction in the breeding gain. Fortunately, it would now appear that such an approach is not, in fact, necessary, since the Doppler effect is, or can be made, sufficiently large to overcome any net positive sodium coefficient characteristic of large FBRs. Once again, the prediction of the Doppler effect is more complicated than in a thermal reactor, where the predominant contribution of temperature broadening is from the increased capture in the lower-energy resolved resonances of U^{238} (see Chapter 3). In the fast reactor, on the other hand, one has a high degree of overlap between resonances of a given isotope and also a considerable overlap between resonances of different isotopes. A further complication is that, while for the thermal reactor the spectrum shape in the resonance region can be assumed to show a $1/E$ dependence, whatever the reactor composition, for fast reactors the spectrum shape in the Doppler region is fairly sensitive to changes in core design.

For a large plutonium-fueled FBR, the overall Doppler coefficient is negative owing to the predominant effect of U^{238}, for which the coefficient is around -5×10^{-6} per kilogram over the temperature range from 300 to 1200 K. A significant negative contribution also comes from resonance broadening in absorbing materials such as steel. The Doppler coefficient for

Pu^{239} is much smaller than that of the U^{238}, while the coefficient of U^{235} is positive, but less than that of U^{238}. If necessary, the overall coefficient of the core can be made more negative by softening the spectrum by adding moderating material, such as BeO. The reason for the large negative coefficient in a full-scale plutonium-fueled FBR can be seen from the spectra shown in Fig. 11.1. In comparison with a small experimental reactor, the full-scale breeder has many more neutrons at energies up to about 20 keV, which contribute most strongly to the U^{238} Doppler effect. While the methods of calculation have been developed to a degree where the Doppler coefficient can be computed to an acceptable accuracy of some $\pm 15\%$, more information is needed on the value at the higher temperatures of 2000 K and over which are likely to be attained in a severe power excursion.

In the absence of the negative Doppler effect, the rate of rise of power in the prompt supercritical regime would be so rapid that any excursion would be terminated only after vaporization of the fuel and consequent explosive disruption of the core, in the course of which very high pressures, in excess of 10^3 atm, could be attained. The negative feedback from a strongly negative Doppler coefficient, however, reduces by a large factor the maximum pressure attained in the excursion, or may even, if large enough, terminate the excursion before fuel vaporization occurs.

Other possible reactivity feedback effects which have to be taken into account arise from expansion of the reactor due to changes in temperature. Axial expansion of the fuel elements normally produces a negative effect, since the reduction in fuel density is more important than the increase in overall core volume. A positive contribution can arise, however, from the phenomenon of fuel element bowing which is due to the radial difference of temperature across a fuel cluster which is subject to lateral restraint at its ends. Since the flux level is higher on the side of the element nearest to the core center, the temperature expansion tends to cause a bowing in that direction, moving fuel inwards and thus increasing the core reactivity. This positive reactivity may be reduced by the provision of mechanical restraints, such as the "leaning post" in the PFR.

A major loss of coolant accident is improbable because of the low operating pressure of a sodium-cooled reactor. There are two basic designs for the liquid metal FBR: the pool and loop types. In the pool version, the core, primary pumps and intermediate heat exchangers are all submerged in liquid sodium within a large tank (Fig. 11.3). In the loop design, only the core is submerged within the sodium tank, the primary circulation pumps and heat exchangers being mounted externally. One of the principal advantages of the pool type is the much greater coolant heat capacity in the event of

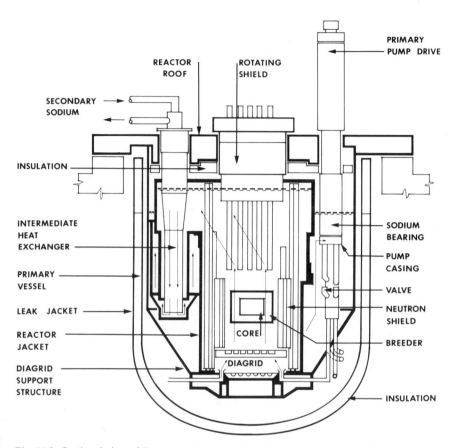

Fig. 11.3. Sectional view of Dounreay Prototype Fast Reactor (courtesy of United Kingdom Atomic Energy Authority).

pump failure, but this is counterbalanced to some extent by the requirement for a larger sodium inventory and the increased difficulty of access for maintenance of components.

The probability of coolant loss can be further minimized by a design, such as the PFR, where a double-walled pressure vessel is supported from the roof, with all penetrations at the vessel top. Any leakage which did occur with this arrangement would be gradual and readily detectable before becoming a problem. Hence the rapid depressurization fault condition which is important in considering the safety of water-cooled and gas-cooled reactors does not really arise for the sodium-cooled fast reactor. A further advantage is that the good heat transfer properties of sodium and the high heat storage capacity possible, particularly with the pool design, remove the necessity for providing elaborate emergency cooling systems.

On the other hand, the high heat rating makes even a minor coolant channel blockage a very serious matter. Such a blockage, due to the detachment of a plate in the Enrico Fermi fast breeder reactor, for example, led to an extensive core meltdown which resulted in the reactor being out of commission for some four years. Progressive propagation of local melting to give an extensive core meltout could lead to the formation of a localized critical volume of fuel, since the quantity of high-enrichment fuel in the core is sufficient to provide several critical masses if rearranged into a more reactive configuration. It is therefore of great importance (a) that the probability of local loss of coolant flow is reduced to an acceptably low value and (b) that in the unlikely event of a major core meltdown taking place, arrangements are made to disperse the molten material and so prevent the occurrence of localized critical masses. Blockage of the coolant channels arising from the passage into the core of debris which may have been detached from components in the primary circuit can be prevented by suitable design of the inlet ports and by the provision of filters. Blockage arising from displacement of, or mechanical damage to, the fuel elements themselves can be prevented by adequate restraints and by extensive monitoring of the coolant leaving the individual assemblies to provide advance warning of damage which could lead to loss of coolant flow. To avoid deterioration of coolant flow as a result of fuel pin expansion caused by irradiation-induced voidage (see Section 5.7) or fuel distortion due to long-term bowing, it is necessary to incorporate a suitable margin against changes in fuel pin geometry in the cluster design. If necessary, fuel clusters in regions of high flux gradient, such as the core–blanket interface, may be rotated through 180° at intervals to prevent accumulation of distortion in one particular direction.

The use of gas cooling, with the consequent high primary circuit pressure compared with a sodium-cooled reactor, increases the probability of a loss-of-coolant accident. With a prestressed concrete pressure vessel however, the only mechanism which could lead to a rapid depressurization is the failure of one of the penetration seals into the vessel. The maximum rate of depressurization can be limited by flow restrictions built into the penetration closures, and the reactivity worth of the helium is typically well below the delayed neutron fraction, so that reactivity transients are not a problem. A reliable emergency heat removal system, independent of the main cooling system, is essential.

Another possible fault in a gas-cooled fast reactor is leakage from the steam generator tubes, leading to an ingress of steam into the reactor core.

The reactivity effect of the steam will tend to be negative, since the enhanced absorption in control rods and fission product poisons arising from the softening of the spectrum more than compensates for the reduction in neutron leakage due to the addition of hydrogen.

Owing to the small thermal capacities of core and coolant in a gas-cooled FBR, the initial rate of temperature rise in the event of a loss-of-flow transient will be greater than in the metal-cooled reactor. The combination of a negative temperature coefficient and the absence of a voidage reactivity effect, however, result in a more rapid negative feedback.

11.4. The Dounreay Prototype Fast Reactor

Following experience with small zero-power reactors designed to investigate the basic physics of the fast reactor system, the United Kingdom Atomic Energy Authority constructed the Dounreay Fast Reactor (DFR) at a site in the North of Scotland. The plant, which started to operate in 1959 at an output of 60 MWt, or 15 MWe, had a variety of objectives, including a study of the problems of handling the liquid sodium coolant and the testing of fast reactor fuel. In addition to its successful role as a fuel test-bed, the DFR has been of value in demonstrating the existence of, and providing information on, materials phenomena such as the swelling and irradiation creep of metals under fast neutron irradiation.

The next stage of the U.K. program was the construction of the 250-MWe Prototype Fast Reactor (PFR), which would permit the testing of plant components of the size appropriate to a full-scale power reactor and also allow the irradiation of full-length fuel elements at the expected level of flux, which is some four times greater than in the DFR. Construction of the reactor started in 1966, with criticality being achieved in 1974 and full-power operation in 1977. The PFR was one of the world's first three large power-producing fast reactors, the others being the French Phenix, of 250 MWe, and the Soviet BN350, of 350-MWe output. All three are of the sodium-cooled type.

The general layout of the reactor core and pressure vessel is shown in Fig. 11.3. Unlike the DFR, the reactor is of the pool type, the core and the primary circuit sodium pumps and heat exchangers all being submerged in liquid sodium within a stainless steel pressure vessel of diameter 12.2 m and depth 15.2 m. The pressure vessel is suspended by its periphery from the concrete roof structure. All the main components of the primary circuit are likewise suspended from the roof and all connections are routed through it, with the consequence that there are no penetrations of the pressure vessel below the level of the top of the sodium. As an additional safeguard against loss of coolant, the pressure vessel is surrounded by a separate steel leak jacket, which is also suspended from the roof.

The reactor core rests on a steel diagrid which in turn is supported by a cylindrical support structure suspended from the roof. The reactor jacket, which surrounds the core and core shielding assemblies, and which is divided into three compartments each containing two of the six heat exchangers, sits on the bottom of the diagrid support structure. The three sodium pumps operate in the space between the support structure and the reactor jacket. Sodium at a temperature of about 400°C is drawn by the pumps from the bulk coolant and delivered to the bottom of the core. After passing upwards through the core, where it is heated to a temperature of 560°C, the sodium flows through the top region of the neutron shield rods and then down through the intermediate heat exchangers before passing into the bulk sodium outside the reactor jacket. The internal surfaces of the reactor jacket are clad with stainless steel insulation packs to minimize the heat transfer from the hot sodium at the core exit to the cooler sodium outside the jacket.

The layout of the coolant circuit and steam-raising plant has already been shown in Fig. 7.2. The intermediate heat exchangers are of the counterflow tube and shell type, with the primary sodium flowing through the tubes. Each pair of intermediate heat exchangers is connected to a secondary loop where the heat is transferred to one of the three 200-MW steam-raising plants, each of which includes reheater, evaporator, and superheater sections. All heat exchangers in the steam plant are vertically mounted shell and U-tube units. Steam at standard conditions of 538°C and 163 kg cm^{-2} pressure is delivered to the turbine, which has a nominal output of 270 MW.

The fuel assemblies for both the core and radial breeder regions consist of pins enclosed within a hexagonal stainless steel shroud which is 142 mm across flats. The pins are supported at intervals by grids attached to the shroud tube. In the core region, each element has 325 pins containing pellets of mixed UO_2–PuO_2 fuel, clad in stainless steel, while the elements in the radial breeder have 85 wire-wrapped pins containing pellets of natural or depleted UO_2. The elements within the core region are subdivided axially, with the mixed oxide in the central 915 mm comprising the core itself and the natural or depleted oxide forming the top and bottom reflector breeder regions of the reactor, each of 228-mm length. In both the core and radial breeder regions, the upper section of the element incorporates the breeder fuel in a 19-pin bundle, the pins being fitted with helical fins to provide a mixing action for the outlet coolant prior to its being sampled for fission product activity. Below the fuel-containing regions of both core and radial breeder elements is a plenum of length 1.19 m, which is needed as a reservoir for the considerable volume of fission product gas generated as a result of the high burn-up of the fuel. The total length of the fuel element is 2.74 m.

The layout of the core is shown in plan in Fig. 11.4. The core is divided into two regions for power flattening purposes, and outside the core and breeder there is a reflector and shield of steel rods and tubes containing graphite. The fuel assemblies are grouped in clusters of six around central leaning posts which act as guide tubes for the absorber rods. The lower support fixtures for the six elements in the cluster are designed so that the elements are forced inwards against the leaning post, thus giving core stability while allowing for element distortion arising from long-term irradiation. The leaning post position at the center of the core is used to accommodate a special safety rod, and there are five tantalum control rods and five boron carbide shutdown rods distributed in other positions throughout the core. The rod operating mechanisms are positioned above the top shield of the reactor.

For the initial fuel loading, the assemblies in the inner and outer regions of the core itself contained fuel having equivalent plutonium contents of 21%

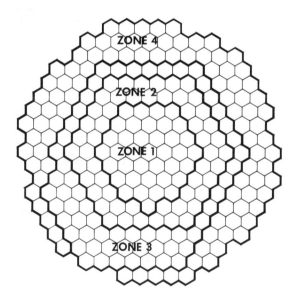

Fig. 11.4. Core layout for Dounreay PFR (courtesy of United Kingdom Atomic Energy Authority). Zone 1, inner core; zone 2, outer core; zone 3, radial breeder; zone 4, breeder reflector.

and 27.5 %, respectively. The target burn-up for a commercial fast breeder is at least 100,000 MW d/t, corresponding to about 10 % heavy atom burn-up. Peak burn-ups of greater than this had been achieved in the earlier DFR, but one of the objectives of the PFR was to demonstrate that the same level of burn-up could be achieved with standard production-line fuel under the different flux conditions expected for a commercial fast reactor. The fuel in PFR has in fact shown an excellent performance, the target of 10 % burn-up having been achieved without failures in the standard fuel. Some experimental fuel pins, containing vibrocompacted fuel, have failed at lower burn-up; if this type of fuel had been equally successful, there would have been advantages from the manufacturing viewpoint, since the preparation of vibrocompacted fuel leads to lower plutonium contamination levels in the production facility.

So far as the nuclear side of the system is concerned, the PFR, like its French counterpart Phenix, has performed extremely well. Owing to the negative power coefficients, both reactors are highly stable, and will run for many hours without the need for control rod adjustments. Tests on the PFR have demonstrated its ability to accept a fault in which all power supplies fail, including the emergency standby systems; these tests show a smooth transition from pumped coolant flow to natural convective circulation without excessive temperature transients. This suggests that it will be possible to design a full-scale LMFBR which would accept the simultaneous failure of all pumps, coupled with a failure to scram, without suffering significant damage over a time long enough to permit long-term remedial action to be taken.

The problems that have arisen at PFR have been almost exclusively on the conventional, or steam-raising, side of the plant. The sodium-to-water steam generators, which were deliberately chosen to be of a type that would be appropriate for a full-scale commercial reactor, developed small leaks due to stress corrosion of welds at the sodium–water interface. Techniques have been developed for locating and repairing the defective welds, and the steam generators have been redesigned so that the welds are above the cover plates, and thus removed from the sodium–water interface. As a further insurance, the tube material for a commercial plant can be changed from stainless to ferritic steel, at the cost of a reduction in the steam temperature obtainable.

Problems

11.1. Radioactivity is induced in the liquid sodium passing through a fast reactor core as a result of neutron capture in Na^{23}. If the transit time of the coolant through the core (t_c) and the time it spends outside the core (t_o) are both very much less than the half-life of Na^{24} (15 h), show that the equilibrium level of activity per cm^3 of the coolant is given by the following expression: (neutron capture rate in Na^{23}) $\times$ (t_i/t_o).

11.2. For a certain fast breeder reactor, the average neutron flux in the core is 9×10^{14} n cm^{-2} s^{-1}. The average capture cross section of Na^{23} in the reactor spectrum is 2 mb. If the sodium spends 2% of the time in-core, calculate the activity per cm^3 of the coolant under equilibrium conditions. (Density of liquid sodium at operating temperature of reactor is 0.81 g cm^{-3}.)

11.3. A 270-MWe fast breeder reactor has a thermal efficiency of 42% and operates at a load factor of 80%. It is fueled with a Pu^{239}–U^{238} mixture, with a total Pu^{239} mass of 1 tonne. If the breeding gain is 0.25 and the ratio of fissions in U^{238} to fissions in Pu^{239} is 0.15, calculate the doubling time of the reactor.

11.4. The sodium flow rate through the core of the U.K. Prototype Fast Reactor is 6400 lb/s and the inlet temperature is 750°F. The thermal efficiency of the reactor is 45%. Calculate the sodium outlet temperature when the reactor is producing its maximum output of 270 MWe. (Specific heat of sodium is 0.30 Btu/lb °F).

12

Safety and Environmental Aspects of Nuclear Reactors

12.1. The Biological Effects of Radiation

In recent years, increasing emphasis has been placed on the safety and environmental aspects of nuclear reactors. In some cases, public opposition has led to long delays in construction and even to the cancellation of projected power plants. While public concern has been expressed over a wide range of aspects of nuclear power, such as the possibility of proliferation of nuclear weapons arising from a move to a plutonium-based breeder reactor program, the principal source of controversy has centered around the potential effects of radiation exposure to the population either from normal operation of reactors or as a result of some postulated accident which could release radioactivity to the environment.

In order to appreciate the magnitude of the potential hazard associated with the release of radioactivity, it is first necessary to consider the effects on the human body of exposure to ionizing radiation. The present section will deal briefly with the biological consequences of radiation exposure and with the international radiation protection standards which have been adopted. Subsequent sections will deal with releases of radioactivity under normal operating and possible accident conditions, and with other environmental aspects of nuclear power stations.

The effects of radiation on living organisms arise from the damage caused to the molecules of their constituent *cells* by the passage of charged particles. The damage can be classified as *somatic* or *genetic*; somatic damage is damage which affects the individual exposed, while genetic damage involves the *gametes*, and can therefore affect future generations.

In order to understand the ways in which radiation affects the body, it is first necessary to look at the structure and function of the *cell*, which is the fundamental unit of life. A typical cell is illustrated in Fig. 12.1. Almost all cells consist of a central *nucleus*, surrounded by the *nuclear envelope*, which divides it from the *cytoplasm*, which in turn is enclosed within the *cell membrane*, which forms the outer boundary of the cell. The cytoplasm, together with organelles such as the *mitochondria*, contained within it, is responsible for the metabolism of the cell, including the buildup of proteins and elimination of waste products. Within the cytoplasm is a complicated network of membranes, known as the *endoplasmic reticulum*, which forms a set of canals for transporting materials around the cell.

The nucleus is responsible for the direction of the metabolic activity of the cell. The key to this activity lies in the *chromosomes*, threadlike bodies containing a nucleic acid, *deoxyribonucleic acid* (DNA), and, more specifically, in the *genes* which are incorporated in the chromosomes. The great majority of human cells, the *somatic cells*, contain 23 pairs of chromosomes. The only exceptions are the *gametes* (sperm cells and ova), involved in reproduction, which contain only half this number. The combination of a sperm and ovum in the fertilization process produces a new cell (the *zygote*) with the required 23 pairs of chromosomes, which forms the basis for the new individual. Further development takes place by *mitosis*, or cell division, a process in which each chromosome duplicates itself prior to the cell splitting into two new cells, each of which emerges with an identical set of 23 chromosome pairs.

The reason that a relatively small amount of energy delivered to the body in the form of ionizing radiation can produce significant cell damage is that the energy transfer to the material takes place by a series of interactions with a few individual molecules, with relatively large amounts of energy being transferred in each interaction. When a charged particle, such as an α or β particle, travels through matter, it interacts with the electrons of the atoms in the matter through an electrostatic interaction, causing the atoms to become excited or ionized. Many of the electrons produced by an initial ionization event are ejected with sufficient energy to produce further excitations or ionizations of the neighbouring atoms. The basic process is therefore one where the kinetic energy originally possessed by the incoming particle is gradually lost in producing excited and ionized atoms. The effect of exciting an atom, by the promotion of one of its electrons to a higher energy level, is to increase the chemical reactivity of the atom, while atoms which have been ionized are even more reactive.

While α and β radiation produce direct excitation and ionization events as described above, a γ ray can only do so by first interacting with an atom of the material to produce a charged particle. There are three types of interaction which can take place. In *photoelectric absorption*, the γ ray transfers all of its energy to an orbital electron which is then ejected from the

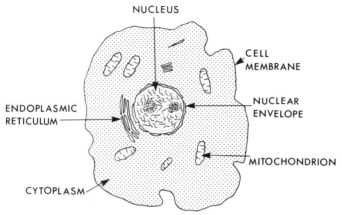

Fig. 12.1. Structure of a typical cell.

atom. In *Compton scattering*, the γ ray is scattered by collision with an electron, and emerges with reduced energy (increased wavelength), the energy lost appearing as kinetic energy of the electron. In *pair production*, a high-energy γ ray interacts with the field of a nucleus to produce a positron–electron pair, some of the energy of the γ being converted into the mass of the two particles and the remainder appearing as kinetic energy shared between the two. In this case, the two charged particles produce ionization and excitation in the tissue as they slow down and, in addition, the positron, on coming to rest, decays into a pair of γ rays, each of approximately 0.51-MeV energy, which can in turn interact to give rise to further ionization and excitation.

The relative probability of each of the three types of γ-ray event depends on the energy of the γ ray. The photoelectric effect predominates at low energies and the Compton at intermediate energies, while pair production can occur only for γ-ray energies in excess of 1.02 MeV (the energy needed to supply the combined rest masses of the two particles).

The spatial distribution of the radiation damage is widely different for the three types of radiation. Alpha particles have a very short penetration (about 35 μm for a 5-MeV particle in body tissue) and leave a short, straight trail of intense ionization behind them. Beta particles have a greater penetration (a few millimeters for a 1-MeV particle) and leave a somewhat erratic track with much lower ionization density. Gamma radiation, on the other hand, can penetrate to considerable depths below the body surface before interacting to produce ionizing events. Fast neutrons also have high penetration, losing their energy mainly by elastic collisions with the nuclei of light elements, particularly hydrogen. In living tissue, about 90% of the energy loss of fast neutrons is by collision with hydrogen nuclei, giving rise to recoil protons which lose their energy by ionizing and exciting the atoms of the material.

The effects of exposure to ionizing radiation will depend on the amount of energy deposited per unit mass of the tissue which is being irradiated, known as the *absorbed radiation dose* (D). For many years, the accepted unit for dose measurement was the *rad*, but this has now been replaced by the *gray*. An absorbed radiation dose of 1 gray (Gy) corresponds to the deposition of 1 J of energy per kilogram of the irradiated material. Commonly used subunits are the milligray (mGy) and the microgray (μGy), which are 10^{-3} and 10^{-6} Gy, respectively. The relation between the gray and the unit it replaces is that 1 Gy = 100 rads. *Absorbed dose rates* are measured in Gy/h, mGy/h, etc.

The damage done by radiation, however, it not simply proportional to the absorbed dose, but depends also on the detailed way in which the energy deposited varies along the path followed by the radiation. For example, an α particle, which gives rise to a dense trail of ionization, produces considerably more biological damage than a β particle which deposits the same energy in the tissue, but spread over a larger volume. This effect is measured by the *linear energy transfer* (L_{∞}), which is the energy deposited by the radiation per unit path length. The greater effectiveness of the high L_{∞} radiation in destroying the cells of higher organisms is related to the number of ionizations required to inactivate the critical part of the cell and the capacity of the cell to repair relatively small-scale damage such as is produced by low L_{∞} radiation.

The relationship between radiation damage and absorbed dose for the different types of radiation is known as the *relative biological effectiveness* (RBE). For a specific type of radiation, the magnitude of the particular biological effect being investigated is compared to that produced by an equal absorbed dose of standard radiation (100-keV γ rays), and the RBE is simply the ratio between the two effects. As a guide to the relative effectiveness of various broadly defined groups of radiation in producing biological damage, one may use the so-called *quality factor Q* which is based on a study of numerous RBE measurements. Typical recommended values of Q are given in Table 12.1.

The biological effect of a given absorbed radiation dose may now be estimated by multiplying the dose in grays by the quality factor. This is called the *dose equivalent* (H). Dose equivalent is measured in terms of a special unit, the *sievert* (Sv), which replaces the older *rem* (1 Sv = 100 rems). The dose equivalent is then given by the relation

$$H(\text{Sv}) = D(\text{Gy}) \times Q \qquad (12.1)$$

The sievert, like the gray, is associated with subunits, the millisievert (mSv), and the microsievert (μSv).

The biological consequences of exposure to ionizing radiation can be divided into *direct effects* and *indirect effects*. In the first of these, the chemical bonds holding together a biologically important molecule, such as a protein or nucleic acid, are broken by the passage of the particle, possibly destroying the function of the molecule. The indirect effect arises when a simpler molecule, such as H_2O, is ruptured, leading to the production of chemically active ions or free radicals, which can in turn migrate through the cell and affect its more complex constituents. For example, the water molecule may break up in any one of the three modes shown below:

$$H_2O \rightarrow H^+ + (OH)^-$$
$$H_2O \rightarrow H^- + (OH)^+$$
$$H_2O \rightarrow H^\cdot + (OH)^\cdot$$

the first two events leading to ion formation and the third to the formation of highly reactive free radicals (the dot indicates the unpaired electron in these). Apart from the possibility of damage to the cell DNA by the migrating ions or free radicals, the recombination of these can lead to the formation of chemical poisons such as hydrogen peroxide.

The key targets for irradiation damage in the cell are the DNA macromolecules of the genes in the chromosomes, which control the functioning of the cell. According to the well-known Watson–Crick model,

Table 12.1. Quality Factor (Q) for Various Types of Radiation

Type of radiation	Q
γ rays (and x rays)	1
β particles	1
Protons	10
α particles	10
Slow neutrons (<1 keV)	3
Fast neutrons (1–2 MeV)	10

the DNA molecule has the structure shown in Fig. 12.2. It is made up of two long chains of smaller molecules twisted around one another to form a double helix. The DNA molecule may be visualized as a sort of spiral staircase, with the supporting sides made up of molecules of a sugar and a phosphoric acid, while the rungs are made up of pairs of the four nitrogenous bases: adenine, cytosine, guanine, and thymine. Adenine in either strand of the helix pairs only with thymine in the opposite strand, and vice versa, while cytosine can pair only with guanine. The two halves of the double helix are held together by hydrogen bonds between the nitrogenous bases. The information required to control the metabolic activity of the cell is provided by the *sequence* in which the bases occur in the DNA composing the genes which are responsible for synthesizing the particular proteins required.

   In the process of cell division, each DNA macromolecule duplicates before the cell divides, so that the daughter cells should have an identical DNA sequence to the parent. The DNA duplication process involves the spontaneous splitting of the molecule into two separate chains by the breaking of the hydrogen bonds linking the nitrogenous bases. Each chain then picks up the required bases, phosphates, and sugars from the material in the nucleus to build on a complementary chain, identical to the one from

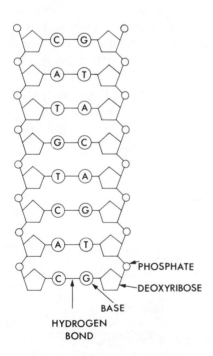

PHOSPHATE

DEOXYRIBOSE

BASE

HYDROGEN
BOND

Fig. 12.2. Structure of DNA molecule. The section shown has been "straightened out" for clarity; in practice, the two long chains are twisted around one another to form a double helix.

which it separated, so that one ends up with two DNA molecules identical to the original.

The identical nature of the product is in principle guaranteed by the requirement that, for example, the only base which can bond to a guanine on the single chain is a cytosine, and vice versa. Occasionally, however, a *mutation* may take place during the duplication, in which a "forbidden" pairing, such as a guanine–thymine, is set up. In general, such mutations are unfavorable; for example, the gene affected may be one which is responsible for the synthesis of a particular enzyme which is critical for the cell. *A lethal mutation* is one where the function of the gene in which the change takes place is so vital that the mutation is rapidly followed by the death of the cell. Even if the mutation is not severe enough to be lethal, the mutated cell is usually disadvantaged to the extent that its progeny will eventually die out.

The effect of radiation exposure is to increase the mutation rate above that normally present. The elimination of mutated cells by immediate or later death means that the short-term effects of the radiation are not serious for the individual unless the dose has been so enormous as to cause the deaths of large numbers of cells, resulting in serious and widespread damage to body organs.

Assuming that the exposure dose has not been high enough to produce death by massive cell damage, certain *delayed effects* may appear over the lifetime of the individual. The most important of these are a general reduction in life span and an increased incidence of cancer. The latter effect, if it occurs at all in the exposed individual, shows up only after a *latent period*, which may be many years in length. The precise mechanism for the carcinogenic action of radiation is still not understood, but it may involve the mutation of a virus normally present in the cell, the mutation being transferred over many generations before the activated virus starts to produce clinically observable changes in the body. The effect seen is one of uncontrolled growth of the cancerous cells, which are no longer subject to the normal body mechanism for growth control. Based on data from survivors of the bombings of Hiroshima and Nagasaki, and on a variety of therapeutic exposures, the incidence of cancer is increased over normal by about one case per 100,000 exposed persons per millisievert, for whole-body exposures exceeding 1 Sv. There is no clear evidence for an increase in the rate of leukemia for exposures below about 1 Sv.

So far, we have been discussing the effect of radiation-induced mutations on the individual exposed. Damage to the *germ cells* which give rise to the gametes can result, following fertilization, in the production of an offspring bearing the mutation. In many cases, the mutation will be lethal, resulting in the immediate death of the zygote. In other cases, however, the mutant may survive the embryonic period, but suffer from more or less severe handicaps such as hemophilia or albinism. Nearly all such mutations are detrimental and result in a shortening of the life span of the individual. The mutation is passed on from one mutant to another through a number of generations, eventually becoming extinct by natural selection.

12.2. Effects of Routine Release of Radioactivity from the Nuclear Industry

12.2.1. Dose Limits for Radiation Exposure

Routine releases of radioactivity result from operations at all stages of the nuclear fuel cycle. In the uranium mining operation, for example, significant amounts of radon gas are emitted from the large tailing piles remaining after the economically recoverable uranium has been extracted from the ore. In the operating reactor, gaseous or volatile radioactive fission products such as xenon, krypton, and iodine can escape through small holes in the fuel cladding and are discharged to the atmosphere in controlled amounts, usually after storage to permit the levels of activity to decay. Radioactive tritium, produced in light or heavy water reactors, may similarly be discharged in gaseous or liquid form. Small quantities of the corrosion products which arise from chemical attack on radioactive parts of the coolant circuit will also leak into the environment. Finally, reprocessing of the highly radioactive fuel, although carried out under the most stringent containment conditions, adds its quota to the total release of activity to the environment.

The most meaningful way of assessing the consequences of the population exposure from normal operation of the nuclear power industry is to compare it with the exposure from the *natural background radiation* due to cosmic rays and the naturally occurring radioisotopes such as potassium-40, carbon-14, and the elements of the uranium and thorium series. The cosmic radiation, a stream of high-energy ionizing particles striking the earth's atmosphere from outer space, contributes an annual dose equivalent of about 280 μSv/yr at sea level, rising by 10 μSv/yr for each 100 m of additional elevation. Some of the exposure from naturally occurring radioelements comes from isotopes which have been absorbed into the body; the bulk of the dose equivalent due to internal sources comes from the element K^{40}, which is widely distributed in body tissue. To the natural background one must now add artificial contributions from sources such as medical x rays and fallout from nuclear weapons tests.

Table 12.2. Average Population Dose Equivalents in the United States from Natural and Artificial Radiation Exposure

Source	Dose equivalent (μSv/yr)
Cosmic radiation	280
γ rays from natural radioelements	260
Internal sources (e.g., K^{40})	270
Medical	380
Weapon fallout	40
Total	1230

The main contributions to the average radiation exposure in the United States are summarized in Table 12.2. It should be noted that the dose from the naturally occurring radioisotopes varies considerably from one location to another, depending on the local geology; in certain regions of India, for example, the soil is so rich in thorium that the annual dose equivalent from this source may be as high as 20 mSv/yr.

The basic guidelines for the exposure of the public to radiation arising from the operation of nuclear power stations are those issued by the International Commission on Radiological Protection (ICRP). While the standards recommended by the ICRP are purely advisory, many countries have established bodies, such as the Environmental Protection Agency (EPA) in the United States, which are responsible for the promulgation of regulations governing standards of radiation exposure. The ICRP recommendations distinguish between two classes of individuals: adults exposed in the course of their occupation and the public at large. Because of the much greater numbers in the second class, and the fact that it includes particularly sensitive categories such as small children, the dose limits for individual members of the public are set at only 10% of the annual doses permitted for radiation workers.

The setting of limits for long-term exposures to very low levels of radiation is difficult because most of the experience of radiation effects in humans has been obtained from the study of individuals who have been exposed to large doses over a short period, such as the survivors of the atomic bomb attacks on Hiroshima and Nagasaki, or to chronic exposures at levels much above the background level, such as watch dial painters who ingested considerable amounts of radium. Studies of long-term exposure of animals to relatively low radiation levels are of doubtful applicability to man because of the strong variation of radiation sensitivity between different species.

In setting radiation limits, it is customary to assume that, for exposure of the whole body to radiation, the effect produced is proportional to the absorbed dose. Since this assumption ignores the possibility of damaged tissue repairing itself during long-term exposure to low radiation doses, it is believed to represent a conservative approach to the estimation of possible damage. Somatic effects where the probability of injury (e.g., the initiation of cancer) is proportional to the dose are known as *stochastic* effects. In contrast to these, an effect such as cataract of the eye is *nonstochastic*; in this case, a threshold dose exists such that no cataract will be induced if the absorbed dose is below this amount.

In order to protect both radiation workers and members of the public from significant radiation injury, the ICRP has made recommendations for the maximum dose equivalent per annum for each of these groups. The most recent recommendations (1977) are shown in Table 12.3. In cases where the whole-body dose has been nonuniformly distributed (e.g., by ingestion of radioactive iodine, which accumulates in the thyroid gland), the dose equivalent to the particular organ is multiplied by a weighting factor (specified by ICRP) to obtain the corresponding whole-body dose equivalent.

Example 12.1. At a certain point in the year, a radiation worker has been exposed to the following whole-body absorbed doses: 10 mGy of γ, 5 mGy of slow neutrons, and 2 mGy of fast neutrons. What further dose equivalent is he permitted before reaching the prescribed annual limit?

Table 12.3. ICRP Recommended Limits for Dose Equivalent[a]

Effects	Dose equivalent limit (mSv/yr)	
	Radiation workers	Public
Stochastic effects:		
Whole body	50	5
Nonstochastic effects:		
Eye lens	150	15
Other tissues	500	50

[a] Based on ICRP Publication 26: "Recommendations of the International Commission on Radiological Protection" (1977).

The dose equivalents for the three types of radiation are as follows:

$$\gamma: H_\gamma = D \times Q = 10 \times 1 = 10\,\text{mSv}$$
$$\text{slow neutrons: } H_{SN} = 5 \times 3 = 15\,\text{mSv}$$
$$\text{fast neutrons: } H_{FN} = 2 \times 10 = 20\,\text{mSv}$$

The total dose equivalent is 45 mSv, so that the maximum allowable increment during the year will be $50 - 45 = 5\,\text{mSv}$.   □

12.2.2. Release of Radioactive Isotopes from Nuclear Power Plants

In order to be of value to the nuclear plant operator, the population dose limits have to be translated into maximum permitted limits for the release of *radioactive nuclides* (*radionuclides*) from the reactor. For each nuclide released, a complicated chain of calculation is necessary to relate the quantity released to the average population dose incurred. One of the first factors to be taken into account is the detailed manner in which the radionuclide is dispersed into the atmosphere or other medium of transmission. The total dose incurred will be derived from material both internal and external to the body. In the former case, it is necessary first to identify all the possible pathways by which the radionuclide can enter the body. These will include the following: inhalation, or ingestion from water supplies; deposition on vegetables used for human consumption; deposition on pasture consumed by animals supplying milk or meat; accumulation in sea, lakes, or rivers where the radioactivity may be concentrated appreciably through the course of complex food chains (e.g., seaweed → crustaceans → fish). The various pathways have to be related to the average inhalation rate of air and the rate of consumption of water and the various foods involved. Once the radionuclide enters the body, the accumulated dose will depend on the type and energy of the radiation, its half-life, the fraction absorbed by each organ and the rate of biological elimination.

As a result of calculations based on a knowledge of the factors above, it is possible to set limits on the discharges of specific radionuclides from the nuclear plant. These limits are referred to as *derived limits*. They may specify simply the maximum allowable rate of release or the maximum permissible concentrations in air, water, or particular members of the relevant food chains.

The principal source of radioactivity in a nuclear reactor arises, of course, from the fission products in the fuel elements. The amount of activity present will depend on the power level of the reactor and the time for which it has been operating. A 1000-MWe PWR at the end of the core life will typically have a total fission product activity level of around 12,000 megacuries (MCi) (4.4×10^{20} Bq). In addition, the fuel will contain some 4000 MCi of actinides (heavy elements formed following neutron capture in the fuel).

Despite the large fission product and actinide inventory of the fuel, the leakage rate from the fuel pins may be kept to a very small value by adequate design and careful quality control in manufacture. The coolant circuits of reactors which are refueled on-load are particularly easy to maintain at a low level of activity, on account of the ease with which defective elements can be removed. In addition to any small escape of fission products from cracks or pinholes in the cladding, some activity will generally arise from slight contamination of the outsides of the fuel pins with fuel while they are being loaded.

In addition to the activity stored in the fuel elements, further radioactive isotopes may be formed by activation of the moderator, coolant, and structural materials by neutron capture. Particulate material arising from corrosion or erosion of the structure by the coolant will be circulated through the core and thus become activated. This can lead to accumulation in cooler parts of the circuit, such as the heat exchangers, or, in a direct cycle plant like the BWR, in the turbine itself.

The half-lives and methods of production of some of the more important fission or activation products are given in Table 12.4. For the BWR, the main gaseous or volatile species include the isotopes of the noble gas fission products krypton and xenon, the isotopes of iodine, the nitrogen isotopes N^{16} and N^{17} produced by neutron activation of the coolant, and tritium. The nitrogen isotopes, on account of their very short half-lives, represent no hazard to the public, but can be responsible for high levels of activity in the turbine and associated plant.

The great majority of the gaseous effluents present in the primary circuit are mixed with relatively large volumes of air when they are removed through the condenser air ejector. The gaseous effluent is delayed long enough to allow the shorter-lived components, such as the nitrogen isotopes, to decay, and is then passed through particulate filters, before being discharged to the atmosphere through a stack, which is typically about 100 m in height. The time taken for the gases released from the top of the stack to reach ground level allows time for further decay, although this and other delays have essentially no effect on the longer-lived fission product gases such as Kr^{85} (10.8-yr half-life).

Under normal conditions, the containment of gaseous fission products in a PWR is simpler than in a BWR since they remain within the nonboiling primary coolant circuit. The relatively small amounts which escape from this

Table 12.4. Characteristics of the Most Important Gaseous or Volatile Fission or Activation Products in Nuclear Reactors[a]

Radioisotope	Half-life	Source
Kr^{83m}	1.86 h	All produced as fission products.
Kr^{85m}	4.4 h	
Kr^{85}	10.76 yr	
Kr^{87}	76 min	
Kr^{88}	2·8 h	
Kr^{89}	3.18 min	
Kr^{90}	33 s	
I^{129}	1.7×10^7 yr	All produced as fission products.
I^{131}	8.05 d	
I^{132}	2.26 h	
I^{133}	20.3 h	
I^{134}	52.2 min	
I^{135}	6.68 h	
I^{136}	83 s	
Xe^{131m}	11.8 d	All produced as fission products.
Xe^{133m}	2.26 d	
Xe^{133}	5.27 d	
Xe^{135m}	15.6 min	
Xe^{135}	9.14 h	
Xe^{137}	3.9 min	
Xe^{138}	17.5 min	
Xe^{139}	43 s	
H^3	12.3 yr	Fission product. Also produced by reaction $B^{10}(n, H^3)Be^8$ in boron added to moderator for control purposes. In HWR, large production from neutron capture reaction $_1H^2(n, \gamma)_1H^3$ in heavy water moderator and coolant.
N^{16}	7.1 s	$O^{16}(n, p)N^{16}$ reaction in oxygen of coolant.
N^{17}	4.1 s	$O^{17}(n, p)N^{17}$ reaction in oxygen of coolant.
O^{19}	29 s	$O^{18}(n, \gamma)O^{19}$ reaction in oxygen of coolant.
Ar^{41}	1.83 h	$Ar^{40}(n, \gamma)Ar^{41}$ in argon present in air or carbon dioxide coolant in gas-cooled reactor.
C^{14}	5730 yr	Produced by following reactions: $N^{14}(n, p)C^{14}$ from nitrogen in fuel, cladding, or coolant; $O^{17}(n, \alpha)C^{14}$ from oxygen in fuel, cladding, or coolant; $C^{13}(n, \gamma)C^{14}$ from carbon in moderator and coolant of graphite-moderated, gas-cooled reactors.

[a] Notations of the type Kr^{83m}, for example, refer to a so-called *metastable* state of the nucleus concerned. Kr^{83} is formed by the beta decay of Br^{83}; this decay may involve a transition either to the ground state of Kr^{83} or to an excited state at 42 keV above the ground state. The excited state decay is very much slower (1.86 hr half-life) than the normal gamma decay from an excited state, which is virtually instantaneous (see Section 1.3.1). The isotope Kr^{83m} is therefore simply an excited state of Kr^{83}, which will decay by gamma emission with a 1.86 h half-life.

circuit are collected and may be stored for periods of a month or more, so that the gaseous waste of a PWR has much lower concentrations of short-lived fission products than the BWR. Under fault conditions, primary coolant may leak into the secondary circuit through defective heat-exchanger tubes, in which case the shorter-lived gases may escape through the condenser air ejector. The content of 12.3-yr half-life tritium in the primary circuit of the PWR tends to be high because of its production through neutron capture in the boron which is added to the coolant for the purposes of reactivity control. Despite this, the gaseous release rates from the indirect cycle plant are so much lower than for the BWR that the gases may be released to atmosphere without the requirement for a high stack.

Liquid wastes in LWRs may be treated by filtration, evaporation, demineralization, and storage to permit decay. The latter technique is less useful than for gaseous effluents, since the shorter-lived isotopes will in any case largely have decayed during their relatively slow passage through the plant. Evaporation is used to increase the concentration of the active wastes, thus facilitating later handling and disposal. For the PWR, tritiated water, formed by the combination of tritium with oxygen, is difficult to remove and is recirculated with the primary coolant.

The activity levels in the primary circuits of heavy water reactors are similar to those in PWRs, except for the larger amounts of tritium formed in the coolant. Release of tritium to the atmosphere, however, is minimized by the fact that conservation of heavy water already requires elaborate arrangements for the recovery of vapor escaping from the coolant or moderator circuits. The air circulated for the ventilation of areas containing high-pressure pipework, for example, is passed through driers which remove the heavy water vapor, and its associated tritiated water. The use of closed-cycle ventilation also ensures that noble gas radionuclides are delayed long enough to permit the decay of short-lived activities.

For gas-cooled reactors using CO_2 as coolant, the main sources of environmental radioactivity are the planned release of the coolant gas and the discharge of air used for ventilation and cooling of plant components. The main contributor to the gaseous activity discharge is 1.8-h half-life argon-41 produced by neutron activation of the argon in the air (1.3% by weight) or in the CO_2 coolant. Liquid effluent activity is contributed by activation of the water used to cool the concrete in prestressed vessel reactors and by contamination of the water in the cooling ponds which are used for storage of fuel discharged from the core.

For the liquid metal fast breeder reactor, of the pool type, the main gaseous release of radioactivity in normal operation arises from leakage of the argon atmosphere above the sodium in the primary circuit. In addition to the noble gases, a very small fraction of any solid fission products released to the sodium from defective fuel elements will eventually accumulate in the argon. Most of the solid fission products, and compounds such as sodium iodide, will subsequently plate out on the colder parts of the primary circuit, but equipment is included in the argon circulating system to remove most of the remaining solid fission products. Before any of the argon is discharged to the atmosphere, it is stored to allow for decay, and the long-lived isotopes, in particular Kr^{85}, are then removed by absorption on beds of activated charcoal.

The releases of radioactivity from nuclear power plants tend to vary markedly, depending on power level, fuel failure rate, and degree of efficiency for the removal of activity from the effluents. The releases of airborne activity from BWRs and PWRs are of the order of 500,000 and 2000 Ci/yr, respectively, while the liquid releases from both types of plant are around 100–200 Ci/yr, mostly in the form of tritium. The general experience in operating nuclear power plants has been that there is little difficulty in keeping routine releases of activity down to levels which result in radiation exposures to the public which are far below the recommended limits. Concern has, however, been expressed about the long-term effects due to buildup of long-lived isotopes such as I^{129}, Kr^{85}, and C^{14}. An evaluation of this hazard, taking account of all stages of the nuclear fuel cycle, will be given in the following section.

12.2.3. Estimated Radiation Doses Due to the Nuclear Fuel Cycle

An assessment of the potential hazard due to the release of radioactive material has to take account of the following parts of the nuclear fuel cycle: (i) mining and fabrication of fuel, (ii) operation of power reactors, (iii) reprocessing of fuel (if desired), and (iv) disposal of wastes.

The following discussion is based on a number of recent studies, the most important of which are those by E. E. Pochin (1976) and by the Study Group of the American Physical Society (1978). Details of these publications are given in the Bibliography.

The most significant hazard from uranium mining is due to radon gas (Rn^{222}) which is released from the mine or the uranium mill tailings. Inhalation of radon leads to irradiation by its α-emitting daughters deposited on the surface of the lung. The quantity of radon emitted from mill tailings and mine ventilation systems is such that its contribution to the natural radon background is not significant except in the proximity of the mine or

Table 12.5. Global per Capita Dose Rate (Sv/yr) at the End of T Years from a Continuous Release of 1 Ci/yr for T Years[a]

T (yr)	Kr^{85}	C^{14}	H^{3b}	H^{3c}	I^{129d}
1	0.5 (-16)	9 (-13)	0.2 (-14)	0.1 (-15)	0.3 (-10)
10	4 (-16)	8 (-12)	0.8 (-14)	0.7 (-15)	2 (-10)
50	8 (-16)	2 (-11)	1 (-14)	1.5 (-15)	7 (-10)
100	9 (-16)	3 (-11)	1 (-14)	1.6 (-15)	8 (-10)
500	9 (-16)	5 (-9)	1 (-14)	1.6 (-15)	10 (-10)

[a] Based on Report to the APS by the Study Group on Nuclear Fuel Cycles and Waste Management [*Rev. Mod. Phys.* **50**(1), Part II (January 1978)].
[b] Release to air.
[c] Release to marine environment.
[d] Dose to thyroid.

tailing pile itself. Problems have arisen where the mill tailings have been used as building fill; reduction of the hazard from tailings to an acceptable level will require the avoidance of this practice and the use of a covering layer of suitably impervious material, such as clay, for the tailings pile.

As mentioned in the previous section, the most significant sources of radiation dose in releases from the operating reactor are the long-lived isotopes, such as H^3 (12.3 yr), C^{14} (5730 yr), Kr^{85} (10.8 yr), and I^{129} (1.7 $\times 10^7$ yr). For the long-lived isotopes, the radiation exposure from a release at some specific instant has to be integrated over very long periods of time. To illustrate the relative importance of the radioisotopes quoted, Table 12.5 gives the global per capita dose equivalents which result from a continuous release of each of the radioisotopes at a rate of 1 Ci/yr for various periods of time. The calculation of exposure doses involves a knowledge of the way in which a radioisotope such as C^{14} (in the form of CO or CO_2 gas) disperses in the global environment after release. This information can be obtained by studying the dispersion of C^{14} produced from nuclear explosions in the atmosphere.

The estimated rates at which these isotopes are released from nuclear power plants are based on operating experience with a large number of reactors. The release rate will depend on the reactor type, power level, turbine and containment design, and population density around the plant. Of the four isotopes, the one whose release rates are most uncertain is C^{14}, since its significance for the long-term radiation dose has been appreciated only rather recently.

The total integrated radiation dose from reactor operation has to be evaluated in terms of two components: the dose to the public and the occupational dose incurred by the plant operating personnel who, in addition to being exposed to higher concentrations of gaseous or volatile emissions, are subject to direct irradiation from sources such as N^{16} in the turbine and from operations such as fuel transfer. The integrated operational dose to LWR operating and maintenance personnel in the United States has been estimated by the APS Study Group to be of the order of 11 man-mSv/MWe yr. The corresponding dose to members of the public is estimated as being about one tenth of this.

It turns out that the greatest potential releases of radioactivity are associated with the introduction of reprocessing plants to recover plutonium and uranium from the spent fuel. This is because, while only a very small fraction of the gaseous or volatile fission product inventory is released in the reactor under normal operational conditions, essentially all of the H^3 and Kr^{85} and a high fraction of I^{129} and C^{14} are released from the fuel during the chemical treatment. The degree to which they are released to the environment depends on the extent to which the plant incorporates retention systems for the various radionuclides. The APS Study, for example, assumes that the standard retention systems are included for I^{129}, but that no provision is made for the more difficult cases of H^3, C^{14}, and Kr^{85}. On this basis, it is found that, after steady state operation of reprocessing plants over a period of 500 yr, 70% of the public dose rate is due to C^{14}, 12% to H^3, and 18% to Kr^{85}. The high fraction due to C^{14} suggests that measures should be taken to identify and minimize the sources of leakage of this radionuclide (as CO_2 gas) from the plants.

The question of waste disposal is important enough to merit separate consideration, and this is dealt with in Section 12.6. For the present, it is sufficient to note that the deep burial method of waste disposal ensures that the dose either to the public or to the radiation workers involved is very much less than that arising from the other stages of the fuel cycle.

The estimated dose commitments for public and occupational exposure are shown in Table 12.6, which is based on the work of Pochin and on the APS Study. Since the contribution of the longer-lived radionuclides, such as C^{14}, will depend on the time period available for their buildup, the dose commitment will be a function of the total time for which the nuclear generating program has been operating; in both the cases considered in Table 12.6, this time is assumed to be 500 yr. The units of dose commitments are man-mSv/MWe yr; if one assumes that at some future time the nuclear generation capacity is at a level of a 1 *kilo*watt per head, the figures quoted will represent the associated *annual dose rate* per head in *micro*sieverts per year. These figures should be compared with the average dose equivalent of

Table 12.6. Estimated Whole-Body Dose Commitments for Various Stages of the Fuel Cycle (Units of Man-mSv/MWe yr)

Operation	Pochin[a]		APS study[b]	
	Public	Occupational	Public	Occupational
Mining and milling	—	<2	<1.4	2
Reactor operation[c]	1	20	1	11
Reprocessing	13.5	20	4	1–6
Total	14.5	∼42	∼6.4	14–20

[a] E. E. Pochin, "Estimated population exposure from nuclear power production and other radiation sources," Nuclear Energy Agency, OECD, Paris (1976).
[b] APS Study Group (see footnote a to Table 12.5).
[c] Routine operation only; radiation dose due to potential accidents will be discussed in Section 12.3.

approximately 1200 μSv/yr from natural background and medical uses of radiation (see Table 12.2).

It is instructive to compare the predicted rate of cancer induction in the public at large, due to this average radiation dose, with the "natural" rates of cancer incidence. Based on the assumption of a linear relation between radiation dose and cancer induction rate, the probability of cancer death per unit of dose absorbed by an individual may be calculated at approximately 10^{-8} per μSv. Taking the average of the two estimates of public dose quoted in Table 12.6, the corresponding risk of cancer death, for an individual member of a population consuming 1 kW per head of nuclear-generated electricity, is about 10^{-7} per year. It is known that the "normal" probability of developing a fatal cancer is approximately 2×10^{-3} per year on average. The additional cancer risk posed by the radiation dose from an all-nuclear economy is therefore about one 20,000th of the normal risk of cancer mortality in the individual.

12.3. Reactor Accidents: Safety Studies on Light Water Reactors

12.3.1. Analysis of LWR Safety

As the majority of power reactors are of the BWR or PWR design, this is the type on which the most extensive safety studies have been made. A 1957 study, *Theoretical Possibilities and Consequences of Major Accidents in Large Nuclear Power Plants* (*WASH-740*), was intended to provide an estimate of the upper limit of the consequences that might be involved in a reactor accident, but did not attempt to link these consequences to the probability of occurrence of the initiating accident mechanisms. A much more detailed study, published by the United States Nuclear Regulatory Commission (NRC) in late 1975, was the *Reactor Safety Study (WASH-1400)*, frequently referred to as the Rasmussen Report, which included a very detailed consideration of accident sequences in light water reactors, using the more recently developed methods of event-tree and fault-tree analysis. An independent study, also published in 1975, was the *Report of the American Physical Society (APS) Study Group on Light Water Reactor Safety* which, although not a review of WASH-1400, proposed some important modifications to the conclusions of the draft version of that report (WASH-1400-D, published in August 1974). The discussion which follows is based very largely on WASH-1400 and the APS report.

In the event of the occurrence of a major accident, resulting in severe damage to the structure of a full-scale power reactor, the principal hazard would arise from the dispersion of the very large quantities of radioactive materials which accumulate over the life of the reactor core. It should be noted that the magnitude of any explosive energy release would be at worst only a very small fraction of that involved in the deliberate explosion of a nuclear bomb, which has to be designed very specifically to achieve a rapid conversion of mass to energy.

The principal sources of radioactivity in the reactor are the *fission products*, supplemented by the transuranic elements, or *actinides*, formed as a result of successive captures of neutrons in the uranium fuel. The fresh fuel loaded into the core is only mildly radioactive (about 300 Ci for the initial core of a typical BWR), but the activity increases steadily over the core life to a value of the order of 1.7×10^{10} Ci just prior to refueling.

The principal components of the stored activity for an 1100-MWe PWR are given in Table 12.7, which is taken from the APS study. The top line applies to the reactor immediately after shutdown following a long period of operation at full power, while the remainder of the table illustrates the rate of decay of activity, as a function of time after shutdown. The activities of the iodine and bromine isotopes, and the noble gases, whose behavior is particularly important in the event of a major dispersal event, are shown separately, and the activity induced by neutron irradiation of the reactor structure and coolant is also listed. The absorption of β particles and γ rays in the fuel and cladding leads to a considerable generation of heat in the core; the last column of the table gives the total thermal power produced in this way as a function of time after shutdown.

The release of a significant fraction of the stored radioactivity requires the breaching of multiple barriers. In the first place, most of the fission products and actinides are embedded in the fuel pellet matrix, and large-scale escape would only occur if the fuel were to melt. The fuel is enclosed in zircaloy cladding and the core is enclosed within the pressure vessel which forms part of a sealed primary circuit. Finally, the whole of the primary circuit is enclosed within a containment structure which is specifically designed to minimize release of radioactivity to the environment.

Melting of all or part of the core could only occur as a result of the cooling capacity dropping significantly below the rate at which power is being produced in the fuel. There are two classes of accidents which could cause this to happen; these are the following:

i. *Transients*, i.e., temporary deviations of the most important reactor operational parameters from their normal values. For example, a sudden positive change in reactivity can cause the power level to exceed the heat removal rate until the automatic reactivity control mechanisms intervene, or a temporary reduction in coolant flow may produce a similar imbalance. The most serious transients are those associated with a loss of off-site and on-site power supplies.

ii. *Loss of coolant accidents* (*LOCA*). A major breach in the pressurized primary circuit, leading to a rapid loss of coolant, will lead to large-scale melting of the reactor core unless emergency cooling water is available to remove the residual shutdown heat. The most serious potential accident for a PWR would be a break in the large-diameter pipe which takes the coolant from the primary coolant pump into the core vessel (see Fig. 12.4). For the BWR, the most serious LOCA would arise from a similar break in the inlet line to one of the recirculation jet pumps (see Fig. 12.5).

In the event of a LOCA, the nuclear reaction in the core would be automatically cut off by the loss of the moderation associated with the coolant, while in the transient case the probability of failure of the shutdown systems is sufficiently low that continued operation at power is not a significant contributor to a core melt situation. For both classes of accident, therefore, the important requirement is the maintenance of a cooling capability sufficient to remove decay heat (see Table 12.7) from the reactor core. The *emergency core cooling systems* (ECCS) for the PWR and BWR are described below.

The general design of both types of light water reactor has already been illustrated diagrammatically in Fig. 7.1. The layout of the main components of the primary circuit of a PWR is shown in Fig. 12.3. The circuit consists of the core vessel itself, together with the four steam generators and main

Table 12.7. Calculated Radioactivity of 1100-MWe PWR at Shutdown and as a Function of Decay Time[a]

Decay time (d)	Radioactivity (MCi)						Total thermal power (kW)
	Iodine and bromine isotopes	Noble gases	All fission products	Actinides	Activation products	Total	
0	1,435	1,240	13,800	3,450	10.6	17,250	225,000
1	265	221	2,890	1,330	9.19	4,230	17,400
5	101	105	1,870	432	8.42	2,310	9,720
15	28.7	29.0	1,280	39.7	7.50	1,330	5,600
30	6.74	4.77	947	9.35	6.40	963	4,060
60	0.494	0.784	656	6.32	4.76	666	2,350
120	0.00282	0.659	401	5.90	2.76	410	1,740
210	0.00000309	0.648	244	5.56	1.36	250	1,100
365	0.00000218	0.630	146	5.17	0.614	152	659
1097	0.00000218	0.553	47.3	4.45	0.324	52.0	204
3653	0.00000218	0.353	17.9	3.27	0.132	21.3	67

[a] From Report of APS Study Group on Light Water Reactor Safety, *Rev. Mod. Phys.* **47**, Suppl. 1 (1975).

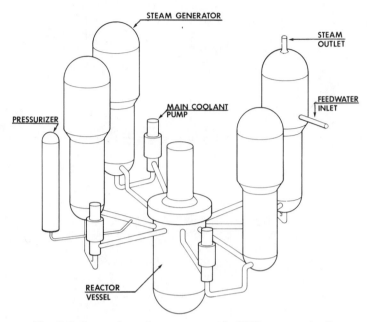

Fig. 12.3. Layout for main components of a PWR pressure circuit.

coolant pumps. The electrically heated pressurizer maintains the primary circuit at a pressure of approximately 2250 psi (157 kg cm^{-2}). Steam from the steam generator outlets passes through the secondary circuit to the turbines.

The emergency core cooling system for the PWR is shown in Fig. 12.4. The primary element is a passive coolant injection system consisting of accumulator tanks containing cool borated water at a pressure of about 650 psi, the boron being included to ensure that the reactor will always remain in the subcritical condition. The accumulators are normally isolated by check valves from the reactor primary circuit. In the event of a breach of the primary circuit, the circuit pressure would drop below that of the accumulators, and the valves between the accumulators and the reactor circuit would automatically open, allowing the emergency cooling water to discharge into the main coolant pipes between the pumps and the reactor vessel ("cold legs").

Two active backup systems are also installed to provide long-term cooling after the accumulators have been emptied. These are the *low-pressure injection system* (*LPIS*) and the *high-pressure injection system* (*HPIS*). The former is required in the event of a large break; it is activated by the drop in pressure and continues to inject coolant from the refueling water storage tank until this supply is depleted (about 30 min). At this point, further cooling is provided by the *low-pressure recirculation system* (*LPRS*), which continually recirculates the water collected in the containment sump back to the core. The high-pressure injection system is needed in the event of a small break in the primary circuit, where the depressurization rate is slow and the circuit pressure remains too high to activate the accumulator system or the LPIS. Like the LPIS, the high-pressure injection system draws its supply from the refueling water storage tank; it discharges into the cold legs at the normal primary circuit operating pressure.

For the BWR, the general design is as shown in Fig. 9.6 and the layout of the emergency coolant system as in Fig. 12.5. As for the PWR, both high- and low-pressure emergency injection systems are provided. The high pressure system operates on an indication of low water level in the pressure vessel following a small circuit break, or on detection of a rising pressure in the

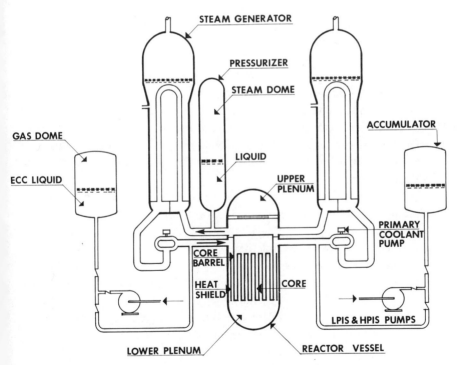

Fig. 12.4. Emergency core cooling system for a PWR [from APS Study Group Report, *Rev. Mod. Phys.* **47**, Suppl. 1, (1975)].

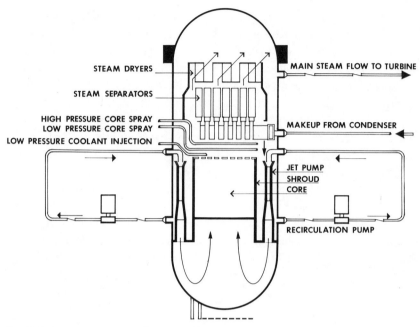

STEAM DRYERS

STEAM SEPARATORS

HIGH PRESSURE CORE SPRAY
LOW PRESSURE CORE SPRAY
LOW PRESSURE COOLANT INJECTION

MAIN STEAM FLOW TO TURBINE

MAKEUP FROM CONDENSER

JET PUMP
SHROUD
CORE

RECIRCULATION PUMP

Fig. 12.5. Emergency core cooling system for a BWR [from APS Study Group Report, *Rev. Mod. Phys.* **47**, Suppl. 1 (1975)].

containment outside the vessel. A turbine-driven pump, operated by steam generated by residual reactor heat, takes water from the condensate water storage tank (later, from the containment suppression pool) and delivers it either as a spray from nozzles located above the core or through the normal feedwater inlets.

In the event that the high-pressure emergency cooling system fails to maintain the reactor water level, relief valves would open to discharge steam into the pressure suppression pool (see Fig. 9.9). The resultant reduction in primary circuit pressure would then operate the low-pressure emergency cooling systems. These consist of the *core spray injection system* (*CSIS*) and the *low-pressure coolant injection system* (*LPCIS*). In the first of these, water from the suppression pool is driven by electrical pumps to a spray header above the core, while the coolant injection system delivers water from the same source into the core through the jet pump recirculation system. The LPCIS employs a separate pumping system from the CSIS, and has enough capacity to protect the core following even a large break in the primary circuit.

As discussed briefly in Chapter 9, the emergency cooling systems described above are supplemented by elaborate containment systems which are designed to minimize leakage to the environment in the event of fission product release from the core. The primary containment for the BWR employs a "drywell" which surrounds the reactor vessel and the recirculation circuit, and is connected through large ducts to a *pressure suppression chamber* containing a large volume of water. The general form of the "bulb-and-torus" and "weir wall" designs of BWR containment have been illustrated in Figs. 9.2 and 9.9. In the event of a major steam release from the reactor vessel, isolation valves would operate to cut off steam flow to the turbines, and the escaping steam would be diverted to the suppression pond, where most of it would be condensed, thus preventing overpressurization of the containment in the event of a major LOCA. This primary containment is surrounded by a secondary containment which essentially consists of a large building designed to provide controlled leakage through high-efficiency air filters.

The containment for the PWR is generally similar, except that a steam quenching system is not necessary because of the high volume within the containment. Cold water sprays or ice condensers are employed, however, to remove heat and reduce the pressure in the containment building and to help retain any radioactivity released from the core. A typical PWR containment has been illustrated in Fig. 9.5.

The probable sequence of events following a major breach of one of the pipes carrying coolant into a PWR (*inlet break*) is described below. The behavior of the important parameters is shown in Fig. 12.6, which is taken from the APS study referred to earlier.

Immediately following the break, the high-pressure coolant is expelled from the rupture (*blowdown period*) causing the core mass flow to decrease rapidly and then reverse. As the pressure drops below the saturation value (about 1500 psi at operating temperature), the remaining liquid in the vessel begins flashing to steam. The accompanying decrease in moderator density causes the nuclear reaction to shut down very rapidly, so that the heat production rate drops to the fission product decay heat level (see Table 12.7). The decrease in heat transfer from fuel to coolant, as a result of the voiding, causes a redistribution of the heat stored in the fuel, with a resultant rise in the temperature of the cladding. Following this initial rapid increase, the cladding temperature will stabilize, or even decrease slightly owing to residual heat removal by the steam still in the core.

When the core pressure drops below that of the EEC accumulators or water storage tanks, the auxiliary coolant starts to be injected into the inlet

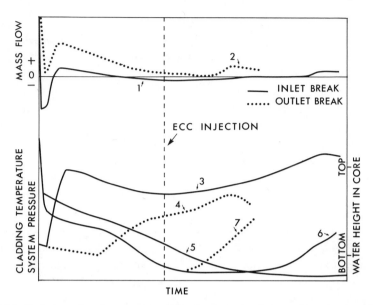

Fig. 12.6. Variation of main parameters following large break in coolant circuit of PWR. Curves 1 and 2 show the core mass flow rates for the inlet and outlet breaks, respectively. For an inlet break, the rapid reduction in flow causes a steep rise in cladding temperature (curve 3); the rise is less rapid for the outlet break (curve 4). The pressure drops continuously for either type of break (curve 5). For an outlet break, the water level starts to rise again (curve 7) shortly after initiation of the ECCS, but for the inlet break (curve 6) the water level rises more slowly owing to steam binding and entrainment of the injected coolant [from APS Study Group Report, *Rev. Mod. Phys.* **47**, Suppl. 1 (1975)].

piping. Initially the entry of the auxiliary coolant is inhibited by steam flow up the downcomer due to backflow through the core and the continued boil-off of liquid in the lower plenum (see Fig. 12.4), but eventually the reduction in steam flow rate as decompression proceeds allows gravity to overcome these inhibiting factors and the lower plenum of the reactor vessel begins to refill.

During the *refill period*, i.e., the time taken for the water in the reactor vessel to reach the level of the bottom of the fuel elements, cooling is particularly limited, since it arises only from steam convection in the core. The temperature of the cladding therefore rises at a rate of approximately 10°C per second. If the cladding temperature exceeds about 650°C, "ballooning" of the fuel pins will occur, owing to the reduced strength of the cladding and the increasing differential between the internal fuel pin pressure and the lowered pressure of the coolant. In the event of the cladding reaching temperatures of 1000°C or above, the zircaloy would start reacting with the steam to form hydrogen and zirconium dioxide, providing an additional source of heating and weakening the structural integrity of the fuel pin due to uptake of oxygen by the cladding. It is important that the refill period be kept short in order to avoid significant steam–zircaloy interaction, since the high temperature of the cladding, coupled with its embrittlement due to oxidation, could lead to serious damage when it is quenched by the coolant rising into the core region.

Once the emergency cooling water rises above the bottom of the reactor core (*reflood period*), steam begins to be generated through contact with the hot elements. The cooling effect of the steam, plus any entrained water, causes a drop in the cladding temperature. In order to maintain an adequate rate of flow of coolant from the ECCS, the incoming water has to displace the steam being generated in the core. For the postulated accident, i.e., a major break in the coolant inlet pipe, the steam and entrained water have to escape by passing through the steam generators and pumps. The frictional pressure drop along this path coupled with the effect of superheating produced by the hot fluid of the secondary circuit, leads to an increase in pressure in the system, a phenomenon known as *steam binding*. The situation would be worsened if any of the steam generator pipes were to fail, releasing steam from the secondary circuit into the depressurized side of the reactor.

The overall result of steam binding may be to reduce the reflood rate by a factor of up to 10 in the worst case, to values as low as 2 cm/s. At reflood rates as slow as this, the fuel temperature rise and resultant damage to the fuel elements would be such that it is not clear that large-scale melting of the core could be avoided. An extensive program of loss-of-coolant simulation tests is currently under way in an endeavor to remove any existing uncertainties in the performance of the emergency cooling systems.

In the event that, following a loss-of-coolant accident, the emergency core cooling system were either to fail to operate or, due to steam binding or some other factor, fail to provide an adequate core reflood rate, the fission product heating, supplemented by the heat produced by the zircaloy–water reaction, would cause melting of the fuel cladding within minutes of the break occurring. The eventual result would be the accumulation of a molten or semimolten core, which would burn through the core support grid and eventually, after a time of the order of one or two hours, through the bottom of the pressure vessel itself. A serious explosion could occur if during this process a large mass of molten core material were dropped by sudden failure of the support grid into a pool of water at the bottom of the pressure vessel. In this event, it would be possible for the vessel to suffer a major disruption and for some core material to be blown out into the containment.

Following the melt-through of the bottom of the pressure vessel, the core would drop to the concrete floor of the containment vessel. Again, some of the core material would be scattered around the containment if the core were to drop into a pool of water which had collected below the pressure vessel. The molten core would melt through the containment vessel in a time which is estimated to be between a few hours and a few days after the initial pipe break leading to the LOCA. Following this, it would continue to tunnel its way into the ground below the containment.

The degree of hazard to the general public arising from a core melt-out incident would depend on the extent and method of failure of the containment building. The most serious failure mode would be the rupture of the containment as a result of a steam explosion caused by the molten core falling into a volume of water in the pit below the reactor, or in the bottom of the reactor vessel, provided that the explosion in the latter case were severe enough to convert the pressure vessel head into a missile capable of damaging the containment building. Failure could also occur by overpressurization due to hydrogen gas from metal–water reactions or CO_2 from concrete decomposed by the molten core. Rupture by gas or steam pressure would be accelerated if the cooling systems designed to remove heat from the containment were to fail. Finally, even if these rupture modes were avoided, the molten core would eventually melt through the floor in a time of the order of a day after the meltdown occurred.

In the event that the only major mode of failure of the containment were by floor melt-through, the escape of radioactivity would be significantly reduced by the filtering action of the soil under the building, though high containment pressure could lead to some bypassing of this filter action through "blowouts."

The quantity and composition of the radioactivity released to the environment following a core melt will depend on the precise sequence of events leading to the accident, and on the extent to which the release is reduced by processes such as wash-out by the fission product removal systems, filtering, and plate-out on surfaces within the containment. The final objective of the safety analysis is to evaluate the risk to the public arising from the possibility of accidents in a nuclear reactor. This involves finding the relationship between the *probability of a given release* of radioactivity and the *magnitude of the damage* to people or property which would be caused by such a release.

The steps in the calculation of this risk are shown in Fig. 12.7. The first step uses the technique known as *event tree analysis*. Starting with a given initiating event, usually a system failure which initiates a potential accident

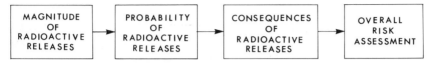

Fig. 12.7. Steps in the analysis of light water reactor accidents.

situation, an event tree is constructed which allows one to calculate the various radioactivity release magnitudes which can arise depending on the success or failure of the safety features of the reactor. A typical event tree, from the WASH-1400 report, is shown in Fig. 12.8; this illustrates the possible sequences of events for a large LOCA, for which the initiating event is a major pipe break. The course of the accident depends on the sequential behavior of the following systems: station electric power, emergency core cooling system, fission product removal system, and containment. These are ordered at the top of the diagram in the appropriate time sequence. At each step in the sequence, the system involved either operates successfully (upper branch of tree) or fails (lower branch). The probability of failure of each system is indicated by the p values shown beside the various branches. For independent events, the probability of occurrence of a given chain is the product of the probabilities of the individual events in that chain, so that the composite probability of each path is as shown on the right of the diagram. [Since, in almost all cases, the failure probabilities are very much less than unity, the probability of success, $1 - p$, is taken as being approximately equal to 1, so that all the upper branches have probability unity.]

In practice, the event tree can be simplified, as shown in the lower part of the figure, by dropping the paths which are automatically eliminated because of the interaction between the safety systems. For example, failure of electric power will inevitably lead to failure of all following systems, since the operation of these depends on the power supply remaining available. Thus there is only a single sequence following power supply failure, leading to core melting with failure of containment.

The second stage of the calculation sequence shown in Fig. 12.7, the determination of the probability associated with each branch of the event tree, uses the method of *fault tree analysis*. A fault tree is the inverse of an event tree in the sense that, instead of enumerating all the possible chains of events following on a single initiating event, one starts with a given failure and sets out the various combinations and sequences of other failures that can lead to the particular failure considered. A typical fault tree, that leading to complete loss of electrical power to the engineered safety features (ESFs)

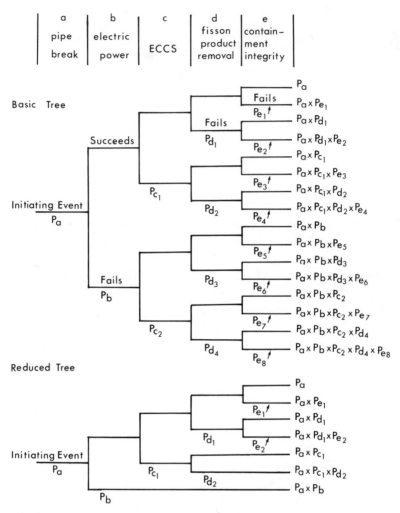

Fig. 12.8. Simplified event trees for a large loss-of-coolant accident (from Reactor Safety Study, U.S. Atomic Energy Commission Report WASH-1400, 1975).

such as the ECCS, also taken from WASH-1400, is shown in Fig. 12.9. The maintenance of this power supply requires both a basic ac supply and a dc supply for operating the control systems which turn on the ac. Thus, failure of either of these systems would lead to loss of the power supply to the ESFs. This fact is indicated in the diagram by connecting the ac and dc failures through an OR gate to the resulting power supply failure. The probability of the latter event is then the sum of the probabilities of the two events below it.

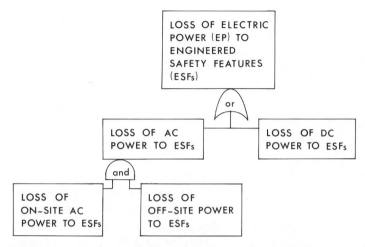

Fig. 12.9. Fault tree for loss of electric power to all engineered safety features (from Reactor Safety Study, U.S. Atomic Energy Commission Report WASH-1400, 1975).

In general, the determination of these probabilities involves extension of the fault tree to lower levels, as is illustrated, for the ac power failure only, in the diagram. Since the ac power can be supplied either by on-site sources (diesel generators) or off-site sources (the electrical grid system), failure of both of these would be needed to produce loss of ac power, and they are therefore coupled through an AND gate to the resultant failure. In this case the probability of failure is given by the product of the probabilities of the two individual failures. In this way, the fault tree may be extended to levels where the failures being considered are those of individual components whose failure probabilities can be estimated from analysis of a statistically meaningful set of failure data, usually based on their use in a wide range of industrial applications. These probabilities are then combined to estimate the probability of the top event in the tree.

A study of the event trees allows one to determine the various containment failure modes and the characteristics of the radioactivity release to the environment following the possible accident sequences. By using the fault tree analysis we can then link each release to its associated probability of occurrence. It is found that many of the possible sequences in the event trees lead to similar types of radioactivity release, with the result that it is possible to summarize the consequences of the PWR accident in terms of as few as nine release categories, each with its corresponding probability. The corresponding figure for the BWR is six categories.

As an illustration, Table 12.8 gives the estimated release fractions into the atmosphere following a particularly severe accident involving core melt in a PWR, coupled with failure of the radioactivity removal systems and rupture of the containment due to hydrogen burning and steam overpressure. The probability of this category of release was estimated as 5×10^{-6} per reactor year. It will be seen that the fractional escape of the various radionuclides considered is strongly dependent on their chemical form; the noble gases (xenon and krypton) would have to be assumed to escape almost entirely, while only a fraction of 1 % of the nonvolatile oxides of metals such as zirconium and plutonium would be expected to escape.

From the fault tree and event tree analysis of the PWR accident, it is possible to identify certain sequences which dominate the probability of the various release categories. It is found that the largest probability of PWR core melt is contributed by the smallest of the three classes of coolant circuit rupture considered, that in which the break has an equivalent diameter in the range of 1.5–5 cm. This arises from the fact that the probability of rupture is higher for the lower break sizes, while the effectiveness of the core protective features does not increase in the same proportion. This smallest LOCA break considered leads to the highest probability of the less severe categories of release, since it is unlikely to result in a containment mode failure other than melt-through of the containment base, with resultant attenuation by ground filtration.

The higher radioactivity release categories are dominated by the so-called "interfacing systems LOCA," where a breach of the reactor coolant

Table 12.8. Radionuclide Atmospheric Release Fractions for PWR Core Melt Followed by Failure of Radioactivity Removal Systems and Failure of Containment Due to Overpressure[a]

Principal chemical groups	Fractional release to atmosphere from core
Noble gases	0.9
Iodines	0.7
Cesiums	0.5
Telluriums	0.3
Alkaline earths	0.06
Volatile oxides	0.02
Nonvolatile oxides	0.004

[a] From WASH-1400, Appendix VI.

system is caused by a failure of the check valves which isolate it from the low-pressure injection system. This type of accident would open a direct path for radioactive release to the atmosphere, bypassing the radioactivity removal systems. It is worth noting that a gross rupture of the reactor pressure vessel itself is such an improbable event that it is a negligible contributor to the probabilities of all categories of radioactive release.

The probability patterns for the BWR are significantly different from those of the PWR. The reliability of the ECC systems is significantly better, so that the probability of LOCA-caused core melts is less than 10^{-6} per year for all release categories and is not a significant contributor to any category. It should be noted, however, that if a LOCA-induced core melt were to occur, the release would be more serious than in the PWR case, since the smaller size of the containment volume leads to failure by overpressurization rather than by melt-through.

For the BWR, the predominant failure mode is by the occurrence of transient events, such as unplanned shutdowns due to failure of off-site power. The transient events, in fact, dominate almost all the release categories. As is the case for the PWR, the probability of a major rupture of the reactor pressure vessel is small enough to make it a negligible contributor.

The probability analysis outlined above is concerned only with potential releases associated with intrinsic failures, i.e., those arising from design inadequacies or operating errors in the plant itself. The WASH-1400 report also considered the possible effects of external events such as earthquakes, tornadoes, floods, and aircraft impacts. In all cases, the probabilities of core melt induced by external events were very much smaller than for the intrinsic sequences previously considered, partly owing to the small probabilities of the events themselves and partly because the reactor structure is designed to withstand events of the magnitude expected.

Having obtained the magnitudes of the classes of radioactivity release, and their probabilities, the next step is to calculate the consequences, in terms of health hazard and property damage, of each release. This depends on the manner in which the radioactivity is dispersed and on the distribution of population around the reactor. The dispersion in turn depends on the type of release, on its duration and elevation above ground, for example, and on the weather prevailing at the time of the accident.

In the WASH-1400 study, a complex calculation was carried out, for a range of weather and population conditions, to give the distribution of radioactivity in space and time for each of the categories of activity release. This was combined with data on the effects of exposure of people to radiation to estimate the health effects for each combination of release, weather, and population distribution (density versus distance). The probability of a specific health effect is then given by the product of the probabilities of the release, of the weather conditions, and of the particular population distribution.

The consequences to health may be expressed in three categories. These are (i) acute fatalities, i.e., deaths which occur shortly after the accident; (ii) acute illness, i.e., people requiring immediate medical attention; and (iii) long-term health effects, e.g., delayed cancer deaths, occurrence of thyroid nodules, and genetic effects, which take place over a period of many years after the accident.

The long-term effects of low dose rates at larger distances are simpler to calculate, since the details of the release and weather conditions are less important. In the draft version of WASH-1400, the simplifying assumption was made that the incidence of cancer and genetic effects was proportional to the total population radiation dose measured in man-Sv. The numerical value for the excess mortality for all forms of cancer induced by low doses of radiation was assumed to lie in the range of 10–13 deaths per million man-mSv. This assumption, based on known data for cancer induction at high dose rates (in the region of several sieverts), probably overestimates the risk at low dose rates since no account is taken of threshold effects or possible repair mechanisms. The final version of WASH-1400 attempted to compensate for this by using dose reduction factors for low-LET radiation at low doses and low dose rates; these had the effect of reducing the assumed number of cancers per unit of radiation dose at low doses by factors of 2.5–5.

In addition to delayed deaths due to whole-body exposure, allowance has to be made for effects on particular organs, such as the lungs and thyroid, which suffer significant β doses due to concentration of inhaled radioactivity in these organs. The estimated mortality rates or lung and thyroid doses, used in the APS study, are shown in Table 12.9. The principal effect of thyroid irradiation, in fact, is to produce benign thyroid nodules, which contribute the main source of long-term illness induced by the accident.

The assumed genetic effects are also shown in Table 12.9. It should be noted that most of the genetic effects are spread over long periods of time following exposure. The figure of 2.5–25 persons with dominant genetic defects over an average of five generations following the million-man-mSv exposure, for example, should be compared with a current incidence of about 2000 per million live births.

For estimating the low-dose exposure associated with the various releases, the APS study used a simple model which gives results which agree well with the more complicated computer analysis of WASH-1400. The

Table 12.9. Some Dose–Effect Coefficients for Delayed Effects (per Million man-mSv) used in APS Report

1. Whole-body effects:	
~13	Cancer deaths
2.5–25	Persons with identifiable dominant genetic defects over an average of five generations following exposure
1.25	Noninheritable congenital defects
0–50	Total extra constitutionally or degeneratively diseased persons over an average of ten generations following exposure
4.2	Spontaneous abortions
2. Lung dose effects:	
0.06–0.16	Cancer deaths per year for the period 5 to at least 27 years following exposure
3. Thyroid dose effects:	
Children under 10,	
0.05–0.30	Cancer cases per year from five years till at least 30 years following exposure (times about 0.04 for mortality during this period)
1.1–5.2	Cases of thyroid nodules per year from five years till at least 30 years following exposure
Adults (persons over 10),	
0.05–0.30	Cancer cases per year from five years till at least 25 years following exposure (times about 0.15 for mortality)

simplification is made possible by the fact that the total low-dose exposure is dominated by the large number of small individual doses hundreds of miles downwind from the accident. By the time that the plume has reached this distance, it can be assumed to have diffused uniformly through the lower "mixing layer" of the atmosphere, which is about 1000 m thick. The radioactivity may be regarded as distributed uniformly within a wedge-shaped plume of depth 1000 m and an opening angle of approximately 0.25 radians, the latter representing a value averaged over a range of atmospheric conditions.

The exposure to an individual is the sum of an internal dose due to inhalation, and external doses arising from the plume itself and from the radioactivity which it deposits on the ground as it passes. The latter can be estimated for the different groups of isotopes by using deposition factors based on experiment. For people fairly near the site, who are likely to be evacuated within a short time of the release, the main dose arises from inhalation, since the whole-body dose due to inhalation is typically an order of magnitude greater than that due to external radiation from the cloud, and the dose from ground deposition is minor provided that the evacuation is reasonably rapid. At large distances, where no evacuation takes place, the integrated dose from ground contamination will be the dominant contributor to the whole-body dose.

As an illustration of the scale of health effects associated with the various categories of accident, Table 12.10 lists the consequences of one of the nine categories of PWR releases, i.e., that where a core melt is coupled with a failure of the radioactivity removal systems, followed by rupture of the containment caused by hydrogen burning and steam overpressure. The numbers quoted are based on the APS report, and in most cases span a considerable range owing to the appreciable uncertainties associated with the assumed dose–effect relationships.

In order to evaluate the impact of the category of accident quoted, the effects must be compared to natural mortality rates in the total population involved over a period of several decades. Thus, according to the APS report, the probability that an individual in the exposed region will die during his natural lifetime of cancer induced by his exposure is, on the average, only one in a thousand (0.1 %). This individual risk would be spread over many years following the assumed accident, with the cancer manifesting itself during the decades following exposure.

The probability of accidents which produce acute fatalities versus the expected number of fatalities, as determined by the Rasmussen study, is shown in Fig. 12.10. The results for the BWR and PWR are shown

Table 12.10. Consequences of PWR Category 2 Radioactivity Release[a]

Cause	Effect
Cancer deaths	
γ radiation from cloud and ground contamination	10,000 cancer deaths
Lung dose from inhalation	600–1600 lung cancer deaths during following 40 years
β-ray-induced thyroid cancer	500–4000 thyroid cancer deaths in following 30 years
Morbidity	
Exposure of thyroid (β and γ radiation)	22,500–300,000 thyroid nodule cases
Genetic	
Total radiation exposure	3000–20,000 genetic defects

[a] From APS Report.

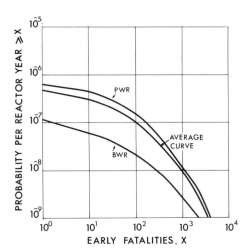

Fig. 12.10. Probability distribution for acute fatalities per reactor year for BWR and PWR (from Reactor Safety Study, U.S. Atomic Energy Commission Report WASH-1400, 1975).

separately, but the differences between the individual curves are much less than the cumulative errors in the estimates and consequently the average curve shown may be considered as representative of either. As an illustration, accidents involving more than 100 fatalities have a probability of about 10^{-7} and would therefore be expected to occur once in ten million reactor years of operation.

The WASH-1400 comparison of the risk of acute fatalities from reactor accidents with fatalities from man-caused and natural events is shown in Figs. 12.11 and 12.12. The reactor accident curve assumes a total of 100 operating reactors. According to the report, the risk of *acute* fatalities is much less for the reactor case than for any other man-caused or natural event with the exception of meteorite impacts. It is important to note, of course, that the acute fatalities constitute only a very small fraction of the long-term fatalities which could be associated with the accident. This is illustrated by Fig. 12.13, also from WASH-1400, which gives the probability per year that an accident at any of the 100 reactors will cause a given number of cancer fatalities *per year* over a 30-yr period after the accident. The *total* fatalities over the 30-yr period would therefore be obtained by multiplying the horizontal axis by a factor of 30.

While the Rasmussen Report is of major significance as the most comprehensive and detailed investigation of water reactor safety yet undertaken, it has been subjected to a great deal of criticism, with respect

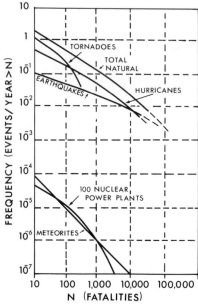

Fig. 12.11. Frequency of natural events with fatalities greater than N (from Reactor Safety Study, U.S. Atomic Energy Commission Report WASH-1400, 1975).

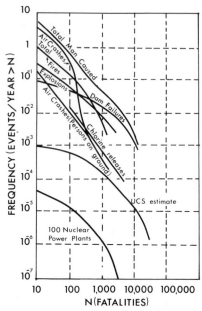

Fig. 12.12. Frequency of man-caused events with fatalities greater than N (from Reactor Safety Study, U.S. Atomic Energy Commission Report WASH-1400, 1975).

both to the treatment of the potential accident sequences and the data base used in calculating accident probabilities. In addition to the APS critique already mentioned, the results have been questioned by, among others, the Union of Concerned Scientists [H. W. Kendall *et al.*, *The Risks of Nuclear Power Plants* (1977)] and by the Risk Assessment Review Group (H. W. Lewis *et al.*, Report to the U.S. Nuclear Regulatory Commission, NUREG/CR 0400, 1979).

One major concern of both groups is the statistical analysis used in WASH-1400. One example is the estimation of the probabilities of "common-cause" failures, i.e., those where a single event produces a common failure in several pieces of equipment. Where the actual safety system is too complicated for a failure probability to be calculated directly, the analysis sets up a series of simple models, each simple enough to permit the calculation of the failure probability for the model, $P(M)$. Next a subjective engineering judgement is made of the probability, $Q(M)$, that the particular model is in fact a correct description of the system. The overall probability of failure, P, is then found by combining the various models and their failure probabilities as below:

$$P = \sum_{M} P(M)\, Q(M) \tag{12.2}$$

Since in practice it would be very difficult to construct an exhaustive set of models, the procedure is simplified by taking only two, an upper bound model, M_U and a lower bound model, M_L. The distribution of the probabilities $Q(M)$ is assumed in WASH-1400 to be log-normal; with the additional assumption that $P(M_U)$ and $P(M_L)$ are symmetrically placed, the value of P is then given by

$$P = [P(M_U)\, P(M_L)]^{1/2} \tag{12.3}$$

An example, quoted in NUREG/CR 0400, is the calculation of the probability of the simultaneous failure of three adjacent control rods in a BWR. M_L is taken as a model where all three fail independently; then $P(M_L) = (P_1)^3$, where P_1, the probability of a single rod failing, is estimated from

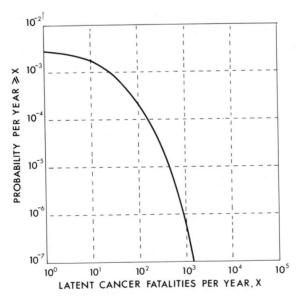

Fig. 12.13. Probability distribution for latent cancers for 100 reactors (from Reactor Safety Study, U.S. Atomic Energy Commission Report WASH-1400, 1975).

experiment to be 10^{-4}. Hence $P(M_L) = 10^{-12}$. M_U is a model where, when one rod fails, there is a 1 % probability that the others will both fail also. Then $P(M_U) = 10^{-2}P_1 = 10^{-6}$. Hence, from equation (12.3), $P = 10^{-9}$.

This method may be criticized on several grounds. In the first place, there is no reason to believe that $Q(M)$ is in fact log-normal in distribution. The lower bound model gives a value too low to be credible, and there is no evidence for the assumption that $P(M_U)$ and $P(M_L)$ in this case are symmetrically placed. Based on a review of this and other features of the statistical techniques used, NUREG/CR 0400 concluded that the uncertainties in the calculated probabilities must be appreciably greater than quoted in the report.

Other criticisms of the report which have been made include the following:

i. The model used for the relation between radiation dose and cancer induction rate was an "absolute" one, where it was assumed that the effect of a given dose is to produce an increment in cancer rate which is independent of the existing rate. An alternative approach is by the "relative" model, where the effect is taken as a multiple of the existing rate. While it is not possible, on the basis of current knowledge, to say which model is more correct, it is possible that the cancer rates used in the Report are an underestimate.

ii. As mentioned above, dose rate reduction factors were used to allow for the possible lower cancer induction rates at low doses and dose rates; these factors are still a matter of dispute and, again, may lead to an underestimate of the mortality rate.

iii. It has been argued, particularly in light of the subsequent Three Mile Island event, that the effects of operator error in increasing the severity of an accident have been underestimated; the Risk Assessment Review Group recommended that further study of human intervention be carried out, to provide more reliable estimates of the consequences, beneficial or otherwise, of operator action.

iv. The report concluded that earthquakes made only a minor contribution to the probability of core meltdown. Subsequent studies have suggested that this may not be true. For example, the effects of an earthquake may be greatly magnified as a result of design and construction errors in a plant. The lack of an adequate data base for predicting the frequency and severity of earthquakes, and the doubtful reliability of the statistical analysis used in the report, have also been cited as reasons for questioning the conclusion that earthquakes produce a negligible contribution to the overall hazard.

v. One of the more controversial analyses was that of the failure of a reactor to shut down ("scram") on demand. Reactor transients which are accompanied by a failure of the shutdown mechanisms are known as anticipated transients without scram (ATWS). The suspect treatment of such events has already been touched on in the earlier comments on common-cause failures. Apart from this factor, the extrapolation of the results of WASH-1400 to a full-scale nuclear industry is inappropriate since the two plants analyzed in the report were not typical with respect to protection against ATWS. The consequences of an ATWS would be particularly serious if the energy release were of sufficient magnitude to cause the pressure vessel to explode in such a way as to rupture the containment. It has been suggested that this could occur in a BWR if the pressure housing of a control rod drive mechanism were to rupture, allowing the rod to be driven out of the reactor by the high-pressure coolant. A control movement blocking device is normally available to limit the movement of a control rod in the event of housing failure, but it is possible that this may also fail owing to improper installation, or be left off accidentally following control rod maintenance.

Another postulated cause of explosive pressure vessel failure is a shutdown failure associated with steam valve closure in a BWR. In the event of a turbine generator failure, the main steam valve is programmed to trip shut, in order to prevent the flow of steam to the generator. The rise of pressure in the reactor then compresses the steam bubbles in the core, leading to an increase in coolant density and thus to an increase in reactivity. The resulting rise in power level will normally be terminated by automatic insertion of the control rods. In the event of a scram failure, however, the power would rise rapidly, even when allowance is made for the negative fuel Doppler coefficient. A backup device is provided in the form of an automatic shut-off of the coolant recirculation pumps, which would once again permit boiling to take place in the core, reducing density and lowering reactivity. It is possible that simultaneous failure of the scram and recirculation shut-off mechanisms would lead to extensive core melting and to a steam explosion which would rupture the pressure vessel and the containment.

vi. It has been argued that the consequences of radioactivity release have been underestimated because the population distribution and meteorological conditions (e.g., prevailing wind) at actual nuclear station sites are inadequately taken into account by the averaging procedures used.

As an example of the divergences from the WASH-1400 analysis, Fig. 12.12 includes the total prompt fatality curve calculated by the Union of Concerned Scientists (UCS) in the review mentioned earlier. The general conclusion of UCS was that the chances of a meltdown accompanied by a large release of radioactivity was about 20 times greater than estimated in

WASH-1400 (one chance in 10,000 per reactor-year compared with one in 200,000) and that the fatality rate was on average about ten times greater.

While the criticisms outlined above imply that the Rasmussen analysis underestimates the hazards of nuclear power plant accidents, there is also evidence which suggests that in some respects the opposite may be true. In particular, recent research indicates that the volatility of iodine may have been greatly overestimated in situations where substantial amounts of water are present. In LWR fuel, the iodine is largely in the form of cesium iodide, which dissolves readily in any water present, leaving only a small concentration in the gas phase. It is this phenomenon that is thought to account for the unexpectedly low release of radio-iodine in the Three Mile Island accident, which is described in Section 12.3.2. Another area in which the analysis may be overconservative is in dealing with the possibility of pressure vessel or containment failure from a steam explosion; recent experimental work suggests that the probability of failure is considerably less than was assumed in the Rasmussen model.

While the Rasmussen Report remains a source of controversy, its significance is probably best summed up in the conclusions of the Risk Assessment Review Group: "Despite its shortcomings, WASH-1400 provides at this time the most complete single picture of accident probabilities associated with nuclear reactors. The fault-tree/event-tree approach coupled with an adequate data base is the best available tool with which to quantify these probabilities." Risk assessment of systems as complex as nuclear power reactors is, by its very nature, a process requiring continual updating and reassessment, using the data from accumulating operational experience to modify and refine the theoretical calculations.

12.3.2. Accident Experience with Light Water Reactors

The two most potentially serious accidents in light water reactors were those of the Browns Ferry BWR in March 1975 and the Three Mile Island PWR in March 1979. The first incident started when a candle being used to check for air leaks started a fire which destroyed electrical cables controlling many of the normal operating and emergency control systems on the Browns Ferry Unit 1 BWR. The damage to the normal and emergency cooling systems was such that the coolant inventory in the reactor was maintained only by resorting to the use of the control rod drive pumps, which were not designed for this purpose. In the event, damage to the reactor core was prevented by improvisation of this kind.

The Rasmussen Report analyzed the Browns Ferry incident in an attempt to establish the probability that a core meltdown could have taken place as a result, for example, of failure to maintain the coolant inventory by the use of the control rod drive pumps or one of the other mechanisms that could have been invoked in the emergency. The analysis led to the conclusion that there was approximately one chance in 300 that the incident could have led to a core meltdown. This calculation has been challenged by UCS, which estimated the probability as being one in 30. Although, as pointed out earlier, the occurrence of a meltdown does not in most cases led to a major public hazard, the realization of the potential consequences of a fire which would disable a high proportion of the control systems has led to design changes and a tightening-up of maintenance procedures in the vicinity of cable runs.

The occurrence which has generated more public concern than any other is the well-known Three Mile Island accident. The reactor involved was the second unit of two located on an island in the Susquehanna River, near Harrisburg, Pennsylvania. The layout of the reactor circuit and ancillary systems is shown in Fig. 12.14. The sequence of events leading to the accident is given below.

1. At 4.00 a.m. on 28 March 1979, the main feedwater pumps tripped; this in itself was not an unexpected or particularly unusual event, and resulted in an automatic shutdown of the turbine; it also caused the emergency feedwater pumps to start up, as they were designed to do. Unknown to the plant operators, however, the block valves on both of the emergency feedwater lines had inadvertently been left closed, so that no water was able to reach the steam generators.

2. Since no feedwater was entering the generators, the reactor coolant temperature increased, causing an expansion which raised the water level in the pressurizer. The pressure increase due to the compression of the steam in the top of the pressurizer caused the pilot-operated relief valve (PORV) to open, and steam and water began flowing out of the reactor cooling circuit through a drain pipe to a drain tank on the floor of the containment building; the circuit pressure, however, was still rising, and this caused a SCRAM which shut down the reactor itself.

3. With the reactor shut down and the PORV open, the pressure in the primary circuit fell. The PORV was designed to close automatically when the pressure dropped to 2205 psi, which it did 13 s after the initial feedwater pump trip. *Although the control room panel indicated that the valve had closed, it had in fact stuck open, so that the coolant was draining off into the reactor's let-down system.* This was to continue for almost two and a half hours before the circuit was eventually sealed by the closure of a backup valve.

4. Owing to the cooling of the core and the loss of coolant through the

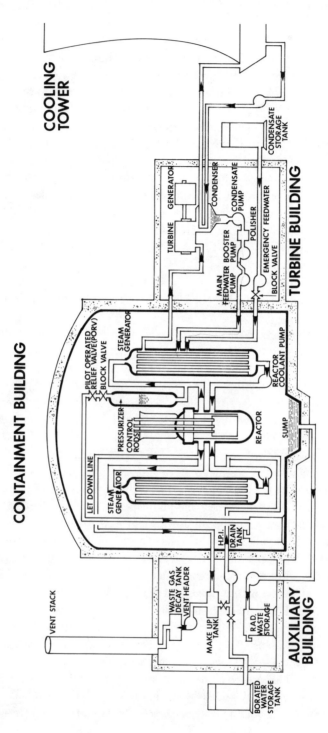

Fig. 12.14. Schematic of the Three Mile Island reactor (from the Report of the President's Commission on the Accident at Three Mile Island, 1979).

PORV, the water level in the pressurizer began to fall. The operators therefore turned on a pump to add water to the system.

5. Because the inflow was now greater than the outflow through the PORV, the pressurizer level began to rise. The rate of rise was increased when the steam generators, starved of feedwater because of the closed block valves, boiled dry, causing the reactor coolant to heat up and expand.

6. Two minutes after the start of the incident, the reactor pressure had dropped to a value low enough to activate the high-pressure injection system (HPIS), which is part of the emergency core cooling system (ECCS). The emergency cooling water began pumping into the reactor at a rate of about 1000 gal/min, and the level in the pressurizer continued to rise.

7. The operators, who had been trained to avoid having the pressurizer filled with water, a condition which would make pressure control difficult and might lead to plant damage, turned one of the HPIS pumps off and reduced the flow from the other.

8. The pressure in the reactor dropped to the saturation point, and steam bubbles began to form in the core. The displaced water moved into the pressurizer, sending its level still higher. The operators, still responding to the perceived need to avoid a water-filled pressurizer, began to drain off coolant through the reactor's let-down system.

9. At this point, it was discovered that no emergency feedwater was reaching the steam generators. The block valves, which had inadvertently been left closed, were opened, restoring the feedwater supply to the generators. The fact that the feedwater had been interrupted for 8 min did not, by itself, lead to significant plant damage, but it added to the confusion of the operators and distracted attention from the serious problem of the stuck-open PORV.

10. At 4.11 a.m., an alarm indicated high level in the sump of the containment building. The drain tank had filled and was now overflowing on to the floor of the containment. From there it was pumped into the auxiliary building. About 30 min later, the pumping operation was stopped, but not before some 8000 gal of slightly radioactive water had been pumped into the auxiliary building.

11. About an hour after the start of the incident, the four reactor coolant pumps began vibrating badly, owing to the fact that the cooling water was mixed with steam produced as a result of the inadequate circulation in the core. The operators reacted by shutting down the pumps, in order to prevent damage to them or to the coolant pipework. The result was to stop all forced flow of water through the reactor.

12. With the coolant continuing to escape through the open PORV, the

top of the core became uncovered and heated up to a temperature where the steam reacted with the zirconium cladding to produce significant quantities of hydrogen, some of which remained within the reactor vessel while the remainder escaped into the containment building.

13. At 6.22 a.m., 2 h 22 min after the PORV had stuck open, the operators closed a backup valve and terminated the release of coolant from the circuit. The damage to the core continued, however, since it was not until another hour had passed that the HPIS was restarted in an attempt to make up for the loss of coolant that had occurred. As much as two thirds of the core appears to have been uncovered, and later calculations indicated that parts of the core reached temperatures in excess of 2200°C, causing extensive fuel damage.

14. From about 6 a.m. onwards, radiation alarms warned of rapidly increasing radiation levels within the containment building, suggesting strongly that serious fuel damage had occurred. At 8.26 a.m., the HPIS pumps were turned on again, at a fairly high rate of flow. It has been estimated, however, that it was not until about 10.30 a.m. that the core was fully covered again. Attempts to establish natural circulation cooling failed because a large bubble of hydrogen gas in the top of the pressure vessel blocked the flow of water.

15. At 11.38 a.m., in an attempt to reduce the pressure in the reactor, the pressurizer valve was opened and the rate of emergency coolant injection was cut sharply. This resulted in a second loss of coolant and an uncovering of the core for an unknown period.

16. An explosion, later identified as being due to hydrogen buildup in the containment building, occurred. The pressure of 28 psi generated by the explosion was insufficient to cause significant damage to the containment.

17. Much confusion had been caused throughout the course of the incident by conflicting estimates of the amounts of radioactive material which had been, or were likely to be, released to the environment. Most of the release which did take place was in the form of fission gases transported through the coolant let-down and makeup system into the auxiliary building and through the building filters and the vent header to the outside atmosphere. The main release arose from a planned transfer of radioactive gases from the makeup tank to the waste gas decay tank, a move arising from the fact that the pressure in the former had risen to a level so high that the water which normally flowed into it for transfer to the reactor cooling system could not enter the tank. This release took place two days after the start of the loss of coolant incident. The question of whether or not an evacuation of the public from the area around the plant was necessary was the subject of

intense debate among the federal, state, and local authorities. In the end, the State Governor issued an advisory recommending that pregnant women and preschool children should leave the region within a five-mile radius of the plant.

18. Three days into the incident, the main concern centered around the possibility of an explosion of the hydrogen *inside the reactor itself*. For the explosion to occur, oxygen would have to be present in sufficient quantity to form an explosive mixture. Initial calculations by the Nuclear Regulatory Commission suggested that the rate of generation of oxygen by radiolysis would be great enough to create an explosive mixture, but this calculation was later found to be in error. In the event, the hydrogen bubble broke up into smaller bubbles which could be more readily eliminated, and the danger of explosion disappeared.

The *consequences* of the Three Mile Island incident can be summarized as follows:

1. Despite serious damage to the plant, the release of radioactive material was relatively small and presented a negligible hazard to the public. Based on the linearity hypothesis of radiation damage, the total number of radiation-induced cancer deaths likely to result from the release was estimated as 0.7. In practical terms, this means that the number of deaths will be zero or almost zero; even in the unlikely event that a few radiation-related deaths were to occur, the number would be far below that which could be detected statistically in view of the normal cancer mortality rate.

2. The full magnitude of the damage to the plant cannot yet be assessed, but it is obviously very extensive. The containment and auxiliary building still hold large quantities of radioactive gases and over a million gallons of contaminated water. The clean-up operation is likely to take several years and to cost in excess of $200 million. Further release of radioactive gases, particularly Kr^{85}, will take place, but it should be possible to control these so that off-site radiation levels do not exceed the NRC limits for routine operation of a nuclear plant.

3. The dramatic nature of the TMI accident, and the frequently sensational media coverage, led to widespread public concern with the safety of nuclear power plants. As pointed out by the Kemeny Commission, the problem had its basis, not in the design of the reactor and its protective systems, but in the actions taken by the operators in response to the emergency. The Report of the Commission criticized the preoccupation of the industry and the regulatory authorities with the design of safety systems intended to give automatic protection against the "large-break LOCA," as compared to the "small LOCA" where the time scale is sufficiently slow to

allow time for potentially misguided operator intervention. It recommended major changes in the structure of the Nuclear Regulatory Commission, the licencing of nuclear power plants, the training of operators, and the organization of emergency plans for dealing with public safety in the event of a major accident. Many of these recommendations have already been put into practice and the end result of the TMI event will be to increase the safety of nuclear plants by rectifying the deficiencies which came to light in the course of the accident.

12.4. Reactor Accidents: Thermal Reactors Other than LWRs

12.4.1. The Gas-Cooled Graphite-Moderated Reactor

No detailed study, comparable to the Rasmussen or APS report, has been produced for thermal reactors other than the light water type. The gas-cooled graphite-moderated reactor is simpler from the safety point of view than the LWR for a number of reasons. There is no longer the possibility of large sudden reactivity changes associated either with rapid changes in the moderator or coolant (comparable, for example, to the PWR cold water accident, where cold unborated water is inadvertently injected into the core from a closed-off coolant loop). Other problems which are avoided include the molten fuel–water and steam-cladding interactions. The loss-of-coolant accident is less severe since the single-phase gas coolant is not subject to a sudden loss of circulation due to change of phase.

One of the problems encountered at an early stage with graphite-moderated reactors operating at a modest temperature was the Wigner energy storage mechanism due to displacement of atoms under neutron bombardment. This phenomenon was responsible for the well-known accident to the air-cooled Windscale No. 1 pile in the United Kingdom, which occurred in October 1957. The incident happened during a graphite "annealing" procedure whereby the reactor was allowed to heat up, with the main blowers switched off, to a temperature at which the stored Wigner energy was released. Unfortunately, owing to a lack of adequate monitoring instrumentation, the release got out of hand, resulting in a graphite fire and the combustion of some of the uranium metal fuel elements. As a result of the accident, the reactor was a total loss and the release of radioactivity was such that a temporary restriction had to be placed on the consumption of milk from the surrounding district. Wigner energy storage is not a problem for the more advanced graphite-moderated reactors, such as the AGR and HTGR, since these operate at graphite temperatures above the range where the Wigner effect is significant.

One of the advantages of the later designs of gas-cooled graphite-moderated reactor is the use of the prestressed concrete pressure vessel (PCPV), which is not subject to the possibility of sudden catastrophic failure as is the steel pressure vessel of the BWR or PWR. The characteristics of the PCPV itself, together with the use of multiple closures of the main penetrations, mean that any realistic failure of a PCPV would be gradual, allowing time for remedial action to maintain cooling. Also, in the event of depressurization, the gas-cooled reactor is not subject to a rapid drop in cooling efficiency due to water flashing to steam. The heat removal capacity of the coolant is approximately proportional to its pressure so that, for the AGR for example, the drop from its normal operating pressure of 40 atm to 1 atm would reduce the heat removal capacity to some 2.5% of normal. The possible depressurization rate from the failure of a PCPV is low enough that the available heat removal capacity should always be in excess of the decay heat generation.

Another possible transient is that due to a loss of power to all the gas blowers, causing complete loss of forced circulation. In this case, natural convection should be adequate to remove the decay heat. Heat removal is aided by the large thermal capacity of the graphite, which acts as a heat sink for the fuel. In the event of a loss of circulation coupled with a depressurization accident, the heat removal capacity of the low-pressure gas is no longer adequate to remove heat by natural convection, and in this case the forced circulation must be restored. The slow depressurization rate of the PCPV allows a reasonable time margin for the institution of emergency cooling.

The above analysis is based on the assumption that the reactor has been shut down as a result of the depressurization, so that only the decay heat has to be removed. If the shutdown system were to fail in an AGR, the negative temperature coefficient of the reactor would not be large enough to prevent melting of some of the stainless steel fuel cladding. The loss of the neutron absorption associated with the steel could initiate a reactivity transient which would result in a large-scale core melt. This event, however, is extremely unlikely on account of the extensive protective instrumentation and reliable shutdown system of the AGR.

Apart from the clad melt mentioned above, it is difficult to find plausible mechanisms for significant reactivity transients in a gas-cooled thermal reactor. The number of control rods tends to be large, so that the maximum reactivity worth of a single rod is usually of the order of 0.1%, in comparison with the 2% or so possible in a light water reactor. Consequently, rapid single-rod removal is not a potential source of serious reactivity transients in the gas-cooled reactor. Simultaneous motoring out of large numbers of rods without reactor shutdown is highly improbable.

For the high-temperature reactor, the mixed graphite–ceramic fuel element and, in particular, the use of coated fuel particles, is an additional safety feature. Excursions to high temperature do not in general lead to severe fuel failure, but tend to produce a temporary increase in the fission product emission rate. Because of the high thermal capacity of the fuel element and the overall negative temperature coefficient, both aided by rapid heat transfer from the dispersed fuel to the graphite, coupled with the large margin between the normal operating temperature and the temperature at which significant fuel damage would occur, relatively large step increases in reactivity can be tolerated, even if the SCRAM system were to fail.

The safety record of the gas-cooled graphite-moderated power reactors in service has been generally very good. For the magnox reactors, the margin between fuel clad temperature and melting temperature is rather narrow, and a channel melt-out due to a coolant flow restriction has occurred (in one of the Chapelcross reactors). Analysis indicates that, following this type of incident, there is no danger of the fault propagating to adjacent channels. While the accumulated operating experience of the AGR and HTGR reactors is very much less than the more than 600 reactor-years cumulative total for the natural uranium graphite systems, the combination of gas cooling and concrete pressure vessel has led to considerable confidence in the safety potential of the more advanced designs.

12.4.2. The Heavy Water Reactor

From the safety viewpoint, the CANDU heavy water reactor has certain similarities to the PWR in that the primary coolant consists of water at high pressure. The CANDU system has not been subjected to the same degree of public analysis as the light water reactors, but on the other hand, it has features which are claimed to make it demonstrably safer than the LWR. One of the most important is the subdivision of the coolant circuit into a large number of pressure tubes, with the largest pressurized component having a diameter of about 0.5 m. The concern about catastrophic failure of a massive pressure vessel is therefore eliminated. The individual pressure tubes are relatively simple components, easily manufactured to high standards of quality control and amenable to a straightforward stress analysis. The containment can easily withstand the effects of the sudden failure of the largest pressurized component.

Emergency core cooling is provided for the CANDU system as it is for the PWR and BWR. In this case, however, the emergency cooling problem is much less severe, partly because of the fact that the CANDU core has almost ten times the volume of a light water reactor core of the same power, and partly because each fuel channel is isolated from its neighbors by a large volume of relatively cool heavy water moderator. Consequently, even the failure of the emergency core cooling system in the event of a circuit depressurization accident would not lead to large-scale fuel melting, provided that the reactor shutdown mechanisms operated properly. The fact that the control rods do not enter the pressurized primary coolant system increases the reliability of their operation, as well as eliminating the possibility of a rod ejection accident. The control rod system is also backed up by an independent mode of shutdown in the form of moderator dumping or injection of poison into the moderator.

While the coolant void coefficient is normally positive, the degree of subdivision of the piping allows any power excursion arising from a loss of coolant accident to be easily terminated by the control rods. An added advantage is that, with on-load refueling, the excess reactivity of the core can be maintained at a much lower level than is required for a light water reactor.

12.5. The Safety of the Fast Breeder Reactor

The move to a fast breeder economy, desirable on the grounds of conserving uranium and minimizing the long-term costs of nuclear power, has perhaps generated more controversy than any other proposed development in the nuclear field. Some of the opposition is due to safety questions associated with the fast reactor itself, and some to the implications of the move to a large-scale plutonium economy.

The controversy has centered on the liquid-metal-cooled fast breeder reactor (LMFBR), since this is the type selected by all the countries involved for at least the initial phase of a fast reactor program. From the safety viewpoint, the main differences between the fast breeder and the thermal reactors considered earlier lie in the use of the sodium coolant. On the one hand, a sodium circuit operates at much lower pressures (up to 10 atm) than does a water or gaseous coolant, thus reducing the consequences of a depressurization accident. In addition, the temperature margin to coolant boiling or burn-out is greater than in the LWR. On the other hand, the potential mixing of molten fuel and sodium following a severe transient could lead to an explosive reaction which is not well understood, but which could conceivably result in very severe damage to the reactor pressure vessel. In addition, the violent reaction of sodium with water could lead to trouble in the event of steam generator leakage.

The high power density of the LMFBR means that even a small local interruption in coolant flow could lead to the meltdown of part or all of the core. In the event of a large-scale core melt taking place, the high enrichment of the fuel raises the possibility of forming a supercritical mass or masses which could materially increase the explosive energy released in the accident.

The sensitivity of the LMFBR to localized interruptions of coolant flow may be seen by considering the sequence of events which led to the partial meltdown of the core of the Enrico Fermi fast breeder reactor in October 1966. This reactor had an output of 200 MWt and was loaded with uranium metal fuel. A late change in the design led to the fitting of thin zirconium liners under the core. While the reactor was being taken up to power, two of the liners were torn off by the coolant flow and swept up into the core, where they blocked the inlets to two of the 105 fuel subassemblies, which melted out because of coolant starvation. The reactor was scrammed before any large-scale melting of the core occurred. The effects of the accident were minimized by the fact that the two subassemblies were in a low-rated region of the core, and in addition the reactor was only at 15 % of full power at the time. On the other hand, if the scram system had failed to operate, a severe core explosion might have taken place because of one of the mechanisms to be discussed later in this section.

An earlier accident to the small fast reactor EBR-1 (November 1955) had resulted in an extensive meltdown of a large fraction of the core. During a set of tests involving a run-up to power at low flow rates, the operator misunderstood an instruction for rapid shutdown as the power reached the safety limit, and instead initiated a slow control rod insertion. The result was a rise of the core center temperature, which caused boiling of the NaK coolant. The excursion was worsened by the existence of a fast positive temperature coefficient, due to inward bowing of the fuel rods. The high temperatures reached caused extensive melting of the core, and the boiling coolant forced the molten central core material outward, causing blockage of the outer coolant channels. The reactivity of the core had been increased by compaction to such an extent that the shutoff rods were no longer adequate to shut the reactor down, and an auxiliary scram system, involving the dropping away of the outer reflector, had to be brought into play to terminate the nuclear reaction. Failure to achieve shutdown would presumably have led to an explosive disruption of the core, although the energy release would have been very modest owing to the small size of the reactor.

Some of the factors which influence the course of an LMFBR accident have already been considered in Section 11.3. These include the high power density typical of the fast reactor (up to a maximum of 300 W/g compared with 40 W/g for an LWR), high plutonium content of the core, low delayed neutron fraction, short neutron lifetime and the reactivity effects of sodium coolant density, core expansion, and Doppler coefficient. Unlike the light water reactor, the LMFBR gains reactivity if the fuel is rearranged into a denser configuration. Thus the consequences of an initial excursion leading to core melt can be very much magnified if the molten fuel is compacted either by gravity or by high pressures generated, for example, by sodium coolant explosion. This is especially serious if the effect of fuel compaction is to drive the reactor rapidly into the superprompt critical condition, since the short neutron lifetime then results in a very rapid rise in power, possibly giving rise to a serious fuel vapor explosion.

The mechanisms which could initiate a large and rapid increase in reactivity include sudden removal of a control rod, rapid insertion of a fuel subassembly, increase in coolant voidage, and the introduction of moderating material into the coolant. For a full-scale LMFBR, the maximum worth of an individual control rod can probably be confined to a value less than the delayed neutron fraction, and elaborate design precautions are in any case taken to eliminate the possibility of sudden rod ejection. The consequences of a fuel subassembly falling into the core while loading can be minimized by ensuring that this operation is only performed with all control rods fully inserted to provide a large shutdown margin. The possibility of a prompt critical excursion due to the sudden injection of a quantity of moderator, such as oil, into the core can be rendered sufficiently remote by careful design, while the isolation of the primary coolant from the steam generators by the intermediate sodium circuit eliminates the possibility of sudden injection of steam into the core.

The sodium voidage coefficient, as mentioned in Chapter 11, is a function of position in the core, being generally positive in the central regions and negative at the periphery. Owing to the large margin to boiling, the creation of voidage by this effect could happen only in the event of coolant starvation to part or all of the core; this possibility will be discussed in detail below. Voidage effects caused by entrainment of cover gas into the sodium have occurred in earlier types of LMFBR, but this effect can be virtually eliminated by careful design.

The accident to the Fermi reactor emphasized the importance of avoiding local coolant starvation due to channel blockage. Filters can be fitted at the coolant inlet to prevent debris from outside the core being swept into the core. Blockages could also occur from debris originating within the core, for example from a failed fuel pin, and a variety of devices, such as thermocouples or flow indicators attached to the fuel subassemblies, may have to be installed to provide warning of flow starvation. These can be backed up by acoustic or fission product detectors. Even if the protective system does not shut the reactor down in time to prevent coolant boiling in the affected subassembly, which would be rapidly followed by fuel melting, the time required for the molten fuel to cause damage to adjacent subassemblies is long enough to make it very unlikely that the reactor would not have been shut down by a signal from one of the numerous protective indicators before any other subassembly could be affected. One has also to consider the possibility that interaction between the molten fuel and coolant within the affected assembly may lead to explosive vapor formation, which could in turn damage other subassemblies. Little of the stored heat will be converted to mechanical energy, however, unless the fuel is very finely dispersed, an unlikely outcome of a meltdown of a single subassembly as a result of flow starvation. More investigation, both experimental and theoretical, is required before the possible magnitude of explosive energy release in the subassembly can be confidently predicted and, in view of the potential consequences of a large release, including possible reactivity transients due to changes in core configuration, this is being vigorously pursued.

Much of the safety work carried out on the LMFBR has been devoted to calculations concerning hypothetical "whole-core accidents" which could involve a rapid excursion into the superprompt critical condition. Such an accident would be a loss of coolant flow, accompanied by a failure to scram. The flow stagnation would lead to boiling of the sodium coolant out of the core, possibly resulting in an increase of reactivity due to the positive sodium coefficient in the central regions of the core. Owing to the continued high power level, coupled with the loss of coolant, the core would melt, the fuel elements slumping into a heap, taking up the space vacated by the sodium. This compaction could lead to a rapid and marked increase of reactivity, which might take the reactor into the superprompt critical regime. The resulting vaporization of part of the core would result in an explosive core dispersion, throwing molten fuel outwards and thereby bringing the nuclear excursion to an end.

The sequence of events following the termination of this initial excursion is still in considerable doubt. It is possible that the dispersed fuel will gradually slump back again in discrete pieces, thereby giving a relatively slow buildup of reactivity and a fairly modest secondary power excursion. On the other hand, it is possible that, in the course of the reassembly under gravity, a single lump representing a large fraction of the original core material may suddenly drop down on top of an assembly which is already almost of critical size, giving a very rapid and very large positive reactivity increase, capable of taking the core well above prompt critical. Another possibility is that the top part of the molten core, on being driven up by the original explosive disassembly, may cause a sodium vapor explosion due to rapid transfer of heat to sodium coolant above the core. This could cause the core material to be driven together into a supercritical mass even more rapidly than by slumping under gravity, and give a more serious secondary explosion.

Much disagreement exists on whether a sodium vapor explosion of the kind postulated above can in fact take place when the molten core mixes with the coolant following the initial explosive dispersal of the core. According to the "superheat theory" of Fauske (see Bibliography), it seems likely that the mixing of molten uranium or plutonium oxide with liquid sodium will not in fact produce a sodium vapor explosion. On the other hand, it is possible that the much more rapid transfer of heat to the sodium from the molten steel of the fuel cladding could either trigger a sodium vapor explosion directly or set up the conditions under which the slower heat transfer from the fuel itself could do so.

In view of the difficulty of calculating the course of a whole-core accident, the design of a modern fast breeder reactor incorporates numerous safeguards to prevent large and rapid increases in reactivity. Phenix and Superphenix, for instance, have a "flared" core, where the fuel assemblies are held at one end but can separate, upon expansion, at the other. This provides a large and rapidly acting negative reactivity coefficient to supplement the Doppler effect. The effect of failure of coolant flow due to loss of power to the circulation pumps can be mitigated by fitting flywheels which increase the coastdown time and allow manual action to remedy a concurrent failure of the shutdown mechanisms.

Early calculations on the magnitude of the energy yields of whole-core accidents suggested that it was possible that explosive releases severe enough to fracture the pressure vessel could occur. More recent calculations, using more sophisticated models and large computer codes, tend to predict much lower energy yields, although the uncertainties are still such that it is impossible to rule out completely an accident sequence which could cause fracture of the vessel. Consequently, any full-scale fast breeder must incorporate numerous protective features designed to reduce the probability of a whole-core accident to a very small value (less than 1 in 10^6 reactor years).

12.6. Long-Term Storage and Disposal of Radioactive Wastes

The question of the ultimate disposal of the radioactive wastes in the fuel discharged from nuclear power plants is one which has generated a good deal of controversy. The very large level of radioactivity involved has tended to encourage exaggerated views of the magnitude of the problem, which can only be satisfactorily treated on the basis of detailed consideration of the possible pathways by which the buried waste can become part of the human environment. Some of the most comprehensive estimations of the potential hazards associated with nuclear waste disposal have been carried out by B. L. Cohen (see Bibliography) and the present section is based largely on this work.

For reference, we consider a light water reactor with a fuel burn up of 33,000 MW d/ton, and an initial fuel enrichment of 3.3 % U^{235}. Over the core lifetime, 25 kg of the 33 kg of U^{235} originally present in each tonne of fuel will have been consumed, together with 24 kg of the U^{238}. In their place, the fuel will contain about 35 kg of assorted fission products, 8.9 kg of plutonium isotopes (mostly Pu^{239}), and 4.6 kg of U^{236}

Following their removal from the reactor, the fuel elements are stored for several months in order to permit the decay of short-lived activities. The subsequent reprocessing involves dissolving the fuel and its cladding in acid and removing some 99.5 % of the uranium and plutonium for recycling. At this stage, gaseous fission products such as krypton and xenon are collected and discharged under controlled conditions. The remaining fission products, actinides, and cladding activation products constitute the "high-level waste" for which satisfactory methods of permanent disposal have to be devised.

The currently favored method for doing this is by incorporating the wastes in a cylinder of borosilicate glass and burying the cylinder in a suitable underground location at a depth of around 600 m. The dimensions of the cylinders are set largely by the requirements for removal of the heat generated by the radioactive decay of the material; the proposed size is about 3 m long by 0.3 m in diameter. Each glass cylinder would be sealed inside a thick casing of stainless steel. The canisters would be buried about 10 m apart, so that the annual wastes from a 1000-MWe plant would occupy an area of some 1000 m^2. In order to keep the temperature of the cylinders low enough to prevent deterioration (embrittlement) of the glass, it is desirable to delay the burial of the waste till about ten years after the reprocessing of the fuel, thereby permitting the heat from the radioisotopes to decay by about an order of magnitude.

About 90% of the fission products are not of concern for longer-term storage, since they have short half-lives (less than four years) or half-lives so long (greater than 4×10^{10} yr) that their activity may be neglected. Over the first few hundred years, the dominant isotopes are Sr-90 (29 yr) and Cs^{137} (30 yr) and their daughters Ba^{137m} and Y^{90}. This is illustrated in Fig. 12.15, which shows the contributions of various isotopes to the overall heat production as a function of time after reprocessing. After 500 yr, the main contributors to the heating are the isotopes of Pu and Am, and beyond about 100,000 yr the daughters of these isotopes, such as Po^{213} and Th^{229}. The predominance of Sr^{90} and Cs^{137} on an *activity* basis in the early years is more marked than would be implied from Fig. 12.15, since the energy release per disintegration from these β-emitting isotopes is considerably less than from the α-emitting actinides.

As an illustration of the methods used in the assessment of the hazards from nuclear waste, we can take the calculation for the isotope Np^{237}, as computed by Cohen. This isotope decays with a half-life of 2.1 million yr by the emission of an alpha particle of energy 4.8 MeV; the principal hazard arising from its ingestion is the induction of cancer of the liver. The first step

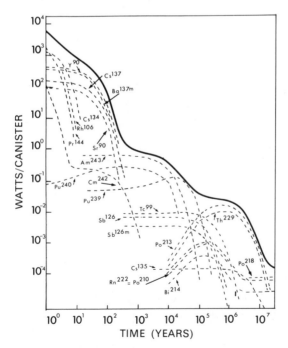

Fig. 12.15. Contributions of various radioisotopes to heat generation in nuclear waste. [From B. L. Cohen, *Rev. Mod. Phys.* **49** (1), 1 (1977)].

is the calculation of the risk of liver cancer due to the ingestion of 1 Ci of Np^{237}. When neptunium is ingested, 1 % is transmitted through the intestinal walls into the blood and of this 45 % is deposited in the liver, from which it is eliminated with a 40 yr half-life. It can thus be calculated that, during this period, the number of alpha decays occurring in the liver from the 1 Ci ingested is equal to 2.0×10^{17}, giving an energy deposition of 1.5×10^5 J, equivalent to an absorbed radiation dose of 8×10^4 gray (Gy).

The cancer risk to the liver is known to be about 3×10^{-2} per Gy, so that the number of cancers induced by the absorbed dose from ingestion of 1 Ci is 2400. From the absorption cross sections of the isotopes involved, it can be calculated that Np^{237} is present in the high level waste at a level of 12 Ci per GWe yr (1 gigawatt, $GW = 10^9$ W). The number of liver cancers corresponding to the ingestion of the Np^{237} in the waste per GWe yr is therefore 27,000. Similar calculations can be done for the cancers induced in other organs, such as the bone and blood; this yields a number of 38,000 for all types of fatal cancer combined.

The results of calculations for the more important isotopes in the wastes are shown in Fig. 12.16; the data are presented in the form of fatal cancer doses per tonne of initial uranium fuel in the reprocessed high-level waste as a function of the time after reprocessing. It is seen that Sr^{90} and Cs^{137} are the dominant isotopes for the first 200 yr, followed by Am^{241} up to about 1500 yr. Am^{243} is the dominant isotope between 1500 and 10,000 yr, after which Np^{237} dominates up to 20 million yr, when the U^{238} decay chain takes over. The hazard from the initial uranium is also shown in Fig. 12.16; it is interesting to note that after 50,000 yr the total hazard from the wastes has decreased to the same level as that from the initial uranium.

What has been done thus far, of course, is simply an assessment of the potential hazard of the isotopes in the high-level waste, i.e., the number of fatal cancers that would be induced if all the waste considered were ingested by a large human population. To convert this into a realistic estimate of the hazard which would actually arise in practice, one has to estimate the probability that the active material will be ingested, given the methods of disposal which are proposed. We are required to estimate the *transfer rate*, or fraction of the waste which will be transferred from its burial site to be ingested by humans. The transfer rate can then be multiplied by the values shown in Fig. 12.16 to obtain the expected cancer fatalities per yr; by integration over time one can then derive the total number of eventual fatalities per GWe yr.

As a first approximation, it is assumed that the probability that an atom of high-level waste is leached out by ground water, carried into a river and

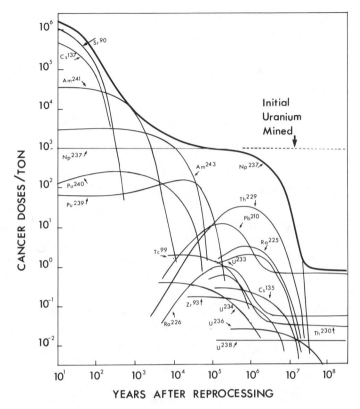

Fig. 12.16. Number of fatal cancer doses per tonne of initial uranium fuel in reprocessed high-level waste as a function of time after reprocessing [from B. L. Cohen, *Health Phys.* **42**, 133 (1982). Used with permission of Health Physics Society].

eventually ingested by a human being is the same as the probability for an atom of an average rock submerged in ground water. There are a number of reasons why this assumption will tend to overestimate the rate of transport of the active wastes to the surface. For one thing, the burial site would be selected for its absence of ground water, both at the time of burial and also, so far as is predictable on geological grounds, for a considerable time in the future. Also, no credit is given for the protective effect of the metal cladding of the glass cylinder; this could be considerable, particularly if some highly corrosion-resistant material, such as titanium, were used.

From analyses of the concentrations of leached-out elements in ground water, it can be calculated that, for elements of similar behavior to the more important waste radionuclides, the fractional rate of removal of the elements from the rock into rivers is less than 10^{-8} per year.

By combining this figure with the probability that an atom in a river is ingested by a human, we can calculate the probability that an atom of stored waste will be ingested. For the United States, the total water flow per year in rivers is 1.5×10^{15} l and the total annual drinking water consumption is 1.8 $\times 10^{11}$ l, and so the probability of a molecule of water in a U.S. river being ingested is about 10^{-4}. Hence the probability of an atom of the buried high level waste being ingested by a human is $10^{-8} \times 10^{-4} = 10^{-12}$ per year.

It is now possible to convert the cancer doses in Fig. 12.16 into the expected number of cancer fatalities per year by multiplying the vertical scale by a factor of 10^{-12}. The total number of eventual fatalities, obtained by integrating over the whole of the time involved, has been estimated by Cohen as about 0.11 eventual fatalities per GWe yr. This calculation assumes that the waste remains buried at 600 m below the land surface, and takes no account of the gradual erosion of that surface that takes place over very long periods of time. The North American continent is being eroded at a rate of 1 m in 22,000 yr, so that the buried waste should reach the surface in 13 million yr if this rate continues. At that time all the remaining waste could be assumed to run off into rivers, increasing its ingestion rate from 10^{-12} per year to a onetime 10^{-4} probability. Inclusion of this effect raises the cancer fatality rate to 0.17 eventual fatalities per GWe yr. This is of the same order as the fatality rate from the dose commitment from the other stages of the fuel cycle, as given in Table 12.6, which may be shown to lead to a cancer fatality rate of about 0.1 per GWe yr.

It can be argued that the integration of the effects of buried waste over time periods of the order of millions of years is unrealistic, in view of the impossibility of predicting the state of technology, or even the existence of a technological civilization, over a very much shorter time scale. Integrations over more realistic time periods, say 100 yr, yield a much smaller fatality rate, of the order of 10^{-4} fatalities per GWe yr, which is far below the rate associated with the wastes from alternative energy sources such as fossil fuel combustion. The hazard from the dominant shorter-lived isotopes, such as Sr^{90}, is in practice negligible because of the difficulty of envisaging mechanisms by which any of these isotopes could reach the surface before decay, in view of the very slow diffusion process involved. Since the longer-term hazard is dominated by the Np^{237}, this effect could be reduced by separation of the transuranic elements from the wastes, to be followed by even more secure isolation of the relatively small quantities of waste represented by these elements.

12.7. The Thermal Effects of Nuclear Power Stations

Like other power plants which operate by the conversion of heat into electrical energy, the nuclear reactor requires a heat sink for rejection of the heat given off by the fuel. For the standard reactor, where the electrical output is produced by a steam turbine, the heat sink consists of the cooling water for the condenser, which is normally obtained from a river or lake or from the sea. The *thermal efficiency* of the plant was defined as

$$\text{efficiency} = \frac{\text{electrical output (MWe)}}{\text{thermal output (MWt)}} \qquad (6.68)$$

The difference between the total thermal output from the fuel and the electrical output is made up of the heat rejected to the heat sink (condenser), plus any heat losses in the reactor circuit, for example from the heat exchangers and the turbine. In practice, these heat losses are small in comparison with the heat rejected to the condenser. The most important factors in determining the thermal efficiency are the temperature of the steam entering the turbine and the temperature of the coolant in the condenser. The efficiency increases as the difference between these two temperatures increases. Since the condenser coolant temperature can vary over only a relatively small range, depending on the ambient temperature of the coolant water, the efficiency depends in practice on the temperature of the steam fed to the turbine.

For most reactors, such as the BWR or PWR, for example, it is not possible because of the limitations on fuel operating conditions to obtain steam temperatures as high as can be produced in a fossil fuel plant. Typical thermal efficiencies for this type of reactor are around 32%, compared with about 40% for a fossil-fueled unit. The lower efficiency means that the plant has to be larger and therefore more expensive than would be possible if an efficiency similar to that of the fossil plant could be achieved. In addition, for a given electrical output, the nuclear plant will discharge more heat to the environment. The effect on the source of the condenser water supply will be more marked still, since effectively all of the heat rejected by the nuclear plant will go to heating the condenser effluent, while in a fossil plant some 10%–15% of the heat is released to the atmosphere through the stack. The reduction of environmental heat loss is one of the incentives in developing reactors, such as the HTGR, which can operate at higher primary coolant temperatures.

In order to avoid adverse effects on the ecology of the river or lake used as a condenser cooling supply, particularly in a climate where the ambient temperature is already high, it may be necessary to adopt one of several techniques for preventing localized overheating of the water source. Among the methods which may be used are the following.

a. *Increase of the flow rate of the condenser coolant:* While reducing the temperature rise, this will involve higher costs in the form of more powerful pumps, larger condenser pipework, etc. One way of avoiding the increase in condenser size is to dilute the condenser effluent with a bypass volume of unheated water, thereby reducing its temperature before discharge to the environment.

b. *Cooling ponds:* These consist of large artificial ponds (about four square miles in area for a 1000-MWe unit) where the discharged water is cooled largely by evaporation.

c. *Cooling towers:* The simplest form of cooling tower is a *natural draft cooling tower* where the discharged water is pumped to the top of a large hyperbolic structure and allowed to flow down over an open lattice where it is cooled by air drawn in by natural convection through holes in the base of the tower. The water not lost by evaporation is recirculated, together with any makeup required. The typical natural convection cooling tower has the disadvantage of being very large, with a base diameter of 100 m or more. It is possible to reduce the size by going to a *forced convection cooling tower*, where fans are used to increase the flow of air. The use of forced convection, however, increases the overall cost of the system compared with natural convection. In regions where the supply of cooling water is very limited, it may be necessary to use *dry cooling towers*, where the condenser coolant is circulated in a closed cycle through a radiator with a high heat transfer surface, cooled by air from large fans. While eliminating evaporation loss, the dry cooling tower is considerably more expensive than the others. It is also the most undesirable from the point of view of noise, but avoids the problems of fog and icing often associated with the wet cooling tower.

In general, the effect of the heat rejected by the condenser will depend very much on the characteristics of the coolant source itself, e.g., its temperature, dissolved oxygen content, and burden of toxic materials. Each site has to be evaluated individually in terms of these and other factors. There are cases where the discharge of warm effluent can even be beneficial. Where a large volume of cold water is available, the growth rates of fish and crustaceans can be increased markedly by the thermal discharge, and research is being carried out on the possibility of using the waste heat from nuclear power stations for fish and lobster farming.

Appendix

Table A.1. Fundamental Constants

Speed of light *in vacuo* (c)	2.9979×10^8 m s^{-1}
Planck's constant (h)	6.6262×10^{-34} J s
Avogadro's number (N_0)	6.0220×10^{23} mol^{-1}
Charge of electron (e)	1.60219×10^{-19} C
Atomic mass unit (u)	1.660566×10^{-27} kg
Mass of electron	9.109534×10^{-31} kg $= 5.4858 \times 10^{-4}$ u
Mass of proton	1.672649×10^{-27} kg $= 1.007276$ u
Mass of neutron	1.674954×10^{-27} kg $= 1.008665$ u
Boltzmann constant (k)	1.3805×10^{-23} J/K

Table A.2. Conversion Factors

To convert	to	multiply by
Btu	joules	1055
Btu	kW h	2.931×10^{-4}
Btu/h	watts	0.2931
Btu/h ft^2	W/m^2	3.155
Btu/h ft^3	W/m^3	10.35
centimeters	inches	0.3937
feet	meters	0.3048
inches	centimeters	2.540
joules	Btu	9.478×10^{-4}
joules	kW h	2.778×10^{-7}
joules	MeV	6.242×10^{12}
joules	MW d	1.157×10^{-11}
kilograms	pounds	2.2046
kg/cm^2	lb/in.^{2}	14.22
kW h	Btu	3412
kW h	joules	3.600×10^6
pounds	kilograms	0.4536
lb/in.^{2}	kg/cm^2	0.0703
meters	feet	3.281
MeV	joules	1.602×10^{-13}
MW d	joules	8.640×10^{10}
watts	Btu/h	3.412
W/m^2	Btu/h ft^2	0.3170
W/m^3	Btu/h ft^3	0.0966

Table A.3. Masses of Selected Particles and Atoms in Atomic Mass Units[a]

Particle/Atom	Mass (u)
Particle	
Proton	1.007276470
Neutron	1.008665012
Electron	5.4858026×10^{-4}
α particle	4.001506180
Atom[b]	
$_1H^1$	1.007825037
$_1H^2$	2.014101787
$_1H^3$	3.016049286
$_2He^3$	3.016029297
$_2He^4$	4.00260325
$_3Li^7$	7.0160045
$_4Be^7$	7.0169297
$_4Be^9$	9.0121825
$_5B^{10}$	10.0129380
$_6C^{12}$	12.00000000
$_7N^{14}$	14.003074008
$_8O^{16}$	15.99491464
$_{43}Tc^{107}$	106.91464
$_{49}In^{115}$	114.903875
$_{49}In^{116}$	115.905257
$_{50}Sn^{116}$	115.9017435
$_{51}Sb^{133}$	132.91521
$_{90}Th^{231}$	231.036298564
$_{90}Th^{232}$	232.038053805
$_{92}U^{234}$	234.040947400
$_{92}U^{235}$	235.043925247
$_{92}U^{238}$	238.050785782
$_{94}Pu^{239}$	239.052157781
$_{94}Pu^{240}$	240.053808657
$_{94}Pu^{241}$	241.056846915

[a] From A. H. Wapstra and K. Bos, *Atomic Data and Nucleus Data Tables, The 1977 Atomic Mass Evaluation*, Academic Press, New York (1977).
[b] Mass given is that of neutral atom.

Table A.4. Microscopic and Macroscopic Absorption and Scattering Cross Sections for the Elements

Element	Symbol	Atomic number	Density (g cm^{-3})	σ_a (b)	σ_s (b)	Σ_a (cm^{-1})	Σ_s (cm^{-1})
Actinium	Ac	89	10.1	515	—	13.8	—
Aluminum	Al	13	2.70	0.230	1.49	0.0139	0.0898
Antimony	Sb	51	6.69	5.4	4.2	0.179	0.139
Argon	Ar	18	—	0.678	0.644	—	—
Arsenic	As	33	5.73	4.3	7	0.198	0.322
Barium	Ba	56	3.5	1.2	—	0.0184	—
Beryllium	Be	4	1.85	0.0092	6.14	0.00114	0.759
Bismuth	Bi	83	9.75	0.033	—	0.00093	—
Boron	B	5	2.34	759	3.6	98.9	0.469
Bromine	Br	35	3.12	6.8	6.1	0.160	0.143
Cadmium	Cd	48	8.65	2450	5.6	113.6	0.260
Calcium	Ca	20	1.55	0.43	—	0.0100	—
Carbon	C	6	1.6	0.0034	4.75	0.000273	0.381
Cerium	Ce	58	6.77	0.63	4.7	0.0183	0.137
Cesium	Cs	55	1.87	29.0	—	0.246	—
Chlorine	Cl	17	—	33.2	—	—	—
Chromium	Cr	24	7.19	3.1	3.8	0.258	0.316
Cobalt	Co	27	8.9	37.2	6.7	3.38	0.609
Copper	Cu	29	8.96	3.79	7.9	0.322	0.671
Deuterium	D	1	—	0.00053	3.390	—	—
Dysprosium	Dy	66	8.54	930	100	29.4	3.16
Erbium	Er	68	9.05	162	11.0	5.28	0.359
Europium	Eu	63	5.25	4600	8.0	95.7	0.166
Fluorine	F	9	—	0.0095	4.0	—	—
Gadolinium	Gd	64	7.90	49000	—	1480	—
Gallium	Ga	31	5.90	2.9	6.5	0.148	0.331
Germanium	Ge	32	5.32	2.3	7.5	0.1015	0.331
Gold	Au	79	19.3	98.8	—	5.83	—
Hafnium	Hf	72	13.31	102	8	4.58	0.359
Helium	He	2	—	<0.05	0.76	—	—
Holmium	Ho	67	8.78	66.5	9.4	2.13	0.301
Hydrogen	H	1	—	0.332	20.436	—	—
Indium	In	49	7.31	193.5	—	7.42	—
Iodine	I	53	4.93	6.2	—	0.145	—
Iridium	Ir	77	22.4	426	14	29.9	0.98
Iron	Fe	26	7.87	2.55	10.9	0.216	0.925
Krypton	Kr	36	—	25.0	7.50	—	—
Lanthanum	La	57	6.17	9.0	9.3	0.241	0.249
Lead	Pb	82	11.35	0.170	11.4	0.0056	0.376
Lithium	Li	3	0.534	70.7	—	3.28	—
Lutetium	Lu	71	9.84	77	8	2.61	0.271
Magnesium	Mg	12	1.738	0.063	3.42	0.00271	0.147
Manganese	Mn	25	7.4	13.3	2.1	1.079	0.170
Mercury	Hg	80	13.55	375	—	15.26	—
Molybdenum	Mo	42	10.22	2.65	5.8	0.170	0.372

(continued overleaf)

Table A.4.—*cont.*

Element	Symbol	Atomic number	Density (g cm^{-3})	σ_a (b)	σ_s (b)	Σ_a (cm^{-1})	Σ_s (cm^{-1})
Neodymium	Nd	60	6.90	50.5	16	1.455	0.461
Neon	Ne	10	—	0.038	2.42	—	—
Nickel	Ni	28	8.90	4.43	17.3	0.404	1.580
Niobium	Nb	41	8.57	1.15	—	0.0639	—
Nitrogen	N	7	—	1.85	10.6	—	—
Osmium	Os	76	22.6	15.3	—	1.095	—
Oxygen	O	8	—	0.00027	3.76	—	—
Palladium	Pd	46	12.0	6.9	5.0	0.469	0.340
Phosphorus	P	15	1.82	0.180	—	0.00637	—
Platinum	Pt	78	21.45	10.0	11.2	0.662	0.742
Plutonium	Pu	94	19.84	1011.3	7.7	50.55	0.385
Potassium	K	19	0.862	2.10	1.5	0.0279	0.0199
Praseodymium	Pr	59	6.77	11.5	3.3	0.333	0.0955
Protactinium	Pa	91	15.37	210	—	8.4	—
Radium	Ra	88	5.0	11.5	—	0.153	—
Radon	Rn	86	—	0.72	—	—	—
Rhenium	Re	75	21.02	88	11.3	5.98	0.77
Rhodium	Rh	45	12.41	150	—	10.9	—
Rubidium	Rb	37	1.532	0.37	6.2	0.00399	0.0670
Ruthenium	Ru	44	12.41	2.56	—	0.189	—
Samarium	Sm	62	7.45	5800	—	173	—
Scandium	Sc	21	2.989	26.5	24	1.06	0.96
Selenium	Se	34	4.79	11.7	9.7	0.427	0.354
Silicon	Si	14	2.33	0.16	2.2	0.0080	0.110
Silver	Ag	47	10.50	63.6	—	3.73	—
Sodium	Na	11	0.971	0.530	3.2	0.0135	0.0814
Strontium	Sr	38	2.54	1.21	10	0.0211	0.175
Sulfur	S	16	2.07	0.520	0.975	0.0202	0.0379
Tantalum	Ta	73	16.65	21	6.2	1.164	0.343
Technetium	Tc	43	11.5	19	—	1.33	—
Tellurium	Te	52	6.24	4.7	—	0.138	—
Terbium	Tb	65	8.234	25.5	20	0.796	0.624
Thallium	Tl	81	11.85	3.4	9.7	0.1187	0.339
Thorium	Th	90	11.72	7.40	12.67	0.225	0.385
Thulium	Tm	69	9.314	103	12	3.42	0.399
Tin	Sn	50	7.31	0.63	—	0.0234	—
Titanium	Ti	22	4.54	6.1	4.0	0.348	0.228
Tungsten	W	74	19.3	18.5	—	1.17	—
Uranium	U	92	19.1	7.59	8.90	0.367	0.430
Vanadium	V	23	6.11	5.04	4.93	0.364	0.356
Xenon	Xe	54	—	24.5	4.30	—	—
Ytterbium	Yb	70	6.97	36.6	25.0	0.888	0.607
Yttrium	Y	39	4.46	1.28	7.60	0.0387	0.230
Zinc	Zn	30	7.133	1.10	4.2	0.0723	0.276
Zirconium	Zr	40	6.506	0.185	6.40	0.00795	0.275

Table A.5. Absorption and Fission Cross Sections for Some Heavy Isotopes Involved in Nuclear Fuel Cycle

Isotope	Atomic number	Abundance (%)	σ_a (b)	σ_f (b)
Th^{232}	90	100	7.40	
Th^{233}	90		1515	15
Th^{234}	90		1.8	
Pa^{233}	91		41	<0.1
U^{233}	92		578.8	531.1
U^{234}	92	0.0057	100.2	
U^{235}	92	0.72	680.8	582.2
U^{236}	92		5.2	
U^{238}	92	99.28	2.70	
U^{239}	92		36	14
Np^{239}	93		45	<1
Pu^{239}	94		1011.3	742.5
Pu^{240}	94		289.5	0.030
Pu^{241}	94		1377	1009
Pu^{242}	94		18.5	<0.2

Table A.6. Westcott g Factors for Some Important Isotopes

Temperature (°C)	U^{233}		U^{235}		Pu^{239}		U^{238}	Pu^{240}
	g_a	g_f	g_a	g_f	g_a	g_f	g_a	g_a
20	0.9983	1.0003	0.9780	0.9759	1.0723	1.0487	1.0017	1.0270
100	0.9972	1.0011	0.9610	0.9581	1.1611	1.1150	1.0031	1.0518
200	0.9973	1.0025	0.9457	0.9411	1.3388	1.2528	1.0049	1.0823
300	0.9987	1.0044	0.9357	0.9291	1.5895	1.4507	1.0067	1.1160
400	1.0010	1.0068	0.9294	0.9208	1.8905	1.6904	1.0085	1.1536
600	1.0072	1.0128	0.9229	0.9108	2.5321	2.2037	1.0122	1.2521
800	1.0146	1.0201	0.9182	0.9036	3.1006	2.6595	1.0159	1.4478
1000	1.0226	1.0284	0.9118	0.8956	3.5353	3.0079	1.0198	1.9026

Table A.7. Properties of Dry Saturated Steam as a Function of Temperature (English Units)[a]

Temperature, T (°F)	Pressure, P (psia)	Specific volume (ft³/lb)		Specific enthalpy (Btu/lb)		
		Sat. liquid v_f	Sat. vapor v_g	Sat. liquid h_f	Evap. h_{fg}	Sat. vapor h_g
32.018	0.08866	0.016022	3302	0.01	1075.4	1075.4
300	66.98	0.017448	6.472	269.73	910.4	1180.2
310	77.64	0.017548	5.632	280.06	903.0	1183.0
320	89.60	0.017652	4.919	290.43	895.3	1185.8
330	103.00	0.017760	4.312	300.84	887.5	1188.4
340	117.93	0.017872	3.792	311.30	879.5	1190.8
350	134.53	0.017988	3.346	321.80	871.3	1193.1
360	152.92	0.018108	2.961	332.35	862.9	1195.2
370	173.23	0.018233	2.628	342.96	854.2	1197.2
380	195.60	0.018363	2.339	353.62	845.4	1199.0
390	220.2	0.018498	2.087	364.34	836.2	1200.6
400	247.1	0.018638	1.8661	375.12	826.8	1202.0
410	276.5	0.018784	1.6726	385.97	817.2	1203.1
420	308.5	0.018936	1.5024	396.89	807.2	1204.1
430	343.3	0.019094	1.3521	407.89	796.9	1204.8
440	381.2	0.019260	1.2192	418.98	786.3	1205.3
450	422.1	0.019433	1.1011	430.2	775.4	1205.6
460	466.3	0.019614	0.9961	441.4	764.1	1205.5
470	514.1	0.019803	0.9025	452.8	752.4	1205.2
480	565.5	0.020002	0.8187	464.3	740.3	1204.6
490	620.7	0.020211	0.7436	475.9	727.8	1203.7
500	680.0	0.02043	0.6761	487.7	714.8	1202.5
510	743.5	0.02067	0.6153	499.6	701.3	1200.9
520	811.4	0.02091	0.5605	511.7	687.3	1198.9
530	884.0	0.02117	0.5108	523.9	672.7	1196.6
540	961.5	0.02145	0.4658	536.4	657.5	1193.8
550	1044.0	0.02175	0.4249	549.1	641.6	1190.6
560	1131.8	0.02207	0.3877	562.0	625.0	1187.0
570	1225.1	0.02241	0.3537	575.2	607.6	1182.8
580	1324.3	0.02278	0.3225	588.6	589.3	1178.0
590	1429.5	0.02319	0.2940	602.5	570.1	1172.5
600	1541.0	0.02363	0.2677	616.7	549.7	1166.4
610	1659.2	0.02411	0.2434	631.3	528.1	1159.4
620	1784.4	0.02465	0.2209	646.4	505.0	1151.4
630	1916.9	0.02525	0.2000	662.1	480.2	1142.4
640	2057.1	0.02593	0.1805	678.6	453.4	1131.9
650	2205	0.02673	0.16206	695.9	423.9	1119.8
660	2362	0.02767	0.14459	714.4	391.1	1105.5
670	2529	0.02882	0.12779	734.4	353.9	1088.3
680	2705	0.03032	0.11127	756.9	309.8	1066.7
690	2892	0.03248	0.09428	783.8	253.9	1037.7
700	3090	0.03666	0.07438	822.7	167.5	990.2
705.44	3204	0.05053	0.05053	902.5	0	902.5

[a] From J. H. Keenan, F. G. Keyes, P. G. Hill, and J. G. Moore, *Steam Tables*, Wiley, New York (1969).

Table A.8. Properties of Dry Saturated Steam as a Function of Pressure (English Units)[a]

Pressure, P (psia)	Temperature, T (°F)	Specific volume (ft³/lb)		Specific enthalpy (Btu/lb)		
		Sat. liquid v_f	Sat. vapor v_g	Sat. liquid h_f	Evap. h_{fg}	Sat. vapor h_g
14.696	211.99	0.016715	26.80	180.15	970.4	1150.5
100	327.86	0.017736	4.434	298.61	889.2	1187.8
200	381.86	0.018387	2.289	355.6	843.7	1199.3
300	417.43	0.018896	1.5442	394.1	809.8	1203.9
400	444.70	0.019340	1.1620	424.2	781.2	1205.5
500	467.13	0.019748	0.9283	449.5	755.8	1205.3
600	486.33	0.02013	0.7702	471.7	732.4	1204.1
700	503.23	0.02051	0.6558	491.5	710.5	1202.0
800	518.36	0.02087	0.5691	509.7	689.6	1199.3
900	532.12	0.02123	0.5009	526.6	669.5	1196.0
1000	544.75	0.02159	0.4459	542.4	650.0	1192.4
1100	556.45	0.02195	0.4005	557.4	631.0	1188.3
1200	567.37	0.02232	0.3623	571.7	612.3	1183.9
1300	577.60	0.02269	0.3297	585.4	593.8	1179.2
1400	587.25	0.02307	0.3016	598.6	575.5	1174.1
1500	596.39	0.02346	0.2769	611.5	557.2	1168.7
1600	605.06	0.02386	0.2552	624.0	538.9	1162.9
1700	613.32	0.02428	0.2358	636.2	520.6	1156.9
1800	621.21	0.02472	0.2183	648.3	502.1	1150.4
1900	628.76	0.02517	0.2025	660.1	483.4	1143.5
2000	636.00	0.02565	0.18813	671.9	464.4	1136.3
2100	642.95	0.02616	0.17491	683.6	445.0	1128.5
2200	649.64	0.02670	0.16270	695.3	425.0	1120.3
2300	656.09	0.02728	0.15133	707.0	404.4	1111.4
2400	662.31	0.02791	0.14067	718.8	383.0	1101.8
2500	668.31	0.02860	0.13059	730.9	360.5	1091.4
2750	682.46	0.03077	0.10717	763.0	297.4	1060.4
3000	695.52	0.03431	0.08404	802.5	213.0	1015.5
3203.6	705.44	0.05053	0.05053	902.5	0	902.5

[a] From J. H. Keenan, F. G. Keyes, P. G. Hill, and J. G. Moore, *Steam Tables*, Wiley, New York (1969).

Bibliography

Chapter 1

A. P. ARYA, *Fundamentals of Nuclear Physics*, Allyn and Bacon, Boston (1966).

A. BEISER, *Concepts of Modern Physics*, 3rd ed., McGraw-Hill, New York (1981).

Brookhaven National Laboratory, *Neutron Cross Sections*, BNL 325, 2nd ed. (1964) and Supplements (1964, 1965, 1966), available from National Technical Information Service, Springfield, Virginia.

W. E. BURCHAM, *Nuclear Physics, An Introduction*, McGraw-Hill, New York (1963).

B. L. COHEN, *Concepts of Nuclear Physics*, McGraw-Hill, New York (1971).

R. D. EVANS, *The Atomic Nucleus*, McGraw-Hill, New York (1955).

H. FRAUENFELDER and E. M. HENLEY, *Subatomic Physics*, Prentice-Hall, Englewood Cliffs, New Jersey (1974).

I. KAPLAN, *Nuclear Physics*, Addison-Wesley Press, Reading, Massachusetts (1955).

R. E. LAPP and H. L. ANDREWS, *Nuclear Radiation Physics*, 3rd ed., Prentice-Hall, Englewood Cliffs, New Jersey (1964).

C. M. LEDERER, J. M. HOLLANDER, and I. PERLMAN, *Table of Isotopes*, 6th ed., John Wiley and Sons, New York (1968).

S. F. MUGHABGHAB and D. I. GARBER, *Neutron Cross Sections*, BNL 325, 3rd ed., Vol. 1 (Resonance Parameters), Brookhaven National Laboratory, Upton, New York (1973).

E. SEGRÉ, *Nuclei and Particles*, Benjamin, New York (1965).

H. SEMAT and J. R. ALBRIGHT, *Introduction to Atomic and Nuclear Physics*, 5th ed., Holt, Rinehart and Winston, New York (1972).

A. H. WAPSTRA and K. BOS, *Atomic Data and Nuclear Data Tables*, The *1977 Atomic Mass Evaluation*, Academic Press, New York (1977).

Chapter 2

E. A. C. CROUCH, *Fission-Product Yields from Neutron-Induced Fission*, Atomic Data and Nuclear Data Tables 19, 417–432, Academic Press, New York (1977).

S. GLASSTONE and M. C. EDLUND, *The Elements of Nuclear Reactor Theory*, D. Van Nostrand Company, Princeton, New Jersey, and New York (1952).

D. R. INGLIS, *Nuclear Energy: Its Physics and Its Social Challenge*, Addison-Wesley, Reading, Massachusetts (1973).

J. R. LAMARSH, *Introduction to Nuclear Reactor Theory*, Addison-Wesley, Reading, Massachusetts (1966).

S. F. MUGHABGHAB and D. I. GARBER, *Neutron Cross Sections*, BNL 325, 3rd ed., Vol. 1, Brookhaven National Laboratory, Upton, New York (1973).

A. M. WEINBERG and E. P. WIGNER, *The Physical Theory of Neutron Chain Reactors*, University of Chicago Press, Chicago (1958).

Chapter 3

G. I. BELL and S. GLASSTONE, *Nuclear Reactor Theory*, Van Nostrand Reinhold, New York (1970).

J. J. DUDERSTADT and L. J. HAMILTON, *Nuclear Reactor Analysis*, John Wiley and Sons, New York (1976).

A. R. FOSTER and R. L. WRIGHT, *Basic Nuclear Engineering*, 2nd ed., Allyn and Bacon, Boston (1973).

S. GLASSTONE and M. C. EDLUND, *The Elements of Nuclear Reactor Theory*, D. Van Nostrand Company, Princeton, New Jersey, and New York (1952).

S. GLASSTONE and A. SESONSKE, *Nuclear Reactor Engineering*, Van Nostrand Reinhold, New York (1967).

P. J. GRANT, *Elementary Reactor Physics*, Pergamon Press, Oxford (1966).

A. F. HENRY, *Nuclear Reactor Analysis*, M.I.T. Press, Cambridge, Massachusetts (1975).

D. L. HETRICK, *Dynamics of Nuclear Reactors*, University of Chicago Press, Chicago and London (1971).

D. JAKEMAN, *Physics of Nuclear Reactors*, The English Universities Press, London (1966).

G. R. KEEPIN, *Physics of Nuclear Kinetics*, Addison-Wesley, Reading, Massachusetts (1965).

J. R. LAMARSH, *Introduction to Nuclear Engineering*, Addison-Wesley, Reading, Massachusetts (1975).

J. R. LAMARSH, *Introduction to Nuclear Reactor Theory*, Addison-Wesley, Reading, Massachusetts (1966).

S. E. LIVERHANT, *Elementary Introduction to Nuclear Reactor Physics*, John Wiley and Sons, New York (1960).

R. V. MEGHREBLIAN and D. K. HOLMES, *Reactor Analysis*, McGraw-Hill, New York (1960).

R. L. MURRAY, *Nuclear Energy*, Pergamon Press, New York (1975).

R. L. MURRAY, *Nuclear Reactor Physics*, Prentice-Hall, Englewood Cliffs, New Jersey (1957).

J. POP-JORDANOV, Ed., *Developments in the Physics of Nuclear Power Reactors*, International Atomic Energy Agency, Vienna (1973).

L. J. TEMPLIN, Ed., *Reactor Physics Constants*, ANL-5800, 2nd ed., Argonne National Laboratory (1963).

J. G. TYROR and R. I. VAUGHAN, *An Introduction to the Neutron Kinetics of Nuclear Power Reactors*, Pergamon Press, Oxford (1970).

L. E. WEAVER, *Reactor Dynamics and Control*, American Elsevier Publishing Company, New York (1960).

A. M. WEINBERG and E. P. WIGNER, *The Physical Theory of Neutron Chain Reactors*, University of Chicago Press, Chicago (1958).

J. WEISMAN, *Elements of Nuclear Reactor Design*, Elsevier/North-Holland, New York (1977).

C. H. WESTCOTT, *Effective Cross Section Values for Well-Moderated Thermal Reactor Spectra*, AECL-1101, 3rd ed., Atomic Energy of Canada Ltd., Chalk River, Ontario (1964).

P. F. ZWEIFEL, *Reactor Physics*, McGraw-Hill, New York (1973).

Chapter 4

S. BANERJEE, E. CRITOPH, and R. G. HART, Thorium as a Nuclear Fuel for CANDU Reactors, *Can. J. Chem. Eng.* **53**(3), 291–296 (1975).

I. P. BELL, Thorium, its Properties and Characteristics, *Nucl. Eng.* **2**(19), 418–422 (1957).

E. CRITOPH, Fuel management in power reactors: Fuel management and refuelling schemes, in *Developments in the Physics of Nuclear Power Reactors*, J. Pop-Jordanov, Ed., International Atomic Energy Agency, Vienna (1973), pp. 201–220.

D. M. ELLIOTT and L. E. WEAVER, Eds., *Education and Research in the Nuclear Fuel Cycle*, University of Oklahoma Press, Norman, Oklahoma (1972).

A. R. KAUFMANN, Ed., *Nuclear Reactor Fuel Elements: Metallurgy and Fabrication*, John Wiley and Sons, New York and London (1962).

R. W. NICHOLS, Uranium and Its Alloys, *Nucl. Eng.* **2**(18), 355–365 (1957).

A. M. PERRY and A. M. WEINBERG, Thermal breeder reactors, *Ann. Rev. Nucl. Sci.* **22**, 317–354 (1972).

T. H. PIGFORD and K. P. ANG, The plutonium fuel cycles, *Health Phys.* **29**(4), 451–468 (1975).

P. SILVENNOINEN, *Reactor Core Fuel Management*, Pergamon Press, Oxford (1976).

Chapter 5

W. V. GOEDDEL and J. N. SILTANEN, Materials for high-temperature nuclear reactors. *Ann. Rev. Nucl. Sci.*, **17**, 189–252 (1967).

R. E. NIGHTINGALE, *Nuclear Graphite*, Academic Press, New York and London (1962).

D. R. OLANDER, *Fundamental Aspects of Nuclear Reactor Fuel Elements*, National Technical Information Service, U.S. Department of Commerce, Springfield, Virginia 22161 (1976).

E. C. W. PERRYMAN, *Nuclear Materials: Prospect and Retrospect*, AECL-5125, Atomic Energy of Canada, Chalk River, Ontario (1975).

E. ROBERTS, J. IORII, and B. ARGALL, Westinghouse pressurized water reactor fuel development and performance, *Nucl. Energy* **19**(5), 335–346 (1980).

J. A. L. ROBERTSON, *Irradiation Effects in Nuclear Fuels*, Gordon and Breach Science Publishers, New York (1969).

M. T. SIMNAD, *Fuel Element Experience in Nuclear Power Reactors*, Gordon and Breach Science Publishers, New York (1971).

M. T. SIMNAD and L. P. ZUMWALT, *Materials and Fuels for High Temperature Nuclear Energy Applications*, MIT Press, Cambridge, Massachusetts (1964).

C. O. SMITH, *Nuclear Reactor Materials*, Addison-Wesley, Reading, Massachusetts (1967).

G. H. VINEYARD, in *Physics and the Energy Problem—1974*, M. D. Fiske and W. W. Havens, Eds., American Institute of Physics, New York (1974), pp. 182–201.

W. D. WILKINSON and W. F. MURPHY, *Nuclear Reactor Metallurgy*, D. Van Nostrand, Princeton, New Jersey (1958).

Chapter 6

P. G. BARNETT, A Correlation of Burnout Data for Uniformly Heated Annuli and Its Use for Predicting Burnout in Uniformly Heated Rod Bundles, United Kingdom Atomic Energy Authority Report AEEW-R463 (1966), available from Her Majesty's Stationery Office, London.

D. BUTTERWORTH and G. F. HEWITT, Eds., *Two-Phase Flow and Heat Transfer*, Oxford University Press, Oxford (1977).

M. M. EL-WAKIL, *Nuclear Energy Conversion*, American Nuclear Society, LaGrange Park, Illinois (1978).

M. M. EL-WAKIL, *Nuclear Heat Transport*, American Nuclear Society, LaGrange Park, Illinois (1978).

G. F. HEWITT and N. S. HALL-TAYLOR, *Annular Two-Phase Flow*, Pergamon Press, Oxford (1970).

J. P. HOLMAN, *Heat Transfer*, 3rd ed., McGraw-Hill, New York (1972).

J. H. KEENAN, F. G. KEYES, P. G. HILL, and J. G. MOORE, Steam Tables (English Units), John Wiley and Sons, New York (1969).
R. V. MACBETH, The burn-out phenomenon in forced-convection boiling, Adv. Chem. Eng. 7, 207–293 (1968).
L. S. TONG, Boiling Heat Transfer and Two-Phase Flow, John Wiley and Sons, New York (1965).
L. S. TONG and J. WEISMAN, Thermal Analysis of Pressurized Water Reactors, American Nuclear Society, Hinsdale, Illinois (1970).
J. WEISMAN, Elements of Nuclear Reactor Design, Elsevier/North-Holland, New York (1977).

Chapter 7

M. M. EL-WAKIL, Nuclear Energy Conversion, American Nuclear Society, LaGrange Park, Illinois (1978).
R. V. MOORE, Ed., Nuclear Power, Cambridge University Press, London (1971).
A. V. NERO, A Guidebook to Nuclear Reactors, University of California Press, Berkeley and Los Angeles (1979).
F. J. PEARSON, Nuclear Power Technology, Oxford University Press, London (1963).
J. G. WILLS, Nuclear Power Plant Technology, John Wiley and Sons, New York and London (1967).

Chapter 8

Central Electricity Generating Board (United Kingdom), Wylfa Power Station, available from CEGB, 825 Wimslow Road, Manchester, England.
K. H. DENT, The Standing of Gas-Cooled Reactors, Nucl. Energy 19(4), 257–271 (1980).
English Electric, Babcock and Wilcox, Taylor Woodrow Atomic Power Construction Company Limited, Hinkley Point–Sizewell–Wylfa Head–A logical development, Nucl. Eng. (April 1965).
L. MASSIMO, The Physics of High Temperature Reactors, Pergamon Press, Oxford (1975).
J. D. MCKEAN, Hartlepool—A milestone in gas-cooled reactors, Nucl. Eng. Int. 14, 724–730 (1969).
National Nuclear Corporation (United Kingdom), Heysham 2/Torness, Nucl. Eng. Int. 26(310), 27–42 (1981).
Nuclear Energy 20(2), Special Issue on Magnox Reactors (April 1981).
L. R. SHEPHERD, The future of the high temperature reactor, J. Br. Nucl. Energy Soc. 16(2), 123–132 (1977).
R. E. WALKER and T. A. JOHNSTON, Fort Saint Vrain nuclear power station, Nucl. Eng. Int. 14, 1064–1068 (1969).
G. L. WESSMAN and T. R. MOFFETTE, Safety design bases of the HTGR, Nucl. Saf. 14(6), 618–634 (1973).

Chapter 9

L. F. DALE, Grand Gulf contributes to growth in the Sunbelt, Nucl. Eng. Int. 25(304), 35–41 (1980).
C. EICHELDINGER, Sequoyah nuclear steam supply system, Nucl. Eng. Int. 16, 850–856 (1971).
General Electric Company, BWR/6, General Description of a Boiling Water Reactor (1980), available from General Electric Company, San Jose, California 95125.

A. L. HEIL, The Sequoyah reactors—Fuel and fuel components, *Nucl. Eng. Int.* **16**, 857–859 (1971).
Westinghouse Electric Corporation, Summary Description of Westinghouse Pressurized Water Reactor Nuclear Steam Supply System (1979), available from Westinghouse Water Reactor Divisions, Pittsburgh, Pennsylvania.
A. ZACCARIA, Advantages of the Mark III Containment, *Nucl. Eng. Int.* **25**(304), 49–50 (1980).

Chapter 10

Atomic Energy of Canada Limited, CANDU Nuclear Power Station, available from Marketing Divison, AECL, 275 Slater Street, Ottawa, Ontario.
Atomic Energy of Canada Limited, CANDU 600 (1979), available from Public Affairs Office, AECL, Sheridan Park, Mississauga, Ontario.
British Nuclear Energy Society, Steam Generating and Other Heavy Water Reactors (Proceedings of a Conference held at the Institution of Civil Engineers, May 1968), available from the British Nuclear Energy Society, Great George Street, London.
J. S. FOSTER and E. CRITOPH, The status of the Canadian nuclear power program and possible future strategies, *Ann. Nucl. Energy* **2**, 689–703 (1975).
J. L. GRAY, The Canadian nuclear power programme, *J. Br. Nucl. Energy Soc.* **13**(3), 227–239 (1974).
Hydro Electric Power Commission of Ontario/Atomic Energy of Canada Limited, CANDU 500 Pickering Generating Station (1969).
W. B. LEWIS and J. S. FOSTER, Canadian Operating Experience with Heavy Water Power Reactors, AECL-3569 (1971), available from Atomic Energy of Canada Limited, Chalk River, Ontario.
H. C. McINTYRE, Natural-Uranium Heavy-Water Reactors, *Sci. Am.* **233**(4), 17–27 (1975).
J. A. L. ROBERTSON, The CANDU reactor system: An appropriate technology, *Science* **199**, 657–664 (1978).

Chapter 11

T. D. BEYNON, The nuclear physics of fast reactors, *Rep. Prog. Phys.* **37**, 951–1034 (1974).
British Nuclear Energy Society, *Fast Reactor Power Stations* (Proceedings of an International Conference held on 11–14 March 1974), Thomas Telford Limited, London (1974).
P. V. EVANS, Ed., *Fast Breeder Reactors*, Symposium Publications Division, Pergamon Press, Oxford (1967).
W. HAFELE, D. FAUDE, E. A. FISCHER, and H. J. LANE, Fast breeder reactors, *Ann. Rev. Nucl. Sci.* **20**, 393–434 (1970).
J. M. LAITHWAITE, The fast breeder reactor at Dounreay, in *Nuclear Power*, R. V. Moore, Ed., Cambridge University Press, London (1971), pp. 152–167.
W. MARSHALL, Some questions and answers concerning fast reactors, *Nucl. Energy* **19**(5), 319–334 (1980).
J. MOORE and J. I. BRAMMAN, Fast reactor development in the U.K., *Nucl. Energy* **20**(1), 15–22 (1981).
Nuclear Engineering International, Creys-Malville nuclear power station, *Nucl. Eng. Int.* **23**, 272, 43–60 (1978).
F. STORRER, Introduction to the physics of fast power reactors, in *Developments in the Physics of Nuclear Power Reactors*, J. Pop-Jordanov, Ed., International Atomic Energy Agency, Vienna (1973), pp. 247–291.

Chapter 12

American Physical Society, Report of the study group on light water reactor safety, *Rev. Mod. Phys.* **47** (Suppl. No. 1) (1975).

American Physical Society, Report of the study group on nuclear fuel cycles and waste management, *Rev. Mod. Phys.* **50**(1), Part II (1978).

V. E. ARCHER, Effects of low-level radiation: A critical review, *Nucl. Saf.* **21**(1), 68–82 (1980).

J. E. BOUDREAU, The mechanistic analysis of LMFBR accident energetics, *Nucl. Saf.* **20**(4), 402–413 (1979).

T. B. COCHRAN, *The Liquid Metal Fast Breeder Reactor*, Johns Hopkins University Press, Baltimore and London (1974).

B. L. COHEN, Hazards from Plutonium Toxicity, *Health Phys.* **32**, 359–379 (1977).

B. L. COHEN, High-level radioactive waste from light-water reactors, *Rev. Mod. Phys.* **49**(1), 1–20 (1977).

F. R. FARMER, *Nuclear Reactor Safety*, Academic Press, New York and London (1977).

H. K. FAUSKE, The role of core disruptive accidents in design and licencing of LMFBRs, *Nucl. Saf.* **17**(5), 550–567 (1976).

J. GRAHAM, *Fast Reactor Safety*, Academic Press, New York and London (1971).

International Atomic Energy Agency, *Environmental Aspects of Nuclear Power Stations* (Proceedings of a Symposium, New York, August 1970), IAEA Vienna (1971).

International Atomic Energy Agency, *Nuclear Power and the Environment*, IAEA Vienna (1973).

H. INHABER, Risk with energy from conventional and nonconventional sources, *Science* **203** (4382), 718–723 (1979).

J. G. KEMENY, Ed., *Report of the President's Commission on the Accident at Three Mile Island*, October 1979, available from the Superintendent of Documents, U.S. Government Printing Office, Washington, D.C. 20402.

H. W. KENDALL, *The Risks of Nuclear Power Reactors (A Review of the NRC Reactor Safety Study, WASH-1400)*, Union of Concerned Scientists, Cambridge, Massachusetts (1977).

R. E. LAPP and H. L. ANDREWS, *Nuclear Radiation Physics*, 3rd ed., Prentice-Hall, Englewood Cliffs, New Jersey (1963).

C. K. LEEPER, How safe are reactor emergency cooling systems?, *Phys. Today* **26**(8), 30–35 (1973).

T. H. MOSS, and D. L. SILLS, *The Three Mile Island Nuclear Accident: Lessons and Implications*, The New York Academy of Sciences, New York (1981).

National Research Council, Advisory Committee on the Biological Effects of Ionizing Radiation, *The Effects on Populations of Exposure to Low Levels of Ionizing Radiation*, BEIR Report, National Academy of Sciences, Washington, D.C. (1972).

T. H. PIGFORD, Environmental Aspects of Nuclear Energy Production, *Ann. Rev. Nucl. Sci.* **24**, 515–559 (1974).

E. E. POCHIN, *Estimated Population Exposure from Nuclear Power Production and Other Radiation Sources*, Nuclear Energy Agency, O.E.C.D., Paris (1976).

A. PORTER, G. A. MCCAGUE, S. PLOURDE-GAGNON, and W. R. STEVENSON, *Report of the Royal Commission on Electric Power Planning, Vol. I (Concepts, Conclusions and Recommendations)*, Royal Commission on Electric Power Planning, Toronto, Ontario (1980).

L. A. SAGAN and R. ELIASSEN, *Human and Ecologic Effects of Nuclear Power Plants*, C. C. Thomas, Springfield, Illinois (1974).

R. D. SMITH, Fast Reactor Safety, *Nucl. Energy* **20**(1), 49–54 (1981).

L. S. TAYLOR, *Radiation Protection Standards*, CRC Press, Cleveland, Ohio (1971).

T. J. THOMPSON, and J. G. BECKERLEY, Eds., *The Technology of Nuclear Reactor Safety, Vol. I, Reactor Physics and Control*, M.I.T. Press, Cambridge, Massachusetts (1964).

T. J. THOMPSON and J. G. BECKERLEY, Eds., *The Technology of Nuclear Reactor Safety, Vol. II, Reactor Materials and Engineering*, M.I.T. Press, Cambridge, Massachusetts (1973).

United Nations Scientific Committee on the Effects of Atomic Radiation, *Ionizing Radiation: Levels and Effects*, United Nations, New York (1972).

United States Atomic Energy Commission, *Theoretical Possibilities and Consequences of Major Accidents in Large Nuclear Plants*, USAEC Report WASH-740, USAEC, Washington, D.C. (1957).

United States Atomic Energy Commission, *The Safety of Nuclear Power Reactors (Light Water Cooled) and Related Facilities*, USAEC Report WASH-1250, USAEC, Washington, D.C. (1973).

United States Atomic Energy Commission, *Reactor Safety Study: An Assessment of Accident Risks in U.S. Commercial Nuclear Power Plants*, USAEC Report WASH-1400 (Rasmussen Report), USAEC, Washington, D.C. (1975).

United States Nuclear Regulatory Commission, *Risk Assessment Review Group Report*, NUREG/CR-0400, U.S. Nuclear Regulatory Commission, Washington D.C. 20555 (1978).

R. E. WEBB, *The Accident Hazards of Nuclear Power Plants*, University of Massachusetts Press, Amherst, Massachusetts (1976).

M. WILLRICH and R. K. LESTER, *Radioactive Waste Management and Regulation*, Macmillan Publishing Company, New York (1977).

R. WILSON, Physics of Liquid Metal Fast Breeder Reactor Safety, *Rev. Mod. Phys.* **49**(4), 893–924 (1977).

Index